DATE DUE

DEMCO 38-296

WAVES AND GRAINS

R

WAVES AND GRAINS

REFLECTIONS ON LIGHT AND LEARNING

Mark P. Silverman

PRINCETON UNIVERSITY PRESS   PRINCETON, NEW JERSEY

Copyright ©1998 by Princeton University Press

University Press, 41 William Street,

n, New Jersey 08540

lom: Princeton University Press,

ester, West Sussex

Rights Reserved

Cataloging-in-Publication Data

verman, Mark P.

ections on light and learning / Mark P.

Silverman.

p.   cm.

Includes bibliographical references and index.

ISBN 0-691-02741-2 (cloth : alk. paper). — ISBN 0-691-00113-8
(pbk. : alk. paper)

1. Physical optics.   I. Title.

QC395.2.S55   1998

97-46555   535′.2—dc21   CIP

This book has been composed in Times Roman

Princeton University Press books are printed on
acid-free paper, and meet the guidelines for permanence
and durability of the Committee on Production
Guidelines for Book Longevity of the
Council on Library Resources

http://pup.princeton.edu

Printed in the United States of America

1   3   5   7   9   10   8   6   4   2
(Pbk.)
1   3   5   7   9   10   8   6   4   2

To Sue, Chris, and Jen

WHO HELPED ME BECOME

A BETTER TEACHER

AND

TO MY GOOD FRIENDS

I. N., A. F., J.C.M., AND A. E.

FROM WHOSE SHOULDERS I SAW THE WORLD

Contents

Preface

A FEW years ago I wrote a popular book of essays, *And Yet It Moves: Strange Systems and Subtle Questions in Physics.*[1] Designed for anyone with scientific curiosity, the book described some of my experiences as a scientist, drawing on projects that ranged widely over the subdisciplines of physics. The parts of those undertakings specifically concerned with fundamental processes of quantum mechanics have been treated more comprehensively in a sequel, *More Than One Mystery: Explorations of Quantum Interference.*[2] In an analogous way, the present book expands in scope and mathematical and experimental detail those of my investigations devoted primarily to optics, except for spectroscopy, which will be published separately.[3]

This is a book of technical essays based upon studies in classical optics that I have made over the past twenty years. In keeping with the activities of a scientist who throughout his career has had little inclination to become a "specialist," the thematic content of these essays is broad; among them are theoretical, experimental, and historical investigations representing virtually all the major elements of physical optics: propagation (in various media), reflection, refraction, diffraction, interference, polarization, and scattering.

The essays explore unusual——indeed, fascinating——questions linking diverse and far-reaching threads throughout the fabric of physics. What, for example, are the explanation and significance of the strange hyperbolic diffraction pattern first reported by Isaac Newton? What is one to make of interference fringes emerging one photon or one electron at a time? How can one magnify small structures a millionfold or more without a single lens or mirror? Can more light reflect from a surface than is incident upon it? What do all living things have in common that might one day render a Star Trek "life scanner" a possibility? How is it possible to look into densely turbid media and see embedded objects ordinarily hidden from view by multiple scattering?

Besides testing or applying important physical concepts, I have long been fascinated with the unfolding of scientific discovery. The field of optics is particularly rich in this regard, and I have included essays (comprising chapters 2, 4, 8, and 10) based primarily upon my own readings of the seminal papers and books of natural philosophers like Newton, Fresnel, Maxwell, and others whose names figure prominently in the exploration of light. These essays not only help give the scientific investigations a historical perspective, but also provide insight into the social dimensions of a life in science.

Finally, I have always been deeply involved in education, as well as in scientific research. There are general lessons that the study of physics, and particularly optics, presents regarding teaching, learning, and the attributes required of a scientist. These, too, I discuss throughout the text and especially in the concluding essays.

To be a physicist is a difficult undertaking, but its intellectual and spiritual rewards are almost beyond imagination to those who have never experienced them. The adventure of discovery is all the more enjoyable when shared with enthusiastic colleagues, and I gratefully acknowledge Professor Jacques Badoz (Ecole Supérieure de Physique et Chimie), Professor Ismo Lindell (Helsinki University of Technology), Professor John Spence (Arizona State University), Mr. Wayne Strange (Trinity College), and Dr. Akira Tonomura (Hitachi Advanced Research Laboratory) for their companionship in some of the investigations described in this book.

I am also thankful to Dr. Trevor Lipscombe, Mathematical and Physical Sciences Editor at Princeton University Press, for his invitation to write this book and for his vigorous support and encouragement, and to my copyeditor, Mr. David Anderson, for his careful attention to detail and his helpful suggestions.

NOTES

1. Mark P. Silverman, *And Yet It Moves: Strange Systems and Subtle Questions in Physics* (New York: Cambridge University Press, 1993).

2. Mark P. Silverman, *More Than One Mystery: Explorations in Quantum Interference* (New York: Springer-Verlag, 1995).

3. Mark P. Silverman, *Probing the Atom: A Study of Fast Beams, Loose Electrons, and Coupled States* (Princeton: Princeton University Press) (forthcoming).

WAVES AND GRAINS

Introduction: Setting the Agenda

> When asked by practical men of affairs for reasons
> which would justify the investment of large sums of
> money in researches in pure science, he was quite
> able to grasp their point of view and cite cogent
> reasons and examples whereby industry and
> humanity could be seen to have direct benefits from
> such work. But his own motive he expressed time and
> again . . . in five short words, "It is such good fun."
> (*H. B. Lemon, on A. A. Michelson*)[1]

FOR MILLENNIA light has exerted an enchantment over the human imagination. What is light? How does it move? How fast does it move? How does it affect matter? Questions like these are, for the most part, no longer mysteries—except, ironically, the first. No one, I suspect, can say what light *is*, any more than what an electron is, or what a quark is. To reply "an electromagnetic wave" or a "massless spin-1 boson" is merely to append labels evocative of two irreducible elements of a dual nature.

Resolving the enigma of light is a dramatic episode in the history of great ideas. As an experimental and theoretical physicist deeply immersed in the interactions and applications of light, I have long been fascinated by the people of this drama—people alive well before I was born and whose acquaintance I would dearly have wished to make. They appeal to me because of their creative genius, experimental skill, articulate expression, incisive wit, personal courage, and, in a few cases at least, their compassion and simple humanity. They have made empirical and conceptual discoveries that I encounter daily in the course of being a physicist and that, in one way or another, have influenced my own scientific pathways and far more modest contributions.

In the inner sanctum of my office/library my eyes have glanced innumerable times at the wall before me to see the grave, yet gentle countenance of Maxwell—a smile, I surmise, beneath the hirsute exterior. Behind me looking over my shoulder is the youthful visage of Fresnel—wistful, distant, perhaps mindful of his fragile health and impending death, his lusterless eyes staring vacantly into the ether whose essence he was unable to penetrate. These two men—major actors in the drama and principal stimuli, in a certain sense, of this book—both died intolerably young: Maxwell at age forty-nine in 1879 and

Fresnel at age thirty-nine in 1827. One can only wonder what further discoveries they might have made had they lived to the venerable age of Newton who took his leave at eighty-five in 1727. The severe and pensive features of Newton, his "mind forever voyaging through strange seas of thought, alone" (in the memorable words of Wordsworth), stare down at me from my right wall. And close beside him is the middle-aged Einstein, a broad, mischievous smile on his face as he steers his bicycle somewhat precariously toward the photographer. I have lived with these "friends" for many years; indeed, quite literally, much of the physics I know I learned directly from them through their writings.

Newton's self-deprecating remark about seeing farther from the shoulders of giants may have been intended sarcastically in his ongoing difficulties with his illustrious contemporary Robert Hooke, but there is surely a deep truth in it. Scientists not only learn from those who have built the foundations of their science; they define themselves in large measure by the company they keep— and for over three decades these colleagues of a distant era have seemed as real to me as those with whom I am in daily discourse by means of electronic mail, fax, and telephone. I thank them, in part, that I am a physicist.

It is with a certain maturity both in age and experience that a scientist comes to see more clearly how his or her work relates to the historical scheme of things. Assuredly at no time in my career have I ever consciously designed a particular project to complement or generalize an investigation by Fresnel or Maxwell. Motivating influences were always more direct. There was the allure of a controversial issue, the practical applications of a new experimental method, the challenge of a difficult mathematical analysis, the delectation of some striking physical phenomenon. And yet, with hindsight, out of this random diversity emerged a pattern. That pattern winds through this volume like a three-stranded rope with intertwining and inseparable strands of theory, experiment, and chronicle.

For a number of years now I have wanted to write a book of essays on different aspects of my studies of light. In one way or another, over the course of a long and varied scientific career, light was either my subject or my tool. To the young medical researcher studying autoimmune disease at a university hospital it was the means of examining fluids essential to life. To the slightly older researcher in a microbiology laboratory, it was the agent that revealed through the microscope the delicate structures of living cells. To the organic chemist attempting to create odd new molecules, it carried the spectroscopic fingerprints of chemical bonds and molecular arrangements. To the atomic physicist probing the limits of the Dirac equation and quantum electrodynamics, the arrival of individual photons spoke tellingly of the interactions that held an atom together. To the materials scientist it signaled new ways to observe the electromagnetic properties of matter. To the quantum physicist it suggested novel experiments for imaging and interferometry with electrons. And to the perpetual student—despite his gray hair—who has throughout his life marveled at the wonders of the natural world, it brought an abundance of colors, patterns, and curious problems to enjoy. Exploring light *is* such good fun, as Michelson said.

Although colored by the many occupational threads that make up the fabric of my experiences, this is nevertheless a book of *physics* essays, addressing virtually all the major themes of physical optics: light propagation, polarization, reflection, refraction, diffraction, interference, and scattering.

I begin in "Following the Straight and Narrow" with an account of light propagation and the various ways of understanding the law of refraction. The accompanying essay, "How Deep Is the Ocean?/How High Is The Sky?" then takes up the theme of geometric optics in the special context of stratified media, in particular, common media compressed by gravity. Here, surprisingly, are examples of the influence of gravity on light in familiar systems close to home rather than more exotic ones reflecting the agency of remote black holes in the deep recesses of interstellar space.

"Dark Spots—Bright Spots" addresses critical experiments on interference and diffraction leading to the recognition of light as a wavelike phenomenon. The essay that follows, "Newton's Two-Knife Experiment," investigates the theoretical underpinnings and experimental details of an unusual example of Fresnel diffraction—a study first performed by Newton and reproduced with modern technology in my laboratory—with an ironical twist of history. In the next essay, "Young's Two-Slit Experiment with Electrons," I look at the striking implications of a modern rendition of an experiment first performed with candlelight by a young Englishman at the turn of the century. Captured on video, the creation of an electron interference pattern one electron at a time illustrates how inadequate "common sense" may be when it confronts fundamental quantum events. The last essay of the second part, "Pursuing the Invisible," discusses the practical applications of interference and diffraction in creating focused, magnified images without the use of lenses. It examines imaging as an interferometric process and the potential advantages of one of the newest types of microscopes based on one of the most ancient forms of image making. Moreover, as the world's simplest electron interferometer employing what is perhaps the brightest beam in science, the device has a special role to play in the exploration of quantum mechanical phenomena.

"Poles Apart" completes the narrative of the wave nature of light by tracing the experimental path to understanding light polarization. In the associated essay, "The State of Light," I give a detailed description of light polarization from the enlightening, yet uncommonly encountered, perspective of a quantum physicist. Besides illustrating a theoretical framework—considerably simpler in my opinion than the prevailing use of four-dimensional matrices—for analyzing optical systems, I discuss as well the operation and application of one of the most versatile, yet "unsung," devices in optics for quantitatively characterizing polarization: the photoelastic modulator.

"The Grand Synthesis" carries the story of light beyond the point of scalar waves to Maxwell's awe-inspiring creation of electromagnetism. Every physics student learns (or at least is taught) electromagnetism—but almost no one, of course, learns it from Maxwell's original papers or *Treatise*. I did, however, and I try to convey in this essay elements of Maxwell's deep insight that have been lost through generations of modern textbooks. The electromagnetic foundation of light provides the basis for understanding light reflection and scattering, the subject matter of the three accompanying essays of this section. "New Twists on Reflection" introduces the theme of chiral asymmetry in nature and discusses the conceptual intricacies and pioneering experiments related to light reflection from materials whose molecules are like a spiral staircase. "Through a Glass Brightly" tells of the theoretical subtleties and ultimate experimental confirmation of an innovative approach, quite different from that of the laser, to amplifying light by reflection. Finally, "A Penetrating Look at Scattered Light" provides a unifying perspective on all the optical processes of the book by regarding them as special cases of light scattering. It also discusses the notoriously difficult problem of light scattering in turbid media (air, fog, clouds, blood, etc.), showing how both chemical and visual information can be extracted by measurements of light polarization.

Although this book is primarily concerned with light, I have included as the last technical essay "Voice of the Dragon," which is actually about sound. The ostensible justification is that the device under study—a child's toy—resembles superficially an unusual and fascinating light source. Another motivation, however—as in Michelson's case—is that the entire undertaking was pleasurable and instructive, and therefore worth relating to others.

The title of this book (aside from evoking an American patriotic song that I often sang to my children when they were very young) has a twofold significance. On the one hand is the obvious allusion to the dual nature of light, which constitutes a central theme throughout the book. But on the other hand, waves also signify in a more general context a disturbance, the "shaking up" of an otherwise unperturbed medium. And this, too, is intended. As one who has long been seriously concerned about contemporary attitudes toward teaching and learning, who believes that there are more appropriate, more effective, more humane ways to do both, I included the three essays "A Heretical Experiment," "Why Brazil Nuts Are on Top," and "What Does It Take . . . ?" as gentle waves in the status quo containing grains of truths culled from a long career of helping students achieve their potential.

NOTES

1. Harvey B. Lemon, from the foreword to A. A. Michelson, *Studies in Optics* (Chicago: University of Chicago Press, 1962), xxi.

Part One

REFRACTION

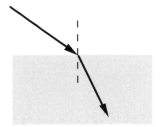

CHAPTER 2

Following the Straight and Narrow

When a hypothesis is true it must lead to the
discovery of numerical relationships that tie together
the most distant facts. When, on the contrary, it is
false, it can represent at best the phenomena for
which it has been imagined ... but it is incapable of
revealing the secret bonds that unite these
phenomena to those of another class.
(Augustin Fresnel)[1]

ANYONE WHO has observed a laser beam passing through a cloudy liquid or smoky room has seen one of the most salient properties of light rays: They propagate rectilinearly (i.e., in straight lines) through a homogeneous medium. This observation is so familiar as to seem almost beneath consideration—and yet from Newton's day to the time of Fresnel, it served as one of the most cogent arguments for a corpuscular or particle description of light. Waves, after all, spread out, as one can readily see by throwing a stone into a pond and observing the ripples.

When light passes from one medium to another, however, the direction of light changes; the rays are "broken" or refracted (from the Latin *frangere*, to break). Newton (1642–1727), having achieved prodigious success in deducing the existence, mathematical properties, and manifold applications of the law of gravity, saw in the phenomenon of refraction another instance of the effects of forces between particles. As a working hypothesis Newton envisaged each color of light to be composed of particles of a particular mass and, possibly, shape. The program of research that he set for himself and "as a farther search to be made by others"—as outlined in a series of some thirty-one queries at the end of his comprehensive and personal treatise, *Opticks*[2]—is in effect to discover the nature of the forces that matter exerted upon the particles of light.

It is worth pointing out that Newton's treatise is remarkably interesting to read. Scientific classics often sit on one's shelf, venerable, sometimes mentioned or referenced, rarely opened and read except by historians. There are reasons for that. The language is antiquated, if not virtually incomprehensible; the mathematical notation or analytical methods obscure; the factual content quite possibly out of date by centuries; the conclusions often incorrect. Try reading Newton's earlier masterpiece, Principia,[3] first published in Latin in

1686. Translations in English are of course available, but one must still be a first-rate geometer to follow Newton's arguments competently and facilely. I suspect that few scientists today are likely to consult the *Principia* to learn mechanics.

By contrast, *Opticks*, which first appeared in 1704, was written in English (although a Latin edition soon followed). Far from being a dry mathematical tome weighted with theorems and proofs, it delineated with obvious loving care Newton's optical experiments—investigations that even today are instructive and enjoyable to read. Moreover, the ordinarily wary Newton expressed himself with greater freedom; he was not averse (as in his study of gravity) to making hypotheses, although these were couched in the less committal form of questions. Nonetheless, these questions permit a penetrating glimpse into what he was thinking, for, as I. B. Cohen pointed out in his preface to *Opticks*, "every one of the Queries is phrased in the negative! Thus Newton does not ask in a truly interrogatory way ... 'Do Bodies act upon Light at a distance ...?'—as if he did not know the answer. Rather, he puts it: 'Do not Bodies act upon Light at a distance ...?'—as if he knew the answer well—'Why, of course they do!'" Whereas the austere and daunting *Principia* concluded an ancient line of inquiry, *Opticks* exposed the exciting phenomena of a new direction of research. Little wonder, then, that experimenters of Newton's day and well over a century later avidly read the book and were much influenced by it.

There is some irony in the fact that Newton's towering reputation was to earn him reproaches—although by that time he was dead and (one would hope) less sensitive to criticism. So deeply had Newton impressed his stamp upon the fabric of science that by the beginning of the nineteenth century most natural philosophers were, like him, partisans of a corpuscular theory of light. When, by the end of that century, a wave theory was firmly established, historians or textbook writers would lament that it was Newton's great authority that held back the progress of a wave theory of light for more than a century.

Such criticism seems to me of dubious validity. That the direction Newton set may have dominated natural philosophy for as long as it did does not necessarily indicate slavish adherence to authority; in fact, the conception of light as a stream of particles acted upon by matter through short-range color-correlated forces actually accounted reasonably well for the phenomena it attempted to explain. Moreover, although a corpuscular theory of light was to become inextricably associated with his name, Newton actually did not find it inconsistent, when necessary, to attribute both particle and wavelike attributes to optical processes. For example, in attempting to understand the pattern of colored fringes in a thin glass plate—a phenomenon eventually to be attributable to wave interference—he wrote:

> As Stones, by falling upon Water put the Water into an undulating Motion and all Bodies by percussion excite vibrations in the Air; so the Rays of Light, by impinging on any refracting or reflecting Surface, excite vibrations in the ...

Medium ... and, by exciting them agitate the ... Body; ... the vibrations thus excited are propagated in the ... Medium ... and move faster than the Rays so as to overtake them.[4]

with the consequence that "every Ray is successively disposed to be easily reflected, or easily transmitted" depending upon whether the motion of the ray is aided or hindered by the overtaking wave. Not a bad try. Indeed, the model bears an uncanny qualitative similarity to a modern picture of photon scattering!

In the opening years of the 1800s when Thomas Young presented his wave theory of interference to a harshly unaccepting community of British scientists, he tried to exploit the ambiguous state of Newton's attitude toward light by pleading that it was indeed the great Sir Isaac himself who first introduced an undulatory theory into optics. The ruse was not very successful, and Young's contributions were actually to find greater appreciation across the Channel among the French—in particular, the remarkably versatile Augustin Fresnel—rather than among the British. But this anticipates matters to come.

By historical accord, it is Christian Huygens (1629–1695) who is usually regarded as the first to propose a wave theory of light, as detailed in his *Traité de la Lumiére* (1690) which predated Newton's own treatise by some fourteen years. Committed Cartesian, Huygens conceived of light as a kind of pressure vibration spreading spherically from each point source throughout an all-pervasive medium, the ether. At each fixed interval of time, the surface tangent to these component wavefronts defined the instantaneous total wavefront—and by outward progression of this envelope Huygens could locate the wavefront at any subsequent time.

It was in extending this mode of reasoning to the more complicated and challenging case of a homogeneous, yet optically anisotropic, medium like calcite that Huygens displayed the full novelty and power of his conception. Here he envisioned the elementary wave surfaces as ellipsoidal, rather than spherical, expanding with different characteristic speeds in the directions of the principal axes (figure 2.1).

As a wavelike description of light, however, Huygens's construction left much to be desired. For one thing, the wave model contained no element of periodicity. Without periodicity, Huygens's waves could not give rise to the attribute of color or to interference phenomena such as the circular fringes Newton saw reflected from a convex lens pressed against a glass slab. For another, it did not even account convincingly for the property of rectilinear propagation. Outwardly spreading spherical waves generate both backward- and forward-moving wavefronts, and the former are not observed. (This was, in fact, a difficulty with which Fresnel, too, would have to contend in developing a comprehensive mathematical theory of diffraction based on Huygens's conception of light; eventually Kirchhoff would provide the final element of the solution, the "obliquity factor.") More damaging still, Newton found a glaring inconsistency in Huygens's argument for constructing anisotropic wavefronts: There

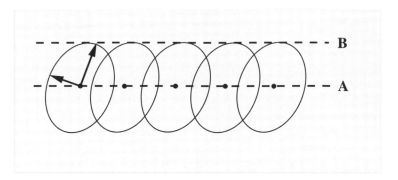

Figure 2.1. Huygens's construction for advance of a plane wavefront in an optically anisotropic medium like calcite with two principal axes. Each luminous point on the initial wavefront *A* produces an ellipsoidal wavelet that propagates with characteristic velocities (arrows) along the principal axes. The new wavefront at *B* is the tangent envelope to all such wavelets.

can be *no* directional dependence of wave speed for "Pressions"—longitudinal vibrations like sound waves—propagating through a uniform medium. Neither Newton nor Huygens apparently entertained the possibility that light could be a transverse vibration.

Although the triumph of the wave theory of light over the particle theory was eventually to be established by Fresnel's sweeping investigations of interference and diffraction, there was, in fact, a far simpler and more direct way in which the two models led to an irreconcilable predictive difference: the speed of light in materials. In this regard, the phenomenon of refraction could have played a key role in the quest to understand the nature of light. History, however, took a different course. Nevertheless, the pertinent physics is instructive.

That light actually moves with a finite speed—no trivial matter to ascertain by direct time-of-flight measurement—had only recently been demonstrated in Huygens's and Newton's day. Indeed, in his treatise Huygens refers appreciatively to "the ingenious proof of Mr. Römer" who determined (~1675) the passage of light to be "more than six hundred thousand times greater than that of sound" by observing the eclipses of Jovian moons at different locations of the Earth about the Sun. Actually, content to have demonstrated that the speed of light was not infinite, Römer, to my knowledge, never gave a concrete numerical value for it; the foregoing figure was deduced by Huygens.

One can regard the eclipses of a moon as a uniform series of time signals sent out by the planet. As the Earth moves away from conjunction (closest approach) with Jupiter, the time required for light to reach the Earth increases. According to Römer the signals were delayed 996 seconds when the two planets were in opposition (farthest approach). Dividing the diameter of the Earth's orbit, known approximately from measurements of solar parallax, by the delay time furnished by Römer, one would obtain a highly respectable value of ~192,000 miles/second (~3.2×10^8 m/s).

It is a much more difficult undertaking, however, to ascertain the speed of light in matter. Although the law of reflection was already known to the Greeks of classical antiquity, the law of refraction was discovered empirically by Willebrord Snel (or sometimes Snell) around 1621—and is consequently referred to as Snel's law in English-speaking countries. Snel died before publishing his discovery, and the law first entered the public record, as far as I know, in the *Dioptrique* of Descartes who is believed to have seen Snel's manuscript, yet not to have acknowledged it. Not everyone believes it; in a French-speaking country—as my Parisian colleagues have gently reminded me—one refers to *la loi de Descartes*.

Without knowing why it was so, Snel observed that the sine of the angle of refraction bears a constant ratio to the sine of the angle of incidence (figure 2.2)—the constant of proportionality depending on the color of the light and nature of the two media. In modern notation the law would read

$$\frac{\sin \phi}{\sin \theta} = \frac{n_1}{n_2} = \text{constant} \tag{1}$$

where n_1 and n_2 are, respectively, the indices of refraction of the incident and refracting media, and θ and ϕ are the angles made by the incident and refracted rays with respect to the normal to the surface (figure 2.2a) and are equivalent to the corresponding angles made with respect to the surface by the wavefronts (figure 2.2b).

Newton knew of relation (1) and made extensive measurements, which he reported in Book 2 of *Opticks*, of the "refractive Power" of substances of different densities. It was Pierre-Simon de Laplace, however, who first derived the law from Newton's corpuscular hypothesis. Published in Book 10 of his *Traité de Mécanique Céleste*, the relation formed the underpinnings of Laplace's treatment of the effects of atmospheric refraction on astronomical observations.

Consider a beam of light originating in air and incident upon the surface of a transparent material such as water or glass. Upon penetrating the surface, the beam bends or refracts toward the normal. This is explicable, according to Newton's ideas, because the molecules of the bulk medium exert a net vertical attractive force on the "molecules" of light. By symmetry there should be no net lateral force if the surface is sufficiently—actually infinitely—extensive. Thus, light moves faster in the refracting medium although the parallel component of velocity remains unchanged. From figure 2.2a, the precise relationship between initial (v_1) and final (v_2) speeds and angles of incidence and refraction is clearly seen to be

$$\frac{\sin \phi}{\sin \theta} = \frac{v_1}{v_2} \quad \text{[particles]}. \tag{2}$$

In a particle theory such as Newton's, the rays of light are effectively identified with a stream of real objects. In Huygens's theory, however, it is the wavefronts

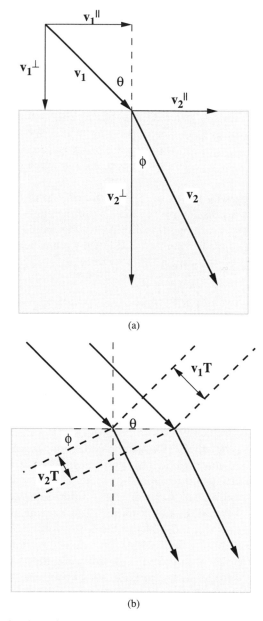

(a)

(b)

Figure 2.2. Refraction of light according to (a) Newton's corpuscular theory and (b) Huygens's wave theory. In (a) the ray is equivalent to the velocity vector of particles of light. Upon moving from medium 1 (air) to medium 2 (water), the particles increase in speed. In (b) a ray is simply the normal to a wavefront. In moving from air to water, the speed of the wave decreases.

that are accorded a physical existence, the rays being merely geometrical sign-posts of beam direction. Figure 2.2b depicts the process of refraction according to Huygens's wave model. Wavefronts (let us say in air) are incident upon the surface, penetrate it, and propagate through the denser medium (for example, water) at a slower speed. In a fixed interval of time T, the distance $v_1 T$ traveled by a wavefront in air is greater than the corresponding distance $v_2 T$ covered by the portion of the same wavefront in water. Sequential wavefronts in water are therefore closer together than in air, and the normal to the wavefront (the ray) must bend toward the normal to the surface.

Although Huygens did not explicitly consider periodic waves, his model readily accommodates them. If T is the period of one oscillation, then the wavelength $\lambda = vT$ represents the distance between points of equal phase on two sequential wavefronts (e.g., the spatial interval between two crests or two troughs). The essential point to recognize is that the oscillation period (or its reciprocal, the frequency $v = 1/T$) depends on the light source and is independent of the various media through which the light propagates. Thus, if $v_2 < v_1$, the wavelength λ_2 in the refracting medium must be *shorter* than the wavelength λ_1 in air, as illustrated in figure 2.2b. From the geometry of the figure, the ratio of the wavelengths is

$$\frac{\sin \phi}{\sin \theta} = \frac{\lambda_2}{\lambda_1} \quad \text{[waves]} \tag{3a}$$

which, upon substitution of corresponding speeds, leads to the reciprocal of Eq. (2)

$$\frac{\sin \phi}{\sin \theta} = \frac{v_2}{v_1} \quad \text{[waves]}. \tag{3b}$$

Equations (2) and (3a, b) are both compatible with Snel's empirical law. To ascertain, however, that the second relation, and not the first, correctly represents the behavior of light requires that one measure directly either the speed or the wavelength of light in transparent bulk matter relative to that in air. Unfortunately, the former could not be done during Huygens's or Newton's lifetimes, or even during the lifetimes of Laplace, Fresnel, and Young. Indeed, by the time the experiment was actually performed—around 1850 by Léon Foucault who compared the speed of light in water to that in air—the issue of light wave versus light particle had been essentially resolved (until the appearance of quantum difficulties after 1900).

Interestingly, the second measurement could well have been performed during Newton's lifetime, for it entailed the use of apparatus already familiar to him, especially in his study of the annular fringes produced by reflection from an air gap of variable width ("Newton's rings"). To my knowledge, however, it was Thomas Young who first reported (~1802) the shortening of wavelengths of visible light in water and inferred from this effect the reduction in light speed.

Young's memoir had little impact. To appreciate its significance one must have already recognized as fundamental the concept of wavelength. But in 1802 the particle theory reigned supreme—and within this framework wavelength had no relevance. Velocity, by contrast, would have been understood by all.

Although the quantum theory of light lies well outside the time frame of the present discussion, I think it is nonetheless useful to relate that the correct law of refraction of light "particles" follows from the conception of light as a stream of photons. Assuming, as did Laplace, that the medium does not affect the lateral motion of the particles, and directing attention to photon linear momentum **p**—a more significant quantity in quantum theory than velocity—figure 2.1a leads to an expression perfectly analogous to Eq. (2):

$$\frac{\sin\phi}{\sin\theta} = \frac{p_1}{p_2} \quad \text{[photons]}. \tag{4}$$

The momentum of a photon, however, is not (as in the mechanics of massive particles) proportional to speed, but reciprocally related to wavelength by de Broglie's famous formula

$$p = \frac{h}{\lambda} \tag{5}$$

in which the constant of proportionality is Planck's consant $h = 6.627 \times 10^{-27}$ erg-sec. Substitution of Eq. (5) into the particle relation (4) leads precisely to the wave result (3a, b) deduced from Huygens's principle. This is but one of many ways in which quantum theory weaves its unifying thread through the seemingly disparate realm of waves and particles.

NOTES

1. Cited (in French) in *Les Cahiers de Science & Vie*, no. 5 (October 1991), p. 1.

2. Sir Isaac Newton, *Opticks: Or a Treatise of the Reflections, Refractions, Inflections & Colour of Light* (New York: Dover, 1952; based on the 4th edition, London, 1730).

3. Sir Isaac Newton, *Principia*, 2 vols. (Berkeley: University of California Press, 1966; based on the 1729 translation by Andrew Motte of *Philosophiae Naturalis Principia Mathematica* [London: S. Pepys, 1686]).

4. Newton, *Opticks*, p. 280.

CHAPTER 3

How Deep Is the Ocean? How High Is the Sky? Imaging in Stratified Media

We saw an ingenious use of this mathematical fact
during our visit to ... China.... [Our friend] placed
the optics of the instrument inside an enclosure and
removed the air. The instrument looks out at the stars
through a horizontal glass window. Because there is
no air inside ... theoretically there will be no effect
of the atmosphere above the glass window, and the
stars will appear in exactly their correct positions.
And, in fact, they do.
(*Aden and Majorie Meinel*)[1]

A STACK of transparent parallel plates produces a virtual image whose location depends sensitively on the angle of viewing. Far from being an idle curiosity confined to a laboratory bench, these "stacks" are everywhere. We breathe in one and swim in another, for gravitational compression of gases and liquids creates stratified media with analogous optical properties. The difference between the actual and apparent locations of objects viewed through the atmosphere and ocean can be significant—and provides a dramatic example of the indirect effect of gravity on light propagation,[2] an influence ordinarily associated with massive stars. Moreover, in a stratified medium of appropriate refractive index gradient, light can move in a closed circular path as if in orbit about a black hole.

3.1 VIEWING THROUGH FLAT LENSES

Although it is doubtless considered ill-mannered to poke a finger in one's tea, this breach of etiquette reveals (with the simplest of apparatus) a striking phenomenon of geometrical optics. The effect is especially marked if the immersed finger is viewed obliquely. The apparent foreshortening of distance that occurs with viewing through a thick refractive medium is a perfect example of optical imaging by a parallel plate or layer.

Most discussions of geometrical optics ordinarily focus on lenses with spherical surfaces. There is, of course, good reason for this: Such lenses are important for historical and practical reasons. Nevertheless, there are interesting things to be learned from the optics of lenses with plane-parallel surfaces.

To the extent that the subject is broached at all in the more than two dozen physics and introductory optics books I have examined, it appears as a standard problem.[3] A coin lies at the bottom of a pond (or a puddle or bathtub of water, etc.) whose depth is x and refractive index is n; what is the apparent depth of the coin as seen from above? The problem is as uninspiring as it is old, for its solution effectively requires only simple substitution into the familiar "lensmaker's" formula applied to lenses of infinite radii of curvature. Phrased as it is, the problem does not stimulate one to discover the enhanced effect of oblique viewing or to explore the implications for stratified continuous media.

Two important aspects of parallel-plate optics are worth mentioning explicitly. First, the location of an image point of a single plate can be determined simply and exactly for arbitrary plate thickness. Second, the analysis of stacked plates with different indices of refraction is not much more complicated than that of one plate. By contrast, the lensmaker's formula for spherical surfaces that one ordinarily encounters in elementary physics is valid only for thin lenses in a paraxial approximation (i.e., light rays close to the optical axis), and analysis of combinations of thick lenses can become algebraically messy indeed.

Perhaps the most prominent feature of the optics of a parallel plate immersed in an ambient medium (e.g., air) is that the incident and transmitted rays are parallel, a consequence readily deduced from Snel's law. A secondary feature, usually negligible for thin plates, is that the refracted ray is laterally shifted upon leaving the plate and reentering the surrounding medium. This has its consequences, however. Insertion of a parallel plate of glass into a converging light beam focuses the beam farther from the converging lens. One might have surmised at first that the plate, being a "lens" with infinite focal length, would have no effect. Moreover, if a parallel plate is sufficiently thick, it can give rise to a surprising effect. Look at both a distant and a nearby object simultaneously through a thick slab of glass or calcite—and turn the slab clockwise and counterclockwise about the vertical axis. The far object appears unmoved, but the apparent position of the near object is markedly altered. The lateral shift does not affect the angular location of an object at infinity but will displace one located at a finite distance.

That the incident and transmitted rays are parallel in the case of multiple layers was well known to Isaac Newton, who drew from that fact a not-so-obvious conclusion regarding imaging by the Earth's atmosphere:

> if Light pass through many refracting Substances or Mediums gradually denser and denser, and terminated with parallel Surfaces, the Sum of all the Refractions will be equal to the single Refraction which it would have suffer'd in passing immediately out of the first Medium into the last. ... And therefore the whole Refraction of Light in passing through the Atmosphere from the highest and rarest Part thereof down to the lowest and densest Part, must be equal to the Refraction

which it would suffer in passing at like Obliquity out of a Vacuum immediately
into Air of equal Density with that of the lowest Part of the Atmosphere.[4]

In other words, to determine the aberration (change in direction) of a light ray
that has entered the Earth's atmosphere—let us say from a distant star—one
need only measure the refractive index of air at the point where the viewing
instrument is located, and this is a relatively easy matter. The aberration is by
no means negligible: about 36 minutes of arc for a star at the horizon. Better
yet, as the Meinels learned in their trip to China, put the instrument into a
vacuum—the medium in which the starlight originated—and there will be no
displacement of the stellar image at all!

The simple analysis that leads to Newton's deduction also provides a basis for
understanding other intriguing aspects of atmospheric optics, such as the origin
of roadway mirages, and the apparent lengthening of the day at the expense
of night from the persistence (by a few minutes) of the Sun's image above the
horizon after actual sunset.

The emphasis of this chapter is on the location of the virtual image produced
by plane-parallel plate lenses singly and in combination. Let us consider first
the apparent depth of a linear object behind an arbitrary number of parallel plates
of arbitrary refractive index. Restriction to normal viewing, as is ordinarily the
case in elementary textbook expositions, would give the mistaken impression
that, if the object is parallel to the plates, then so also is the image. This is not
necessarily true for oblique viewing.

The optics of discrete stacked plates can be readily generalized to apply to
continuous media, and I will take up the cases of an isothermal gas and liquid
stratified by gravitational compression. The examples are of practical interest
and are particularly suitable for instructional purposes because they entail—in
contrast to the sort of formula-substitution problem previously cited—a synthesis of elementary ideas from different domains of physics: optics, electrodynamics, mechanics, and thermodynamics. Moreover, these examples dramatize
the fact—often unrecognized and therefore initially surprising—that the gravity
of the Earth, rather than of some exotic object like a neutron star or black hole,
can play a significant role in the location of optical images.

3.2 LOCATING VIRTUAL IMAGES

The geometrical configuration upon which the following discussion is based is
summarized in figure 3.1 for the case of two parallel plates of thicknesses d_1
and d_2, and respective refractive indices n_1 and n_2. The object to be viewed,
an arrow of length a parallel to the interfaces, is located a distance d_0 to the
left of plate 1 in a medium (e.g., air) of refractive index n_0. It is viewed by an
external observer a distance d_e to the right of plate 2, again in a medium with
index n_0.

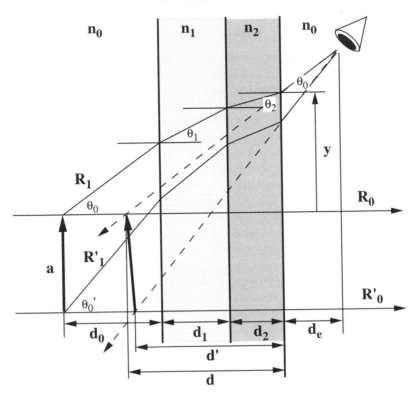

Figure 3.1. Refraction of rays through a stacked array of parallel plates. The location of the virtual image of the arrow tip (base) is determined by the intersection of ray R_0 (R_0') and the backward extension of ray R_1 (R_1'). The figure illustrates the case $n_2 > n_1 > n_0$.

Now consider a light ray R_1 emitted from the arrow tip and incident on the first plate an angle θ_0 to the normal (i.e., horizontal) line. Snel's law

$$n_0 \sin \theta_0 = n_i \sin \theta_i \tag{1}$$

then leads to the angle of refraction θ_i at the interface $i = 1$ or 2, and it is easy to demonstrate that the ray emerges from the final plate at the initial incident angle θ_0. The point of emergence, as already remarked, is displaced—for example, downward if $n_0 < n_1, n_2$.

A ray R_0 emitted by the arrow tip perpendicular to the plates passes through the stack undeviated. The point of intersection of this ray with the backward extension of R_1 locates the virtual image of the tip at a distance d from the front surface of plate closest to the observer (which is plate 2 in the present case). Strictly speaking, a fan of rays (such as would be produced by a realistic light source) in the plane of the figure and in a perpendicular plane leads to closely separated virtual images. This weak astigmatism, however, vanishes for normal

viewing. A detector with large depth of field, such as the human eye, would be largely insensitive to the effect for arbitrary angle of viewing—and so we will not consider parallel-plate astigmatism further.

From the geometry delineated above, the image distance d is given by

$$d = y \cot \theta_0 \tag{2}$$

where

$$y = y_0 + y_1 + y_2 = d_0 \tan \theta_0 + d_1 \tan \theta_0 + d_2 \tan \theta_2 \tag{3}$$

is the vertical distance between the tip of the arrow and the point of emergence of ray R_1 at the rear surface of plate 2 (interface between medium 2 and the air). By using Snel's law one can express the tangent functions in Eq. (3) in terms of the incident (and directly measurable) angle θ_0. Doing so, and generalizing the resulting relation to cover an arbitrary number m of plates, leads to the desired virtual image distance

$$d = d_0 + \sum_{i=1}^{m} \frac{d_i \cos \theta_0}{\sqrt{(n_i/n_0)^2 - \sin^2 \theta_0}}. \tag{4}$$

It is worth pointing out explicitly here that the dependence of the image location on the angle of ray incidence is a true refractive effect and not merely a geometrical consequence of oblique viewing, for the apparent distance d reduces to the actual distance $d = d_0 + d_1 + \cdots + d_m$ irrespective of the angle θ_0 when all refractive indices are the same (n_0).

In the frequently treated example of normal incidence $\theta_0 = 0$, Eq. (4) reduces to

$$d = d_0 + n_0 \sum_{i=1}^{m} \frac{d_i}{n_i} \tag{5}$$

and reproduces the familiar result $d = d_1(n_0/n_1)$ in the simplest case of an object lying in the left interface of a single refracting plate.

The virtual image of an extended object, as a result of its dependence on the angle of incidence, is not simply a projection of the object onto a displaced parallel plane. Look again at figure 3.1; a ray R_1' emitted from the base of the arrow must be incident upon the first surface at a larger angle θ_0' in order to intersect ray R_1 at the aperture of the detector, here taken to be the pupil of the eye. The apparent location d' of the virtual image of the base, determined by the intersection of the normally emitted ray R_0' and the backward extension of R_1', is less than d for $n_i > n_0$ $(i = 1, 2, \ldots, m)$.

From Eqs. (2) and (3) and the comparable geometric relation

$$d' + d_e = (y + a + d_e \tan \theta_0) \cot \theta_0' \tag{6}$$

one can derive an implicit relation between θ_0' and θ_0 that, for objects of small extension, leads to the following displacement δ between the virtual images of

the tip and base:

$$\delta = d' - d = \frac{aD \cos^2 \theta_0}{d + d_e - D \sin \theta_0 \cos \theta_0} \tag{7}$$

where D is the rate at which the apparent tip location varies with viewing angle

$$D = \left| \frac{\partial d(\theta)}{\partial \theta} \right|_{\theta=\theta_0} = \sum_{i=1}^{m} \frac{d_i \sin \theta_0 [(n_i/n_0)^2 - 1]}{[(n_i/n_0)^2 - \sin^2 \theta_0]^{3/2}}. \tag{8}$$

For the configuration of figure 3.1, where the top of the arrow lies closest to the eye and $n_2 > n_1$, the base of the arrow appears closest to the front surface ($d' < d$). This distortion vanishes for normal viewing since $D(\theta_0 = 0) = 0$.

3.3 IMAGING IN A STRATIFIED GAS

Let us consider an isothermal ideal gas in a uniform gravitational field as an analytically tractable model of an atmosphere without thermally induced refractive index gradients. (The real atmosphere is of course more complicated—but this book is not a treatise on meteorology!) The gas compresses under its own weight, thereby leading to variations in pressure, density, and refractive index that depend on height. From the perspective of geometric optics, the atmosphere is an infinite stack of infinitesimally thin plates.

From the ideal gas equation of state and the barometric pressure relation[5]

$$\frac{dp}{dz} = \rho(z)g, \tag{9}$$

where g is the acceleration of gravity near the Earth's surface, one can derive the dependence of mass density $\rho(z)$ on height z (measured from ground level $z = 0$)

$$\rho(z) = \rho(0) \exp\left(\frac{-z}{h_c}\right). \tag{10}$$

The characteristic height

$$h_c = \frac{RT}{Mg}, \tag{11}$$

determined from the universal gas constant $R = 8.31$ J/K-mol, absolute temperature T, and formula weight (actually mass) M of gas, is a rough measure of the vertical extension of the atmosphere. (At a height $z = h_c$ the density falls to $1/e = 0.37$ its value at ground level.) For the Earth's atmosphere—comprising about 75% N_2 and 25% O_2—h_c is approximately 8.8 km, which is roughly the cruising height of a passenger jet. The validity of relation (10) requires that the acceleration of gravity a_g remain close to g over the range of heights of

interest. This is a reasonable assumption here. From Newton's law of gravity it is not difficult to show that at the height h_c the fractional departure from g is $(g - a_g)/g \cong 2h_c/R_E \sim 3 \times 10^{-3}$ where $R_E \sim 6400$ km is the radius of the Earth.

The index of refraction n of a gas is simply related to density by an expression of the form

$$n - 1 = K\rho \tag{12}$$

where K is a constant depending on the light frequency and atomic structure. Later, in chapter 13, I will discuss the physics underlying this expression from the perspective of light scattering. (As a historical aside, it is of interest to note at this point that François Arago and Jean-Baptiste Biot—two French physicists of whom I shall say more soon—jointly investigated, at the suggestion of Laplace, the refractive index of the components of air with the intention of using refractive power as a means of chemical identification. The two collaborators were to become bitter enemies—the first a partisan of the wave theory, and the second a defender of the corpuscular theory.) Substitution of the mass density, Eq. (10), into Eq. (12) leads to the refractive index

$$n(z) = 1 + \beta \exp\left(\frac{-z}{h_c}\right) \tag{13}$$

where $\beta = K_\rho(0) = n(0) - 1 \ll 1$. For air at sea level one has $n(0) - 1 \sim 3 \times 10^{-4}$.

To someone at ground level looking up at angle θ_0 (to the vertical) at an object at height h, the apparent height, deduced from Eq. (4) applied to a continuous medium (with $d_0 = 0$), is

$$d = \int_0^h \frac{\cos\theta_0 \, dz}{\sqrt{(n(z)/n_0)^2 - \sin^2\theta_0}}. \tag{14}$$

Changing the integration variable from z to $v = \exp(-z/h_c)$ leads to a form found in integral tables. Then, following a little algebraic rearrangement, one obtains the expression

$$d(h) = \frac{n(0)\cos\theta_0}{\sqrt{1 - n(0)^2\sin^2\theta_0}}\left[1 - \frac{h_c}{h}\ln\left(\frac{Q_1}{Q_2}\right)\right] \tag{15}$$

in which

$$Q_1 = n(0)(1 - n(0)\sin^2\theta_0) + n(0)\cos\theta_0(1 - n(0)^2\sin^2\theta_0)^{1/2}, \tag{16a}$$

$$\begin{aligned}Q_2 = \; & n(h) - n(0)^2\sin^2\theta_0 \\ & + \{(n(h)^2 - n(0)^2\sin^2\theta_0)(1 - n(0)^2\sin^2\theta_0)\}^{1/2}.\end{aligned} \tag{16b}$$

For Eq. (15) to be meaningful the denominator must not vanish. The viewing angle is therefore limited by the inequality

$$\sin \theta_0 < \frac{1}{n(0)}. \tag{17}$$

This is not particularly restrictive, for air at sea level θ_0 must be less than 88.6°. Thus, the expression for $d(h)$ is applicable to nearly the full range of viewing.

There is a subtle, but physically important, point relating to the form of Eq. (15), which can be written as the sum of two terms. With Eq. (4) in mind, it is tempting to interpret the first term—which does not depend on h_c or, equivalently, on the gravitational acceleration g—as the effect of a uniform layer of gas with index $n(0)$ on the refraction of light originating in a vacuum. The second term, in which h_c appears, would then represent the refractive influence of gravity. One might think that the first term is always by far the larger, and that the location of the virtual image effectively results from refraction at a "boundary" between the vacuum and the atmosphere at ground level. This picture calls to mind the remark of Newton cited earlier (which concerned the net angular deviation of a light ray in air). This interpretation, however, is not strictly valid.

The gas is not a uniform layer, but has an effective thickness, as indicated in relation (13), of the order of h_c (or, at most, a few tens of h_c) that can be much smaller than the object distance h. Even for purposes of analogy there is no refractive boundary separating the object and the gas; both object and observer are immersed in the gas, which is a continuous stratified layer. Had the gas actually been an optically homogeneous layer, no refractive effect would have resulted at all. That is, in the limit that g approaches zero, the image must be located at the object ($d = h$), and *not* at the position given by the first term of Eq. (15).

Examination of the integral in Eq. (14) shows immediately that d reduces to h when g vanishes. This is not so apparent, however, in the integrated expression Eq. (15) in which h_c becomes infinitely great and the argument of the logarithm approaches unity in the limit of vanishing g. Careful evaluation of this limit nevertheless yields, as it must, the expected result. The physical significance of these remarks is that the gravitational compression of the gas is responsible for the entire difference between object and image locations.

With regard to the title of this chapter, let us examine the case $h \gg h_c$ for which Eq. (15) reduces to

$$d(h \gg h_c) = \frac{n(0) \cos \theta_0}{\sqrt{1 - n(0)^2 \sin^2 \theta_0}}$$

$$\times \left\{ 1 - \left(\frac{h_c}{h}\right) \ln \left[\frac{n_0}{2} \left(1 + \frac{\cos \theta_0}{\sqrt{1 - n(0)^2 \sin^2 \theta_0}} \right) \right] \right\}. \tag{18}$$

For overhead viewing ($\theta_0 = 0$) one obtains the simple expression

$$d(h \gg h_c) = n(0)h - n(0)h_c \ln\{n(0)\}. \tag{19}$$

(It is to be noted that once a reduction for $h > h_c$ is made, one can no longer take the limit $g \to 0$.)

How high then *is* the sky? More to the point, actually, is the deviation, $d - h$, of the image from the true object location. Suppose, as a reasonably distant object, one views the Moon, which is at $h \sim 3.8 \times 10^5$ km from the Earth.[6] The index of refraction of the atmosphere at that distance is as close to unity as one is likely to achieve on Earth. Looking at the Moon overhead would result in an apparent increase of its distance by 114 km contributed principally by the first term in Eq. (19); the second term contributes 2.6 m. At $10°$ above the horizon the overestimate of the Moon's distance would be nearly 3840 km (of which the second term in Eq. (15) contributes 46 m).

The perceptual error is an overestimate, rather than an underestimate, because at any level above ground the effective n_i/n_0 in Eq. (4) is less than unity. One can see this geometrically in figure 3.1 where the observer's eye would now be placed in the medium with the darkest shading corresponding to the slab of largest refractive index. In other words, n_2 in the figure would correspond to $n(0)$. Backward extension of the portion of ray R_1 in this region would intersect the horizontal ray R_0 at a point to the *left* of the arrow.

Are these effects significant? The answer of course depends on what they are compared with. It is thought-provoking to note, however, that the above differences between d and h resulting from refraction in a gravitationally compressed atmosphere—that is, the net effects as well as the contributions of the separate terms in Eq. (15)—lie well within the precision with which the Earth-Moon distance can be determined. Lunar laser ranging by time-of-flight measurements employing the Apollo 11 retroreflector[7] can locate the Moon to better than 3 cm with an expectation of eventual further reduction to 1 mm.[8]

Suppose that, instead of the Moon, one looks at an airplane flying at the altitude $h = h_c = 8.8$ km. From the exact relation (15) one can deduce net deviations $d - h$ of 0.9 m and 73.8 m, respectively, for overhead viewing and viewing at $10°$ above the horizon. Of interest, however, is that the term in Eq. (15) specifically dependent on g contributes to the same order of magnitude, but with opposite sign, as the term suggestive of a uniform layer. For overhead viewing these contributions are 2.6 m and -1.7 m.

3.4 IMAGING IN A STRATIFIED LIQUID

The extent to which a liquid of volume V and mass density ρ is compressed isothermally is given by the bulk modulus B,

$$B = -V \left(\frac{dp}{dV}\right)_T = \rho \left(\frac{dp}{dr}\right)_T. \tag{20}$$

Combining Eq. (20) and the barometric pressure equation (9) leads to a depth dependence of the density expressible in the form

$$\rho(z) = \frac{\rho(0)}{1 - z/d_c} \tag{21}$$

where the characteristic depth (analogous to Eq. [11]) is

$$d_c = \frac{B}{g\rho(0)}. \tag{22}$$

For water, the preceding physical parameters have the values $B \sim 2.0 \times 10^9 \, N/m^2$, $\rho(0) = 1.0 \times 10^3 \, kg/m^3$, and $d_c \sim 204$ km.

It is worth noting that, whereas the "height of the sky" in the previous example can either greatly exceed (for the Moon) or be approximately equal to (for an airplane) the characteristic height of the atmosphere, the "depth of the ocean," whose mean is about 5 km (and which reaches a maximum of about 11.5 km),[9] is considerably smaller than the characteristic depth of water given by Eq. (22). Thus, for the range of depths z that will be of relevance here, the term z/d_c will be small, and one can write the liquid density as

$$\rho(z) \sim \rho(0)\left(1 + \frac{z}{d_c}\right). \tag{23}$$

How does compression affect the index of refraction in this case? The mean electric field of a light wave in a liquid or solid ordinarily differs from the local electric field at an atomic or molecular site. This difference is taken into account in the Lorentz-Lorenz equation for the index of refraction

$$\frac{n^2 - 1}{n^2 + 2} = A\rho \tag{24}$$

(which reduces to relation [12] for a gas upon setting $n \sim 1$). A is a constant referred to as the atomic refractivity; for water, $A \sim 0.21 \, cm^3/g$. Combining Eqs. (23) and (24) leads to the expression

$$n(z)^2 \sim n(0)^2\left(1 + \frac{\gamma z}{d_c}\right) \tag{25}$$

for the depth dependence of the refractive index; the constant

$$\gamma = A\rho(0)\left[\frac{1 - 2A\rho(0)}{1 - A\rho(0)}\right] \tag{26}$$

is approximately 0.15 for water. Equation (25) was obtained by expanding the exact expression for n^2 to first order in $A\rho(0)(z/d_c)$ (which attains a value of ~ 0.01 for a maximum depth of $z = 10$ km). In contrast to the case for a gas, expansion in $A\rho(0)$ (~ 0.21 for water) does not lead to an accurate value for the index of refraction.

Substitution of relation (25) into the general formula (14) for a continuous stratified medium and evaluation of the integral between the limits 0 and z leads to the apparent depth

$$d = \frac{d_c \cos \theta_0}{\gamma} \ln \left[\frac{1 + \gamma(z/d_c) + \sqrt{\cos^2 \theta_0 + 2\gamma(z/d_c) + \gamma^2(z/d_c)^2}}{1 + \cos \theta_0} \right] \quad (27)$$

which, with retention of terms of first order in γz, reduces to the simpler expression

$$d \sim \frac{d \cos \theta_0}{\gamma} \ln \left(1 + \frac{\gamma z}{d_c \cos \theta_0} \right). \quad (28)$$

The restriction $\gamma z/d_c < \cos \theta_0$ poses a limit (for water 5 km deep) of about 89.8° on the maximum angle for which Eq. (28) is valid. Expanding the natural logarithm in Eq. (28) to first order leads to an apparent shortening of the actual depth by

$$d - z \sim \frac{-gz^2}{2d_c \cos \theta_0}. \quad (29)$$

The deviation from true depth in the preceding calculation is totally attributable to the effect of gravitational compression of the water. As discussed in the previous section, this quantity must vanish in the limit that g approaches 0 (or d_c approaches infinity). Thus, to an observer whose eyes are just below the upper surface, the bottom of a transparent ocean 5 km straight down would appear about 9.2 m closer; an object on the ocean bottom viewed at 45° would appear about 18.3 m closer to the surface.

Of course, if the observer views the ocean bottom from the air, then the refraction at the air-water interface foreshortens the apparent depth to an even greater extent. In that case, an essentially uniform 5 km layer of water with refractive index 1.33 leads (by the results of section 3.2) to an apparent depth of only 3.8 km.

3.5 ORBITING RAYS OF LIGHT

Although a ray of light traveling through the Earth's atmosphere is bent, the degree of curvature is always much greater than that of the Earth's surface. Even for a ray entering the atmosphere parallel to the Earth's surface, the radius of curvature of the ensuing optical path will be approximately seven times that of the Earth's surface. Nevertheless, it is possible to conceive of an atmosphere with just the right refractive index gradient so that a ray can travel completely around the planet—an exotic proposition, but no more so, perhaps, than that of a black hole.[10] Many years ago, in fact, the proposal was made by someone named Schmidt that the Sun was such a body—and that the bright disc we see is an optical illusion![11]

In Schmidt's model the Sun is a ball of gas whose density decreases from the center outward attaining at a certain radius—which defined what he called the "critical sphere"—a value such that light rays within that spherical surface had the same radius of curvature. The rays that reached our eyes, and that therefore defined the size of the Sun's disc, were just those escaping in a narrow range beyond tangency. Rays leaving the surface of the critical sphere more obliquely would pass into space without reaching us.

Much has been learned about the Sun since then, including the origin of its luminosity, but the idea of light constrained to move in circular paths by virtue of simple geometric optics, rather than the esoteric laws of general relativity, is an intriguing one well worth examining. How must the refractive index vary if orbiting light rays are to result? The answer is in fact quite simple.

The fundamental equation that locates a point $\mathbf{r}$ on a light ray in a medium of refractive index $n(\mathbf{r})$ is

$$\frac{d}{ds}\left(n\frac{d\mathbf{r}}{ds}\right) = \nabla n \tag{30}$$

where s is the length of the ray as measured from a fixed point through which it passes. This equation, whose derivation will be left to the appendix, follows straightforwardly from the definition of a light ray as a trajectory orthogonal to the wavefront.[12] In particular, the gradient of the wavefront yields a vector in the direction $d\mathbf{r}/ds$ (a unit vector since $d\mathbf{r} \cdot d\mathbf{r} = ds^2$) with magnitude n.

As a quick test of the plausibility of Eq. (30), consider the simplest case of an optically uniform medium, $n = $ constant. Then $d^2\mathbf{r}/ds^2 = 0$, and the solution for the trajectories is a straight line (with two constants of integration to be specified).

The case of a medium with spherical symmetry is also simple. The refractive index $n(r)$ depends only on the radial distance $r = |\mathbf{r}|$; all rays are plane curves, and it is not difficult to show that along each ray

$$nr \sin \phi = \text{constant}. \tag{31}$$

In the above relation ϕ is the angle between the position vector $\mathbf{r}$ of a point on the ray and the tangent to the ray at the same point (as shown in figure 3.2). If the ray is to trace out a circular path, ϕ must be $90°$, whereupon it follows from Eq. (31) that the index of refraction must take the form

$$n(r) = \frac{\text{constant}}{r}. \tag{32}$$

One can verify that an inverse radial dependence of the refractive index leads to circular rays of light by solving directly the fundamental equation (30), which,

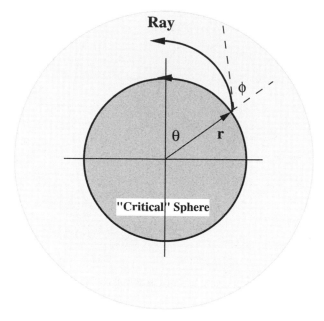

Figure 3.2. Rays in a medium of spherical symmetry.

for spherical symmetry, reduces to

$$\frac{dr}{d\theta} = \frac{r}{a}\sqrt{n^2 r^2 - a^2}.$$
(33)

In Eq. (33) θ is the polar angle of the ray (whose vertex is at the origin $r = 0$), and a is the constant in relation (32). Since r does not vary with θ for a circular orbit, the left-hand side of Eq. (33) vanishes, and one arrives again at Eq. (32).

Suppose, then, that one had a huge ball of gas extending at radius R to the vacuum of space where $n(R) = 1$. Then circular orbits of light should occur at the apparent solar radius $R_S < R$ if the refractive index at that radius were $n(R_S) = R/R_S$.

The foregoing example raises a simple question that, at first thought, may seem perplexing. Since the circular ray is always moving in the direction in which the refractive index does *not* vary, why does the ray bend? In fact, the same question could be raised with respect to any ray moving in a stratified medium perpendicular to the optical gradient. Yet refraction occurs.

In directing attention to rays of light, it is sometimes easy to forget their correct significance—that rays are merely normals to wavefronts. The answer to the question is that refraction occurs because the wavefront, not the ray, extends into the medium and experiences the variation in refractive index. The different parts of the wavefront move at different speeds; the wavefront changes

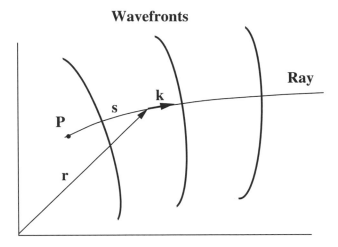

Figure 3.3. A ray propagating in direction $\hat{\mathbf{k}}$ at point $\mathbf{r}$ in a medium of nonuniform refractive index.

direction; the ray bends. A simple idea, yet one that nineteenth century partisans of the corpuscular theory found so difficult to accept.

APPENDIX: THE RAY EQUATION

Figure 3.3 illustrates schematically the passage of wavefronts through a medium of nonuniform refractive index n. Any trajectory normal to the wavefronts constitutes a light ray, such as the one shown in the figure whose location relative to the (arbitrary) coordinate system is marked by the coordinate vector $\mathbf{r}$. The variable s denotes distance along the ray measured from some fixed point P.

Upon covering distance $\mathbf{dr} \cdot \hat{\mathbf{k}}$ in the direction of propagation $\hat{\mathbf{k}} = \mathbf{dr}/ds$, a wave of vacuum wavelength λ incurs the phase change

$$d\varphi = \frac{2\pi}{\lambda} n\mathbf{dr} \cdot \hat{\mathbf{k}} = nk_0\mathbf{dr} \cdot \hat{\mathbf{k}} \qquad (A1)$$

where the constant k_0 is termed the "wavenumber." Comparison of Eq. (A1) with the general result of vector calculus

$$d\varphi = \nabla\varphi \cdot \mathbf{dr} \qquad (A2)$$

leads to the following expression for the phase gradient:

$$\frac{\nabla\varphi}{k_0} = n\hat{\mathbf{k}} = n\frac{\mathbf{dr}}{ds}. \qquad (A3)$$

Taking the derivative with respect to s of both sides of Eq. (A3) results in the following chain of steps:

$$\frac{d}{ds}\left(n\frac{d\mathbf{r}}{ds}\right) = \frac{d}{ds}\left(\frac{\nabla\varphi}{k_0}\right)$$

$$= \nabla\left(\frac{\nabla\varphi}{k_0}\right)\cdot\frac{d\mathbf{r}}{ds} = \nabla\left(\frac{\nabla\varphi}{k_0}\right)\cdot\frac{1}{n}\left(\frac{\nabla\varphi}{k_0}\right)$$

$$= \frac{1}{2n}\nabla\left(\frac{|\nabla\varphi|^2}{k_0^2}\right) = \frac{1}{2n}\nabla(n^2)$$

$$= \nabla n \qquad\qquad\qquad\qquad\text{(A4)}$$

which leads to the ray equation (30).

NOTES

1. Aden and Marjorie Meinel, *Sunsets, Twilights, and Evening Stars* (New York: Cambridge University Press, 1991), p. 13.

2. M. P. Silverman, "Some Thoughts on Imaging by Parallel Plates and Gravitationally Stratified Media," *Eur. J. Phys.* 11 (1990): 366–71.

3. F. A. Jenkins and H. E. White, *Fundamentals of Optics*, 4th ed. (New York: McGraw-Hill, 1976), p. 43.

4. I. Newton, *Opticks* (New York: Dover, 1952), p. 273.

5. C. Kittel and H. Kroemer, *Thermal Physics* (San Francisco: Freeman, 1980), pp. 125–26.

6. Over a distance this large ($h/R_E \sim 0.6$), one should in all rigor take account of the inverse square spatial dependence of a_g in deriving the refractive index $n(z)$, which will, in any event, fall off to unity at the Moon, and Eq. (15) still probably renders a reasonable value for d.

7. J. E. Faller and E. J. Wampler, "The Lunar Laser Reflector," *Scientific American* 222, no. 3 (1970): 38.

8. D. C. Morrison, *Science* 246 (1989): 447.

9. W. Wertenbaker, *The Floor of the Sea* (Boston: Little, Brown, 1974), p. 9.

10. In the gravitational field (the "Schwarzschild field") of a nonrotating, spherically symmetric neutral black hole of mass M, light can move in a closed circular orbit of radius $3GM/c^2$, where G is the universal gravitational constant. The orbit, however, is not stable. For details see H. C. Ohanian, *Gravitation and Spacetime* (New York: W. W. Norton, 1976), p. 303.

11. R. A. Wood describes Schmidt's theory of the Sun in *Physical Optics* (New York: Macmillan, 1919), pp. 83–84. No published reference to the work is given, nor is the author's first name mentioned.

12. M. Born and E. Wolf, *Principles of Optics*, 4th ed. (New York: Pergamon, 1970), p. 121.

Part Two

DIFFRACTION AND INTERFERENCE

CHAPTER 4

Dark Spots—Bright Spots

But as yet no one has sufficiently determined the
motions of the rays near the bodies proper, where
their inflection occurs. The nature of these motions
. . . provides that aspect of diffraction that is of the
most consequence to deepen, because it contains the
secret of the physical mode by which the rays are
inflected and separated into diverse bands of unequal
directions and intensities.
(*Paris Academy of Science, 1818*)[1]

THE PURSUIT of physics is a passion. Some, like Isaac Newton, undertake it out of religious conviction that the laws of Nature reveal the handwork of their Creator. Some, like Etienne Malus, having witnessed the atrocities of war, undertake it as a refuge from the troubles of a harsh world. Some, like Augustin Fresnel, undertake it in search of discovery and recognition. All are insatiably curious; all are persevering. Nature does not yield secrets to the timid.

The controversy over the nature of light exposes more than new laws of physics. It is a laboratory of social dynamics where human emotions—from pride of discovery to anger and jealousy of defeat—are played out. The direction of one's research, the conceptual framework for organizing one's observations, may depend as much on the contingencies of personal interactions as on the outcome of experiments. And yet despite human frailties and transient personal concerns, it is a wondrous fact that objectively verifiable principles eventually emerge, that this marvelous science of physics continually progresses toward a distant truth.

By the beginning of the nineteenth century, the conviction that light was a particle phenomenon held sway both in Britain and on the Continent, especially in France. It was not that the doctrine held together firmly and consistently; on the contrary, so flexible was it that by the stretch of imagination one could manage to explain almost anything.

The events that overturned the particle hypothesis relate to two conceptually distinct aspects of the nature of light. On the one hand, there are phenomena like interference and diffraction that characterize waves in general. On the other, there are phenomena directly concerning the polarization of light— in effect, the vector nature of the electromagnetic field—that are specific to light waves. Athough investigations of these two facets of light are neither mutually exclusive nor chronologically separable, it is nonetheless expedient to consider here waves proper, and to postpone discussion of polarization to part 3.

As the fact does not not seem to be widely appreciated, let us recognize explicitly that Newton's optical investigations embraced not only the composition and dispersion of colors—which the reader probably already knows—but nearly *all* light phenomena including those that we would today regard as hallmarks of classical wave behavior. Apart from observation of his eponymous rings, he knew that long thin objects displayed shadows with colored fringes, as reported by Francesco Grimaldi in the mid–seventeenth century, and he had in his own studies gone well beyond Grimaldi. Newton also examined in scrupulous detail—and I will discuss these experiments further in chapter 5—the passage of light through a wedge-shaped aperture and observed the intricate pattern of hyperbolic curves thereby produced. Indeed, if conceptual advances inexorably follow from accumulations of data—the familiar caricature of scientific method often taught in schools—then Newton lacked nothing to construct a wave theory of light. But he didn't. Instead, he regarded all such deviations from rectilinear propagation as the manifestation of forces exerted on light particles by the edges of material objects.

The first serious challenge to this Newtonian perspective came from another Englishman, the physician Thomas Young. Born in 1773 nearly fifty years after Newton's death, Young, like Newton, had far-ranging interests among which were archaeology, medicine, and foreign languages. Indeed, his biographers frequently describe him as a prodigy, invariably pointing out that by the time he was a teenager he had studied, apart from the classical languages Latin, Greek, and Hebrew, and the languages French and Italian of a cultured European, other fairly exotic tongues such as Chaldean, Persian, Syriac, and Turkish. I am perhaps less inclined than Young's biographers to see the achievement as an example of higher cognitive powers, for my Swiss relations had traditionally to learn five languages from childhood on merely to converse with their compatriots—and I have, myself, studied eight. His pioneering investigations in physics, however, are impressive.

Young was the master of what I would call "binary interference"—all the instances where two wavefronts alone suffice for explanation of an optical phenomenon. It was Young, for example, who first gave a correct explanation of Newton's rings as the superposition and interference of waves reflected from the upper and lower surfaces of a thin film of air between two glass surfaces of variable separation, as shown in figure 4.1. The two reflections, however, occur under different circumstances: the upper wave (ray 1) originating in the optically denser medium (glass) and reflecting from air, the lower wave (ray 2) originating in the optically rarer medium (air) and reflecting from glass. There is—in addition to the continuous retardation that accrues from propagation

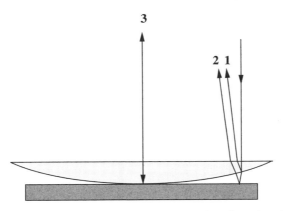

Figure 4.1. Newton's rings experiment. Incident light is reflected at either the upper surface (ray 1) or lower surface (ray 2) of the air film between the plano-convex lens and flat glass plate. The superposition of these two waves produces bright and dark fringes depending on the optical pathlength difference and reflective phase shift. The center of the pattern (ray 3) is dark if the reflective phase shift is 180°.

along optical paths of different length[2]—a discrete 180° difference in phase between these two waves.

An important consequence of this relative phase is that, at the point where the convex lens contacts the flat glass slab—that is, where the air gap is of almost zero thickness and no retardation from propagation occurs—the two reflected waves (ray 3) undergo complete destructive interference. The center of the annular pattern is *black*. Young recognized the reason for this seemingly baffling effect and devised a beautiful experiment in support of it. Using a lens of crown glass and a plate of flint glass with a film of cassia oil between them, he produced a system in which the upper and lower waves both reflect from optically rarer media, for the oil had a refractive index intermediate between the two types of glass. Under these circumstances waves interfere constructively where the optical pathlength vanishes, and the annular pattern displays a *bright* spot at the center.

With the same mode of reasoning, Young could account for the interference colors seen in soap bubbles, and the fact that the thin soap films eventually turn black at their thinnest sections shortly before bursting.

In his investigations of diffraction from slender objects, Newton had attributed the resulting fringes to the action of a force on those rays (i.e., streams of light particles) that just skirt one edge or the other of the object. To test this hypothesis, Young repeated Newton's experiments using a thin wire as the diffractor and mounting an opaque mask at one side. If Newton were correct, fringes should be seen on the unobstructed side of the wire—but in fact the entire pattern of fringes vanished. Interpreted in terms of binary interference, the

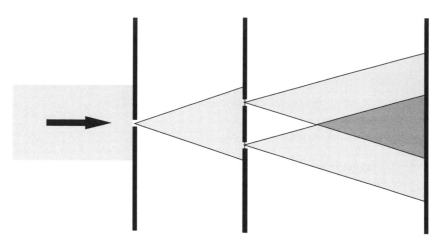

Figure 4.2. Young's two-pinhole interference experiment. Sunlight filtered through a pinhole produces a coherent beam which is subsequently split by two close-lying pinholes and recombined at a distant screen. In the region of overlap interference fringes are observed.

explanation is clear: Remove one of the waves, and there can be no interference, hence no fringes.

Could one, however, use the wave theory quantitatively to *predict* the locations of fringes and not merely account for their existence? To this end Young conceived of an experiment that has since become practically synonymous with binary interference.[3] "Young's two-slit experiment" is one of the archetypical physical models, a tool for studying interference conceptually and practically not only in optics but all throughout quantum physics.[4] (In a later chapter I will discuss the subtleties of Young's experiment performed with electrons.) Young's procedure was simple and straightforward. In an otherwise darkened room sunlight was admitted through a pinhole in an opaque screen, the diverging cone of light falling upon two other pinholes in a second screen nearby as illustrated in figure 4.2. Cones of light spread from each of these pinholes, and, where they overlapped on a third distant viewing screen, dark and light bands were produced. Covering one of the holes with a finger—like masking one side of a diffracting wire—Young could make the pattern of fringes vanish.

From the perspective of a corpuscular theory, these effects were baffling: How could the addition of light to light lead to darkness? Young explained that the dark bands are the loci of points for which the optical pathlengths from the pinholes to the viewing screen differ by an odd number of half-wavelengths. The two waves then overlap 180° out of phase and cancel each other by destructive interference. Similarly, the bright bands comprise those points for which the two waves overlap in phase, their optical paths differing by an integral number of wavelengths. From the simplicity of the configuration Young could calculate the expected locations of intensity maxima and minima and found agreement with his experiment. Finally, knowing the distance between pinholes and the fringe locations for a specific color, he could deduce the wavelength corresponding to that color: for example, 0.7 μm for red and 0.4 μm for violet.

Many physicists are of the impression, I have found, that Young's experiment was a sort of *experimentum crucis* that decided the scientific world in favor of a wave theory of light. In truth this was hardly the case. Young published his results in 1804, but his presentation was not convincing. Despite the precision with which the measurements were apparently made, he failed to describe his experimental procedure carefully or even in some cases to provide the quantitative data that he had recorded. Nor did he make clear how he arrived at his theoretical relations. Young's work did receive notice, however—a scathing critique from Lord Henry Brougham, physicist and future chancellor of England.

Manner of presentation aside, the very nature of the two-slit experiment undermined Young's conclusion. Since the fringe pattern was again produced by diffraction—that is, the light had skirted the edges of an object—the fact that two streams of light could destroy one another at a point might very well reflect the modifications impressed upon light particles by matter rather than the more suspect notion that light was a wave.

It was Augustin Fresnel who first performed (~1816) an ingenious experiment in which interference was made to occur in total absence of diffraction. Fresnel accomplished this by illuminating with a point source two mirrors slightly inclined to one another, as illustrated in figure 4.3. The reflected light from the mirrors overlapped at a distant screen, giving rise, as in Young's experiment, to interference fringes. As the source was shielded so that it could not illuminate the screen directly, the Fresnel "two-mirror" experiment is another example of binary interference. Indeed, one can analyze the experiment quite simply by ignoring the mirrors themselves and considering each of the interfering beams to have originated from a virtual point source behind each mirror. The two virtual sources emit light coherently since they are images of the same real source, and the pattern of fringes, therefore, should be that due to the interference of light from two luminous points. The subtlety and beauty of this experiment are nicely captured in Fresnel's own words (rendered into English): "only the vibration theory could give one the idea of this experiment, and . . . it is sufficiently hard to perform that it would be nearly impossible to discover it by accident."

Fresnel, as the reader will discover (if it is not already obvious), is a central figure and something of a hero in this book. Pathetically all too human in his desperate desire to distinguish himself in the world of science, his ambitions are the ambitions of all of us who do research, write papers, and seek recognition. As a young man trained in engineering, he first turned his attention to industrial chemistry but learned to his chagrin that what he thought was original work was anticipated by others. Disappointed, he later immersed himself in the wave

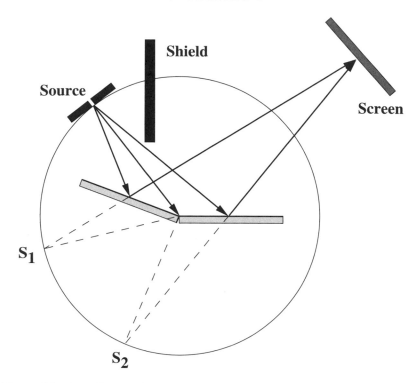

Figure 4.3. Fresnel's two-mirror experiment. Light from a point source reflects from two mirrors and overlaps at a distant screen. The rays appear to originate from virtual sources S_1 and S_2 behind the mirrors. Real and virtual sources lie on a circle centered about the mirror junction. A shield blocks direct illumination of the screen.

theory of light, guided and encouraged by François Arago—one of very few wave enthusiasts in the Paris Academy—who helped publicize his work both in France and abroad. One interesting footnote of history: Fresnel, who would eventually provide overwhelming support for the wave nature of light, was the son of an architect employed by the Duc de Broglie whose descendant, Louis de Broglie, would predict the wave nature of electrons.

In his early studies Fresnel (like Newton and Young) investigated the diffraction of light from thin objects and attempted to account for the locations of fringes with a binary model of interference. For example, within the geometrical shadow of a long narrow diffractor he assumed that only rays propagating from the two edges of the object contribute to the pattern. Correspondingly, within the illuminated region to each side of the shadow he attributed the fringes to interference between a direct ray from the light source and a ray from the closest edge. I have at times used such procedures myself in the study of both light and electron diffraction, and it is surprising how well they work for so drastic a simplification of the actual state of affairs. Nevertheless, Fresnel's binary ray theory was neither consistent nor unified and could not in general be applied to situations—like the illuminated region behind a diffracting aperture—where many rays clearly superposed at each point.

Directing his attention to the three-dimensional spatial distribution of fringe maxima and minima, Fresnel deduced what he considered one of the most remarkable consequences of the wave theory of diffraction. Indeed, it is a striking result that I have rarely seen mentioned in the vast number of general physics and optics textbooks in which diffraction patterns are usually shown only in a single plane a fixed distance from the diffractor. What, however, is the locus of points traced out by a fringe as the distance of the viewing screen from the diffractor is varied? One might think—based on familiar treatments of diffraction within (what we now call) the far-field or Fraunhofer approximation—that the variation is linear. Fresnel recognized that the fringes do not propagate in a straight line, but along a hyperbola whose foci, in the case of fringes outside the geometric shadow of a diffracting wire, are the point source and one of the edges. Furthermore, he devised a simple, yet precise, micrometer to demonstrate this fact.

Why a hyperbola? Recall that within a binary ray model of interference there are always just two point sources. They may be, for example, two pinholes in Young's experiment, or a point source and an edge point in Fresnel's treatment of diffraction from a wire. The locus of all points equidistant from two given points (the sources of light) is a perpendicular plane bisecting the line joining the two points. Thus, the principal (zeroth-order) maximum in the fringe pattern is spatially distributed in a plane lying midway between the two sources. The first-order maximum is the locus of all points in space so situated that their distances to the two sources differ by one wavelength. This, however, is the condition that defines a hyperboloid of revolution, a surface generated by the movement of a point whose distance from two fixed points (the foci) is a constant. Likewise, the second-order maximum will be spatially distributed on the surface of another hyperboloid for which the constant difference is twice the wavelength. In this way the interference fringes form in space a family of confocal hyperboloids—and the intersection of these surfaces with a plane passing through the two foci will be a system of hyperbolas.[5]

Under conditions characteristic of most interference experiments, where the source points are very close and the viewing screen far removed, the hyperbolic variation is practically linear. To discern the actual fringe variation with the apparatus available to Fresnel required great patience and experimental skill. Fresnel, like Newton, was not only a mathematically adept theorist, but a superb experimentalist.

The hyperbolic fringe variation with distance from the source carries a major conceptual implication. If light, as Newton construed it, were truly a beam of particles deviating from rectilinear propagation as it passed by the edges of an object, then once well beyond the short-range influence of matter it must resume its linear motion. After all—by Newton's Second Law—if there is no force, there can be no acceleration. But, to the extent that one regards the fringe maxima as locations of the light particles in space, these particles are moving forward along a hyperbolic trajectory in the absence of any discernible force (ignoring the negligible effect, if any, of the surrounding air).

Alas for Fresnel! Both Newton and Young knew of the hyperbolic motion. Indeed, Newton, in trying to resolve this difficulty was forced to introduce the existence of an ether—a major inconsistency in a theory purportedly describing the motion of particles in empty space.

Unable to read Latin or English, and having little access to scholarly journals, Fresnel was unaware particularly of Young's work—and subsequently learned with bitter disappointment from Arago, upon the latter's return from a trip to England, that his researches, so arduously and painstakingly undertaken, were regarded as largely unoriginal. What scientist today could fail to sympathize with Fresnel's disheartened reply: "I have decided to remain a modest engineer . . . and even to abandon physics, if circumstances require it. I would resolve to do so the more easily that I now see it is a stupid plan troubling oneself to acquire a small bit of glory."[6] Perhaps it might have been some consolation to Fresnel to know that Newton himself had to contend with such challenges: "I see," wrote Newton in regard to his work on telescopes communicated to the Royal Society in 1672, "a man must either resolve to put out nothing new, or become a slave to defend it."[7]

Fortunately, Fresnel did not abandon physics. First, his innate curiosity would not permit it, and second, the pursuit of a "small bit of glory" is too strong a human need to be denied. When reputable scientists pretend such, it is in their dotage after they have already attained their measure of recognition.

In any event, encouraged by Arago, Fresnel's spirit revived, whereupon he set about with resolute determination to go beyond anything Young had done; he would create a complete theory of diffraction that accounted not only for the locations of fringes, but their intensities as well. In the ensuing work Fresnel changed his diffracting element from a narrow object to an open slit and ultimately was led to take cognizance of what he must certainly have initially recognized as the mainstay of a wave theory, but had hitherto largely ignored: Huygens's principle. The light that constitutes a diffraction pattern does not arrive as rays from two cleverly, but nevertheless arbitrarily, chosen points. Rather, every unobstructed point that receives light from the source can act as a source of secondary emission—and the task was to determine how one sums the infinite number of wavefronts that converge at a point on the viewing screen.

Fresnel, it should be mentioned, was an *ancien élève* (graduate) of the Ecole Polytechnique in Paris, a school noted for its comprehensive mathematical instruction. In applying himself to this problem, Fresnel, in marked contrast to his predecessors, availed himself not so much of geometry, but of analysis. Having figured out by a trigonometric decomposition how to sum two waves of arbitrary relative phase, he had surmounted the principal obstacle to creating an integral theory of wave propagation and diffraction. The final expression at which he arrived in terms of "Fresnel integrals" resembles closely the relation found in nearly all optics books today—with one important exception. Fresnel did not formulate correctly the inclination factor, the factor that modulates the wave amplitude as a function of angle with respect to the forward direction. The factor he deduced (the cosine of this angle) worked very well for small diffractors and remote viewing screens, but it did not vanish for backward propagation.

In 1817 the Paris Academy launched a competition for the essay best accounting for the diffraction of light. With the exception of Arago, the committee responsible for the event consisted exclusively of partisans, like Laplace and Biot, of the particle hypothesis. The very wording of the thematic content to be judged was framed in Newtonian terminology; indeed, the task of each participant was to deduce mathematically the movement of *rays* in their passage near material bodies.

Fresnel, as one might imagine, was not initially enthusiastic about entering—his whole direction of research having apparently already been ruled out by the wording. Nevertheless, urged on again by Arago, he composed a lengthy paper summarizing his philosophical approach, his methods, and his results. It is an amusing irony of history that Siméon-Denis Poisson—another graduate of the Polytechnique noted for his broad theoretical contributions to physics and mathematics, and a staunch advocate of the corpuscular theory—noted a glaring inconsistency in Fresnel's theory. Applying this theory to an opaque circular screen, Poisson deduced the (to him) ludicrous result that the center of the shadow *(doit) être aussi éclairé que si l'écran n'existait pas* (must be as brightly illuminated as if the screen did not exist). Arago performed the experiment in advance of the committee's decision, and the bright center—which history records as Poisson's spot—showed up as predicted.

Fresnel, his relentless efforts finally recognized, received the prize[8]—but Biot, Poisson, and others remained unshaken in their particle convictions.

NOTES

1. Cited in J. Z. Buchwald, *The Rise of the Wave Theory of Light* (Chicago: University of Chicago Press, 1989), p. 170.

2. The optical pathlength of a wave through a uniform transparent medium is nL, where L is the geometric pathlength and n is the index of refraction. If the medium is nonuniform, then the optical pathlength is determined from the integral $\int_0^L n(s)\,ds$.

3. For historical accuracy it should be noted that this experimental design actually originated with Grimaldi around 1665. He performed the two-pinhole experiment using

sunlight but with a separation between pinholes that greatly exceeded the transverse coherence length of the light, and thus no fringes were observed. I discuss coherence length in more detail in chapter 6.

4. See M. P. Silverman, *More Than One Mystery: Explorations in Quantum Interference* (New York: Springer-Verlag, 1995).

5. Consider two vertically aligned pinholes separated by an interval d. Let r_1 and r_2 be, respectively, the separations of the top hole and bottom hole from a distant image point P defined by coordinates y (vertical displacement from the incident forward direction of the light beam) and z (distance of viewing screen from the apertures). Then for fixed wavelength λ and integer n the relation $r_2 - r_1 = n\lambda$—which defines a hyperbolic locus of points P in the y-z plane—locates a specific fringe maximum. In Cartesian form, one obtains

$$\frac{y^2}{(n\lambda/2)^2} - \frac{z^2}{(n\lambda/2)^2[(d/n\lambda)^2 - 1]} = 1$$

where $d > n\lambda$. To good approximation the points of minimum light intensity also trace out a hyperbolic trajectory as z is varied.

6. Cited in Buchwald, *Wave Theory of Light*, p. 138.

7. Cited in E. T. Bell, *Men of Mathematics* (New York: Simon and Schuster, 1937), p. 108.

8. There were but two contestants; the identify of the second is not known with certainty.

CHAPTER 5

Newton's Two-Knife Experiment:
The Hyperbolic Enigma

And when the distance of their edges was about the
400th part of an Inch, the stream parted in the
middle, and left a Shadow between the two parts.
This Shadow was so black and dark that all the Light
which passed between the Knives seem'd to be bent,
and turn'd aside to the one hand or to the other.
(Isaac Newton)[1]

ABOUT A CENTURY before Young's celebrated two-slit experiment, Isaac Newton quantitatively investigated the diffraction of light from a wedge-shaped aperture but failed to understand the implications of his findings. I have repeated the experiment using a laser light source with pinhole spatial filter and charge-coupled device (CCD) camera, and reexamined its theoretical basis by means of the scalar diffraction theory created by Fresnel and improved upon by Kirchhoff.[2] Both the far-field shadow region and near-field directly illuminated region reveal aesthetically striking images. The images are deducible from the mathematical analysis, but their interpretation is still subtle and best elucidated by an alternative and less widely known perspective of diffraction—ironically, a model of diffraction first suggested, and later repudiated, by both Young and Fresnel.

5.1 A GREAT DISCOVERY UNPURSUED

Every physicist has undoubtedly heard of Thomas Young's celebrated "two-slit experiment," widely regarded (rightly or wrongly) as the *experimentum crucis* of the wave nature of light and the archetypical configuration for demonstrating superposition and interference of matter waves as well. By contrast, I have encountered few physicists who have ever heard of Newton's "two-knife experiment"—which predates the publication of Young's results (∼ 1802) by over a century. Newton, of course, never called his experiment by this name; I did. But labeling aside, I have seen no optics text or monograph, elementary or advanced, in which is mentioned, let alone analyzed, the intriguing problem that captured Newton's attention some three centuries ago. And yet, in some ways, this is an extraordinary problem.

On a number of occasions I have described the experimental setup to physics students and professional physicists and asked them to predict the outcome, that is, the pattern of light that one might expect to see (and that Newton, using sunlight, reported in astonishing detail)—only to be met with shoulder shrugs and blank stares. Although, once revealed, the pattern is amenable to a certain heuristic interpretation, further scrutiny reveals that this apparent simplicity is deceiving. Indeed, Newton's two-knife experiment represents a fairly difficult system to analyze generally and does not fit neatly into the elementary categories by which diffractive phenomena are ordinarily classified. Perhaps that is one reason why so little mention is made of it.

There is also a certain irony in the fact that Newton himself—embarrassed, I suspect, by the implications of his own painstakingly obtained results—virtually ignored this experiment when he summed up the properties of light and the directions for further inquiry in the concluding chapter of his treatise *Opticks*.[3] Yet in its precision and thoroughness, the "two-knife experiment" (actually a series of experiments) provided as conclusive a quantitative demonstration as one could wish of the wave nature of light. Newton, however, did not pursue this line of thought—for, despite public reticence on the subject, he long favored a corpuscular interpretation of optical phenomena—and it was ultimately Augustin Fresnel who created around 1816 the theoretical framework for understanding these beautiful experiments.

5.2 PASTEBOARD, KNIVES, AND PITCH

In the domain of optics Newton's name usually conjures up thoughts of the prismatic decomposition and recomposition of white light, the first scientific inquiries he presented before the Royal Society. Unfortunately, they were among the last as well, for the conflict over priority that developed with Robert Hooke discouraged Newton from publishing his subsequent researches for many years thereafter. Nevertheless, Newton, notorious for beginning many a manuscript and completing but few, has left in *Opticks* a comprehensive account of his studies of light.

Of the extensive chronicle of experiments concerning "Reflections, Refractions, Inflections, & Colours of Light" (as the subtitle reads), there is a remarkable collection of investigations—to me the most interesting and conceptually significant of the three-volume opus—that I will designate (for reasons soon apparent) the "two-knife experiment." These investigations fall under Newton's category of "inflections," which one would refer to today as diffraction.

"Grimaldo has inform'd us," Newton began the Third Book of *Opticks*, "that if a beam of the Sun's Light be let into a dark Room through a very small hole, the Shadows of things in this Light will be larger than they ought to be if the Rays went on by the Bodies in straight Lines, and that these Shadows have three parallel Fringes . . . of colour'd Light adjacent to them." Thus was Newton aware of the diffraction of light and had, himself, in preliminary experiments investigated the "Shadows" cast by "Hairs, Thred, Pins, Straws, and such like slender Substances." But Newton was nothing if not tirelessly thorough, and his goal was not merely to observe an interesting phenomenon, but to nail down the mathematical law that governed it. "My Design," he inscribed at the beginning of his treatise (1704 edition), is not to explain the Properties of Light by Hypothesis, but to propose and prove them by Reason and Experiment."

To determine the quantitative relation governing light propagation near the presence of matter—and hence the "forces" that particles of light were subject to—Newton ultimately devised an experiment whose description is best given in the designer's own words: "The Sun shining into my darken'd Chamber through a hole a quarter of an Inch broad, I placed at the distance of two or three Feet from the Hole a Sheet of Pasteboard, which was black'd all over on both sides, and in the middle of it had a hole about three quarters of an Inch square for the Light to pass through. And behind the hole I fasten'd to the Pasteboard with Pitch the blade of a sharp Knife."

Examining the light that passed over the edge of the knife and fell squarely onto a parallel sheet of white paper some three feet distant, Newton observed streams of faint light "shoot out . . . into the shadow like the Tails of Comets." Unable to see the faint streams clearly because the direct light of the Sun was too bright, he then punctured a small hole in the paper to let pass a fraction of the sunlight, which he viewed upon a black cloth. Newton could then see the streams of light clearly. There was unmistakably a "leak" of illumination beyond the border of what ought to have been the totally dark shadow of the blade.

Besides the odd fact that light had apparently passed over the edge and into the region of the knife's shadow, the projected image displayed another curious feature. Above the projected edge of the knife—in the region where light was free to pass unimpeded and where one might consequently have expected to see a uniformly white image—the shadow was fringed with alternating bright and dark parallel bands—further confirmation of Grimaldi's and Newton's own prior investigations with slender objects.

Up to this point Newton had observed the now familiar phenomenon of Fresnel diffraction from an effectively infinite straight edge, which the reader will find in almost any sufficiently advanced optics book. Characteristic of the straightedge diffraction pattern is (1) the monotonic decrease toward zero of the light intensity with increasing penetration into the shadow region, and (2) the diminishing contrast of the fringe pattern with increasing penetration into the region of direct illumination. It is useful to note that edge diffraction patterns of this kind can now be seen not only with light, but also with electrons,[4] and provide essential information about the intrinsic size, brightness, and coherence of newly developed highly coherent subnanometer-size electron sources.

Since the precise location of the fringes was the key to discerning how light moved through space, Newton modified his apparatus by adding another knife. And here the interesting part begins: "I caused the edges of two Knives to be ground truly strait, and pricking their points into a Board so that their edges might look towards one another . . . I fasten'd their Handles together with Pitch. . . . The knives thus fix'd together I placed in a beam of the Sun's Light, let into my darken'd Chamber through a Hole the 42nd Part of an Inch wide, at the distance of 10 or 15 Feet from the Hole."

Before pursuing the matter further, the reader might be curious to know how Newton determined precisely that the diameter of the hole was 1/42 (0.024) of an inch. That is a difficult measurement to make—if one focuses attention on measuring the *hole*. I have on a number of occasions posed the question to various students and teachers, but no one ever arrived at a satisfactory solution. For Newton, it was easy. He had punctured the hole in a piece of lead with a pin and therefore simply laid twenty-one pins side by side to find that they measured $\frac{1}{2}$ inch. As an experimental scientist with access to sophisticated equipment that Newton never dreamed of, I am continually amazed by his ingeniously simple solutions to experimental problems. Quantitative observation was his hallmark, and in assessing the various distances that played a role in these diffraction experiments Newton recorded lengths to a precision of at least a thousandth of an inch.

Figure 5.1 shows a silhouette to scale of the diffraction aperture created by the two knives fixed with pitch so that the edges of the blades, at 4 inches from their point of intersection, were separated by $\frac{1}{8}$ inch, giving a wedge angle of 1.79° or 0.031 radian. Viewed on a sheet of white paper normal to the beam "at a great distance from the knives"—which Newton subsequently specified as about 9 feet—the projected image revealed a remarkable pattern that he described as follows: "When the Fringes of the Shadows of the Knives fell perpendicularly upon a Paper at a great distance from the Knives, they were in the form of Hyperbola's."

Newton's sketch of the pattern in *Opticks* is reproduced in figure 5.2. Lines CA and CB in the figure represent boundaries drawn upon the paper parallel to the knife edges and between which all the light would fall if it passed between the edges without diffraction. Where the shadows of the blades were widely separated, a given fringe ran parallel to the nearest edge, as perhaps expected (even if not understood) from the previous experiment with a single blade. But where the blade separation narrowed and eventually vanished, the fringe curved hyperbolically away from its parallel course, crossing (according to Newton) other fringes in the pattern until it penetrated deeply the shadow region of the opposite edge.

The hyperbolic shape was no crude estimate. By measuring to within a reported thousandth of an inch the deviation of individual fringes for different edge separations—the white paper all the while a fixed distance from the

Figure 5.1. Computer-generated silhouette of Newton's wedge aperture formed by two knives at an angle of 0.031 radians.

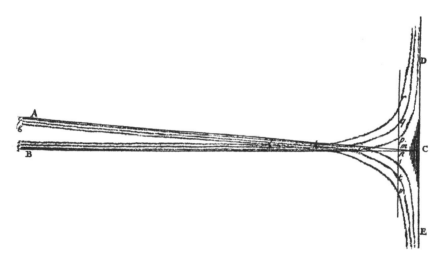

Figure 5.2. Newton's sketch of the hyperbolic fringes from a wedge aperture. Heavy lines CA and CB are the geometric shadows of the knife eges.

knives—Newton demonstrated that each fringe followed a hyperbolic curve precisely. Do not confuse this hyperbolic relation with the one discussed in the previous chapter that later occupied the attention of Young and Fresnel. That hyperbola resulted from noting the locations of a particular fringe at increasing distances from the diffracting aperture or object. By contrast, the conspicuous pattern of hyperbolas in the two-knife experiment was projected onto a viewing screen perpendicular to the light beam at a fixed distance from the knife edges.

Why a family of hyperbolas? I will examine the theory of this experiment in the following section, but first let us look at a modern reproduction of the wedge diffraction experiment that I undertook with my American colleague

Wayne Strange. The original motivation was to provide optical simulations of electron diffraction in a new type of electron microscope, discussed in detail in chapter 7.

Since no sunlight shines into the "darken'd Chamber" of my basement laboratory, we used as light source the 544 nm green line or 633 nm red line of a helium-neon laser outfitted with a 20 μm diameter pinhole spatial filter. The wedge aperture was created by bolting to an aluminum frame the razor-edged blades of two paint scrapers. Measurement showed the wedge angle to be 0.0238 radians, rather than 0.031, but this is close enough to Newton's value. By means of a translation stage with micrometer adjustment, the aperture could be swept across the detecting surface of a CCD camera employed for precision recording of the near-field diffraction pattern in the illuminated region between the blades. To record the far-field pattern, the image was projected about a meter (in the absence of the CCD surface) onto a translucent screen and photographed from the rear by a conventional single-lens reflex camera. Sometimes the camera lens was removed and the image projected directly onto the film.

As Newton accurately described, the far-field shadow region does indeed show a family of hyperbolic curves, as illustrated in the photograph of figure 5.3. Imperfections in the knife edges, which were not "ground truly strait," produced the irregularities in the fringes. In fact, introducing a single nick in a polished blade at a measured distance from the pointed end can usefully serve as a calibration mark for vertical scanning. The flared end of the image corresponds to the vertex end of the aperture. The image in figure 5.3 was recorded at a distance of $\sim$ 1 m from the aperture. To enhance the apparent fringe contrast, we also dispensed with the translucent screen and recorded the image directly on film at a distance from the aperture of approximately 1 cm. Except for demagnification, the same pattern was produced. Since only the scale, and not the shape, of the pattern varied with distance after a fraction of a centimeter from the plane of the aperture, I will continue to refer to this as the far-field image.

The muted black and white photograph does not do justice to the sharply defined, brilliant green profusion of hyperbolas that we saw; it is really quite a pretty phenomenon. Why a family of hyperbolas?

5.3 DIFFRACTION AT A WEDGE

There have been many treatments, both in textbooks and in the research literature, of light diffraction by a parallel-edged slit.[5] Because the edges are parallel—and usually assumed to have infinite extension—the problem is basically one-dimensional; the diffraction pattern in any plane perpendicular to the edges of the slit is the same as in any other perpendicular plane. It is as

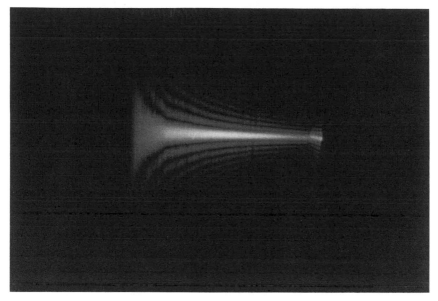

Figure 5.3. Far-field laser diffraction pattern of a wedge aperture (vertex angle of 0.0238 radian) showing a family of hyperbolic curves (recorded at $z = 1$ m on Kodak Tri-X film, asa 400; exposure time 1 sec). The vertex of the wedge is on the reader's left. Irregularities in the fringes are due to nicks or grind marks in the knife edges.

if the image on the viewing screen were a composite of "slices," each slice formed projectively by only those light rays lying in the corresponding cross section through the slit. This symmetry makes it relatively easy to determine the locations of the fringe minima by a simple "two rays at a time" geometric construction—and, indeed, one could derive an analytical formula for the entire pattern by a geometrical method only a little more elaborate.

The problem of diffraction from a wedge-shaped slit, however, is in general considerably more difficult. Lacking the Cartesian symmetry of an aperture with parallel sides, the intrinsically two-dimensional diffraction integral

$$\Psi(P) = -i\frac{\psi_o}{\lambda} \int_\Sigma \frac{e^{ik(R+r)}}{Rr} K(\varphi)\, dS \tag{1}$$

constructed by Fresnel and improved by Kirchhoff does not reduce a priori to the product of one-dimensional integrals. In Eq. (1) ψ_o is the amplitude of an incident monochromatic spherical wave (of wavelength λ and wave number $k = 2\pi/\lambda$) centered on a point source S a distance R_0 from the aperture; $\Psi(P)$ is the diffracted wave at the observation point P.[6] The integration is over all unobstructed points of the plane Σ of the diffracting aperture, as shown in

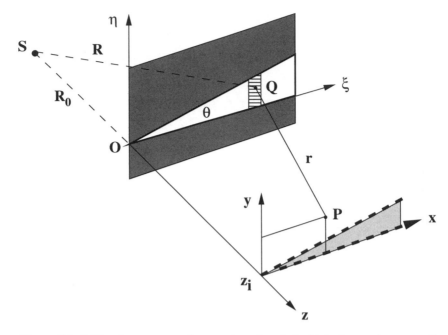

Figure 5.4. Diffraction geometry for a wedge of angle θ. Light from point source S illuminates aperture point Q which contributes to the image point P. Heavy dashed lines in the image plane denote the geometric shadow of the edges between which is the directly illuminated area (shaded region).

figure 5.4; R and r are, respectively, the variable distances of a point Q on Σ from the fixed source S and image point P.

$K(\varphi)$—where φ is the angle (not shown in the figure) between the coordinate vector $\mathbf{r}$ from Q to P and the outward normal to the aperture—is the obliquity function (derived by Kirchhoff) that takes account of the angular dependence of the light scattering. For an incident spherical wave $K(\varphi) = (1/2)(1 + \cos\varphi)$, in which $\varphi = 0$ is the forward direction of the light beam. Since $K(\pi) = 0$, the obliquity function eliminates the spurious back-reflected wave of the original Huygens-Fresnel theory. If the diffraction geometry is such that φ remains small ($\varphi < 1$), then $K(\varphi)$ is approximately unity. This is the condition assumed by Fresnel—and virtually everyone else after him who wished to obtain tractable analytical formulas.

Equation (1) is a mathematical statement of Huygens's and Fresnel's conception that illuminated points on an aperture serve as secondary sources of spherical waves that superpose at the screen to form the diffraction pattern. The factor $-i$ represents a 90° phase shift ($-i = e^{-i\pi/2}$) between the incident and diffracted waves; this relative phase plays no role in the experiments I am discussing, but its existence can nevertheless be revealed by a split-beam interference experiment employing holography.[7]

Let us consider first a wedge with two sides of length L making an angle $\theta \ll 1$. L is very much larger than any other length characterizing the light such as the wavelength, the longitudinal coherence length (in effect, the size of a "wave packet" emitted by the source), or the diameter of the illuminated region of the aperture. Coordinates $(\xi, \eta, 0)$ locate a point in the aperture plane where $\eta = 0$ defines the lower edge and $\eta - \xi\theta = 0$ defines the upper edge (see again figure 5.4). A point on the image plane, a fixed distance z from the aperture plane, is assigned coordinates (x, y, z).

It is assumed at the outset that the source-to-aperture distance and aperture-to-screen distance are large compared with the wavelength, and that both vary sufficiently slowly for points over the aperture that they are effectively constant (with $R \sim R_0, r \sim z$) in the denominator of Eq. (1). The phase of the incident spherical wave is assumed to vary sufficiently slowly over the aperture that one can approximate the incident wave by a constant amplitude $\Psi_o = (\psi_o e^{ikR_0})/R_0$. The r-dependence of the phase factor in Eq. (1) cannot be neglected, however, since the product kr can be large. Last, the diffraction pattern on the screen is assumed to lie essentially in the forward direction so that the obliquity factor is always approximately unity. Under these conditions the Fresnel-Kirchhoff integral (1) takes the form

$$\Psi(P) = -i\Psi_o \frac{e^{ikz}}{\lambda z} \int_0^L \int_0^{\xi\theta} \exp\left[i\frac{k}{2z}\{(x - \xi)^2 + (y - \eta)^2\}\right] d\eta\, d\xi \quad (2)$$

where the argument of the exponential was obtained from the binomial expansion

$$r = [z^2 + (x - \xi)^2 + (y - \eta)^2]^{1/2} \sim z\left[1 + \frac{(x - \xi)^2}{2z^2} + \frac{(y - \eta)^2}{2z^2}\right]. \quad (3)$$

The wedge angle θ, which appears in the upper limit of the η-integration, is small enough that no distinction need be made at present between the angle, its sine, or its tangent. For the wedge orientation of the figure, however, it is $\tan\theta$ that would appear exactly.

The different ways to approximate the integral in Eq. (2), which cannot be evaluated analytically in closed form, constitute the various diffraction models commonly encountered. For example, expansion of the quadratic terms in the argument of the exponential with subsequent neglect of contributions of order ξ^2, η^2 leads to the expression for Fraunhofer diffraction that ordinarily determines the far-field diffraction pattern (although, depending on geometry and wavelength, "far" may need to be very far indeed). The Fraunhofer diffraction pattern is equivalent to the spatial Fourier transform of the aperture function, a correspondence that lies at the heart of optical information processing by such techniques as spatial filtering. Retention of the quadratic phase terms constitutes the more general Fresnel diffraction.

Upon transforming the integration variables,

$$u = \sqrt{\frac{k}{2z}}(y - \eta), \tag{4a}$$

$$v = \sqrt{\frac{k}{2z}}(x - \xi), \tag{4b}$$

and taking the wedge to be of infinite length ($L \to \infty$)—a reasonable simplification since the actual illumination in the two-knife experiment does not extend to the extremity of the wedge—one can put Eq. (2) into the form

$$\Psi(P) = -i\frac{\Psi_o}{\pi}e^{ikz}\left[\int_{-\infty}^{v_0}e^{iv^2}\,dv\int_0^{u_0}e^{iu^2}\,du - \int_{-\infty}^{v_0}e^{iv^2}\int_0^{w_0+\theta v}e^{iu^2}\,du\,dv\right] \tag{5}$$

where the terms in the upper limits are defined by

$$u_0 = \sqrt{\frac{\pi}{\lambda z}}y, \tag{6a}$$

$$v_0 = \sqrt{\frac{\pi}{\lambda z}}x, \tag{6b}$$

$$w_0 = \sqrt{\frac{\pi}{\lambda z}}(y - \theta x). \tag{6c}$$

Were it not that the upper limit in the rightmost integral of Eq. (5) contains the integration variable v, the diffracted wave could have been written exclusively in terms of a standard Fresnel integral

$$F(x) = \int_0^x e^{it^2}\,dt. \tag{7}$$

It is perhaps more common to see the integral defined with a factor $\pi/2$ in the exponent, but I prefer the simpler form for numerical evaluation. The following properties of $F(x)$ are not difficult to demonstrate and will be useful in the ensuing discussion:

$$F(-x) = -F(x), \tag{8a}$$

$$F(x) \sim \sqrt{\frac{i\pi}{4}} + \frac{e^{ix^2}}{2ix} \quad (|x| > 1) \tag{8b}$$

and, as a limiting case of Eq. (8b),

$$F(\infty) = \sqrt{\frac{i\pi}{4}}. \tag{8c}$$

For both Newton's experiments and the modern rendition described in the previous section, in which the wavelength is on the order of 5×10^{-5} cm, the factor $\kappa \equiv (\pi/\lambda z)^{1/2}$ is sufficiently large, whether the detector is placed close to the aperture ($z \sim 2$ cm; $\kappa \sim 170$ cm^{-1}) or reasonably far from the aperture ($z \sim 200$ cm; $\kappa \sim 17$ cm^{-1}), that u_0, v_0, and w_0 exceed unity for nearly all observation points—in the shadow regions as well as illuminated area—not directly at the geometric projection of the edges. Examining the double integral in Eq. (5) in light of the assumption that θ is a small angle, one sees that the major contribution to the integral is generated within the range $0 < u < w_0$. Although v may become so large that θv is no longer a small quantity, the phase factor $\exp(i v^2)$ in the integrand then oscillates so rapidly that its overall contribution for $u > w_0$ is small. Thus, the wave function (5) can be recast in the form

$$\Psi(P) = -i \frac{\Psi_o}{\pi} e^{ikz} \left[\left(\sqrt{\frac{i\pi}{4}} + F(v_0) \right) (F(u_0) - F(w_0)) - H(v_0, w_0, \theta) \right] \quad (9)$$

where the last term in brackets is defined by

$$H(v_0, w_0, \theta) = \int_{-\infty}^{v_0} \int_{w_0}^{w_0 + \theta v} e^{i(u^2 + v^2)} \, du \, dv. \quad (10)$$

In obtaining Eq. (9), I have used Eq. (8a) to integrate v over the portion of the range from $-\infty$ to 0, as well as the fact that v_0 is always greater than 0 (since there is no image at negative values of the longitudinal coordinate x). The transverse coordinate y, and hence u_0, can be positive or negative depending upon whether one examines the shadow region of the lower or upper blade, respectively.

Since $u_0 = 0$ and $w_0 = 0$ denote the edges of the wedge projected onto the observation plane, the first term within the bracket of Eq. (9) looks as if it were a linear superposition of two waves—one scattered from each knife edge—modulated by a function of the longitudinal distance from the vertex (specified by v_0 or x). Neglecting the function H—an omission that subsequent calculation and comparison with measurement will show to be well justified for the conditions of the Newton two-knife experiment—note that the diffraction pattern in any vertical plane ($x = $ constant) is effectively the same as that of a parallel-edge slit of corresponding fixed width $d = x\theta$. [For an infinite parallel-edge slit, one applies Eq. (9) with $H = 0$, $w_0 = \kappa(y - d)$, and $F(v_0 = \infty) = \sqrt{i\pi/4}$.] I will refer to this as the projective approximation. Considering that a laterally *asymmetric* aperture is broadly illuminated by a coherent wavefront from all points of which (according to the mathematical formalism of Fresnel-Kirchhoff) spherical wavelets emerge, this projective characteristic of the image is by no means an obvious result, and I will discuss this feature again later.

By a judicious transformation from Cartesian to plane polar coordinates one can, to good approximation, reduce the wedge double integral H to an expression containing only Fresnel single integrals. Leaving the details of this procedure to the original article (see note 2), I will simply record the final result:

$$H(v_0, w_0, \theta)$$

$$\sim \frac{e^{iw_0^2}}{2i\,w_0} \left[e^{-i(w_0\theta)^2} F(v_0 + w_0\theta) - F(v_0) - \frac{\sqrt{i\pi}}{2} \left(1 - e^{-i(w_0\theta)^2} \right) \right]. \quad (11)$$

Numerical evaluations of H by computer confirm that Eq. (11) holds reasonably well for $(v_0, w_0) > 1$ and $\theta < 1$. The restriction on θ is not severe for it encompasses fairly sizable angles. For example, $\theta = 0.5$ radians (with $\sin\theta \sim 0.48$) corresponds to a wedge of $28.6°$, an angle greatly exceeding that employed by Newton.

5.4 THE FAR-FIELD SHADOW

The shadow region of the image plane, as shown in figure 5.4, is defined by $y < 0$ and $y - x\theta > 0$. Let us examine the far-field pattern in the dark region above the upper knife edge for which both u_0 and w_0 are greater than 0. Because the factor κ is in general large for visible light and macroscopic distance z, the preponderant contribution to the image occurs for $w_0 > 1$, in which case the approximation (8b) can be employed to estimate the pertinent Fresnel integrals. In what follows, I assume $\theta < 1$ and neglect H.

The argument of each Fresnel integral in Eq. (9) being positive, the difference $F(u_0) - F(w_0)$ reduces to

$$F(u_0) - F(w_0) = \frac{1}{2i} \left(\frac{e^{iu_0^2}}{u_0} - \frac{e^{iw_0^2}}{w_0} \right). \quad (12)$$

Substitution of Eq. (12) into Eq. (9), factorization of the resulting expression to obtain explicitly terms that depend only on relative phases, and use of the identity $i^{-3/2} = -e^{i\pi/4}$ lead to the waveform

$$\Psi(P) \sim -\frac{\Psi_o}{2\sqrt{\pi}} e^{i\Omega} \left[1 - \frac{e^{i(v_0^2 + \frac{\pi}{4})}}{2\sqrt{\pi}\,v_0} \right] \left[\frac{1}{u_0} - \frac{e^{i(w_0^2 - u_0^2)}}{w_0} \right] \quad (13)$$

where Ω is an unimportant global phase. Upon evaluating the difference $w_0^2 - u_0^2$ (with neglect of a term proportional to $\theta^2 x^2$), one finds that the light intensity

$I(P) = |\Psi(P)|^2$ at image point P takes the form

$$I(P) \sim \frac{I_o}{4\pi^2} \left(\frac{\lambda z}{y^2}\right) \left[1 - \frac{\sqrt{\lambda z}}{\pi x} \cos\left(\frac{\pi x^2}{\lambda z} + \frac{\pi}{4}\right)\right]$$

$$\times \left[1 + \left(\frac{y}{y - \theta x}\right)^2 - 2\frac{y^2}{y(y - \theta x)} \cos\left(2\pi \frac{xy\theta}{\lambda z}\right)\right]. \quad (14a)$$

Deep within the shadow region, the transverse image coordinate y can greatly exceed the wedge width θx at the longitudinal distance x from the vertex; the second bracketed expression in Eq. (14a) then simplifies further, leading to

$$I(P) \sim \frac{I_o}{4\pi^2} \left(\frac{2\lambda z}{y^2}\right) \left[1 - \frac{\sqrt{\lambda z}}{\pi x} \cos\left(\frac{\pi x^2}{\lambda z} + \frac{\pi}{4}\right)\right] \left[1 - \cos\left(2\pi \frac{xy\theta}{\lambda z}\right)\right]. \quad (14b)$$

In arriving at Eqs. (14a, b), a negligibly small term $1/(4\pi v_0^2)$ was dropped in the first bracketed expression. Although x and y appear in denominators, expression (14a) cannot become singular, for the assumed conditions do not permit the values $x = 0$ or $y < \theta x$. It is worth emphasizing in any event that the general relation (9) remains finite for all image points $P(x, y, z)$.

Examining Eqs. (14a, b), one sees that the argument of the (small-amplitude) cosine term in the first bracket (the longitudinal modulation) undergoes rapid oscillation, whereas the argument of the (larger-amplitude) cosine term in the second bracket varies much more slowly because θ is small. The observable fringe pattern, therefore, is attributable almost exclusively to the second bracket, which is characteristic of two-slit interference. One may think of the process as the superposition and interference of waves scattered from the two knife edges (for there is no direct incident wave in the shadow region). To good approximation the fringe minima occur when the argument of the cosine is an integer multiple of 2π, or

$$xy\theta = n\lambda z \quad (n = 1, 2, 3, \ldots) \quad \text{[Minima]}. \quad (15a)$$

Correspondingly, fringe maxima are given approximately by

$$xy\theta = (n - \tfrac{1}{2})\lambda z \quad (n = 1, 2, 3, \ldots) \quad \text{[Maxima]}. \quad (15b)$$

Expressions (15a, b) represent a family of hyperbolas with asymptotes $x = 0$, $y = 0$. In figure 5.5 are shown two surface graphs of the far-field pattern calculated directly from Eq. (9) with numerical evaluation of the Fresnel integrals. The first (figure 5.5a) surveys the total diffraction pattern including the directly illuminated area (which, as in the experiment with laser light, appears bright, diffuse, and structureless); the second (figure 5.5b) gives a detailed view of just the shadow region of the upper knife, thereby avoiding the illuminated region and showing the hyberbolic fringes with greater contrast.

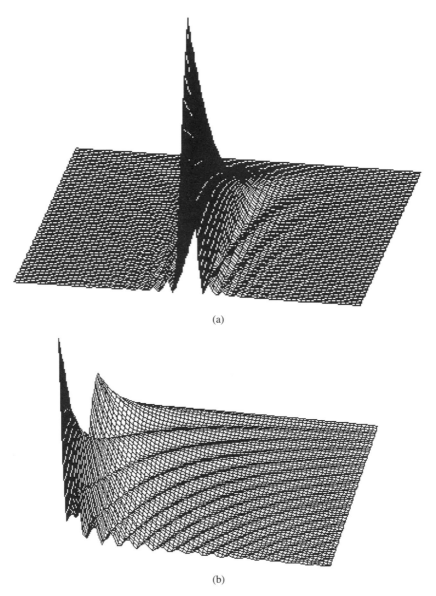

(a)

(b)

Figure 5.5. Surface graphs of the far-field diffraction pattern showing hyperbolic fringes. Scan region (in cm) (a) ($0 \leq x \leq 5$; $-1 \leq y \leq +1$) shows full symmetry, but low fringe contrast; (b) ($0 \leq x \leq 5$; $0.2 \leq y \leq 1.5$) shows heightened fringe visibility in absence of directly illuminated area. Plot parameters: $\lambda = 633$ nm, $\theta = 0.031$ radian, $z = 200$ cm.

It is instructive to consider how much of the Fresnel-Kirchhoff integral (2) needs to be retained in order to deduce the pattern of hyberbolic fringes. For sufficiently large image distance z (on the order of several meters for visible light and $x <$ wedge length $L \sim 10$ cm) such that

$$\frac{x^2\theta^2}{2\lambda z} \ll 1 \tag{16}$$

the quadratic contribution $k\eta^2/2z$ to the phase in Eq. (2) can be neglected (Fraunhofer approximation). The integral over η is then easily performed and leads to the waveform (with subscript F for Fraunhofer/Fresnel)

$$\Psi_F(P) = \frac{\Psi_o e^{i\Omega}}{2\pi^{3/2}} \left(\frac{\sqrt{\lambda z}}{y}\right) \left[A_1 e^{-iv_1^2} - A_0 e^{-iv_0^2}\right] \tag{17}$$

where

$$v_1 = \sqrt{\frac{\pi}{\lambda z}}(x + y\theta) \tag{18}$$

and

$$A_0 = \int_{-\infty}^{v_0} e^{iu^2}\,du, \qquad A_1 = \int_{-\infty}^{v_1} e^{iu^2}\,du. \tag{19}$$

(Note, too, that v_1, unlike w_0 in Eq. [6c], does not correspond to the geometric shadow of the knife edge.) Since image points of interest in the shadow region ordinarily satisfy $x \gg y\theta$ for small θ, one can set $A_1 \sim A_0$. The resulting intensity distribution

$$I_F(P) = \frac{I_o}{4\pi^3} \left(\frac{2\lambda z}{y^2}\right) |A_0|^2 \left[1 - \cos\left(2\pi\frac{xy\theta}{\lambda z}\right)\right]$$

$$= \frac{I_o}{\pi^3} \left(\frac{\lambda z}{y^2}\right) |A_0|^2 \sin^2\left(\frac{\pi xy\theta}{\lambda z}\right) \tag{20}$$

is equivalent to Eq. (14b) upon use of Eq. (8b) to evaluate the longitudinal Fresnel integral.

Except for the Fraunhofer condition (16), no restriction has been placed on the lower limits of the x and y coordinates in Eq. (20). The integral is not singular but leads to a maximum value as $y \to 0$. This hybrid result—longitudinal Fresnel diffraction, lateral Fraunhofer diffraction—reproduces the correct far-field shadow pattern but does not give an accurate description of the near-field image in the directly illuminated region.

At this point the reader may be inclined to believe that the fringe pattern in the shadow region is principally a consequence of Fraunhofer diffraction from a narrow slit, albeit one with variable width. This is not really so, however; variation in curvature of the wavefront along the longitudinal dimension of the wedge is a critical component, as will be seen next.

For a wedge of finite length L, the corresponding image distance beyond which one would expect the quadratic phase contribution $k\xi^2/2z$ to be negligible is given by

$$\frac{L^2}{2\lambda z} \ll 1. \tag{21}$$

In general, this condition cannot be met for visible wavelengths and "tabletop" image distances. For example, for a wedge 10 cm long illuminated by light of wavelength $\sim 5 \times 10^{-5}$ cm, condition (21) would require $z > \sim 10$ km. In the case of a very short wedge, or upon use of a lens to simulate a remote image plane, the far-field diffraction pattern would be given by the Fourier transform of the aperture function (as pointed out in the preceding section). In that case all wavefront curvature (contributions of ξ^2 and η^2 to the phase) is neglected, and to within an unimportant global phase factor the waveform becomes (with FT for Fourier transform)

$$\Psi_{\mathrm{FT}}(P) = \frac{\Psi_o L}{2\pi y} \left[\frac{\sin q_1}{q_1} e^{-iq_1} - \frac{\sin q}{q} e^{-iq} \right] \tag{22}$$

where

$$q = \frac{\pi L x}{\lambda z}, \qquad q_1 = \frac{\pi L}{\lambda z}(x + y\theta). \tag{23}$$

The corresponding intensity $I_{\mathrm{FT}}(P)$ is

$$I_{\mathrm{FT}}(P) = \frac{I_o L^2}{4\pi^2 y^2} \left[\left(\frac{\sin q_1}{q_1}\right)^2 + \left(\frac{\sin q}{q}\right)^2 - \frac{2 \sin q_1 \sin q}{q_1 q} \cos\left(\frac{\pi y\theta L}{\lambda z}\right) \right] \tag{24a}$$

which, for $x > y\theta$, simplifies to

$$I_{\mathrm{FT}}(P) = \frac{I_o L^2}{\pi^2 y^2} \left(\frac{\sin\left(\frac{\pi L x}{\lambda z}\right)}{\frac{\pi L x}{\lambda z}} \right)^2 \sin^2\left(\frac{\pi y\theta L}{2\lambda z}\right). \tag{24b}$$

The true Fraunhofer diffraction pattern represented by Eqs. (24a, b) does *not* show hyperbolic fringes. Rather, $I_{\mathrm{FT}}(P)$ vanishes for field points P such that y and x independently satisfy

$$y = \frac{2n\lambda z}{L\theta}, \qquad n = 1, 2, 3, \dots \quad \text{(for all } x\text{)}, \tag{25a}$$

$$x = \frac{n\lambda z}{L}, \qquad n = 1, 2, 3, \dots \quad \text{(for all } y\text{)}. \tag{25b}$$

In other words, for a fixed image distance z, the fringes in the Fraunhofer pattern constitute an orthogonal rectilinear grid of lines $y = \text{constant}$, $x = \text{constant}$.

5.5 THE NEAR-FIELD ILLUMINATED REGION

Returning to Eq. (9), we again make the projective approximation (neglect H) and limit the following discussion to the directly illuminated region defined by $y > 0$ and $y - \theta x < 0$—or, equivalently, $u_0 > 0$ and $w_0 < 0$. Figure 5.6 shows surface graphs of the near-field pattern beginning with the high "summit" (figure 5.6a) in the vicinity of the vertex and evolving into a craggy range of peaks and valleys (figure 5.6b) as the image widens. Over the longitudinal domain surveyed ($0 \leq x \leq 2$ cm) lateral Fresnel fringes fill the illuminated region. The figures were obtained by exact numerical evaluation of the Fresnel integrals in Eq. (9).

For image points sufficiently distant from the geometric shadow of the knife edges that $|w_0| > 1$, Eq. (8b) can again be used to evaluate the Fresnel integrals analytically. In this case, the arguments have opposite signs (see Eq. [8a]), and the difference $F(u_0) - F(w_0)$ becomes

$$F(u_0) - F(w_0) = \sqrt{i\pi} + \frac{1}{2i}\left(\frac{e^{iu_0^2}}{u_0} + \frac{e^{iw_0^2}}{|w_0|}\right). \tag{26}$$

Now the diffracted wave takes the form

$$\Psi(P) \sim \Psi_o e^{ikz}\left[1 - \frac{e^{i(v_0^2+\frac{\pi}{4})}}{2\sqrt{\pi}\,v_0}\right]\left[1 + \frac{1}{2\sqrt{\pi}}\left(\frac{e^{i(u_0^2+\frac{\pi}{4})}}{u_0} + \frac{e^{i(w_0^2+\frac{\pi}{4})}}{|w_0|}\right)\right]. \tag{27}$$

If we drop, as before, a small term proportional to $1/v_0^2$ and phase contributions of order $x^2\theta^2$, Eq. (27) leads to the the near-field intensity $I(P)$:

$$I(P) \sim I_0\left[1 - \frac{\sqrt{\lambda z}}{\pi x}\cos\left(\frac{\pi x^2}{z\lambda} + \frac{\pi}{4}\right)\right]$$

$$\times\left[1 - \frac{\sqrt{\lambda z}}{\pi}\left(\frac{\cos\left(\frac{\pi y^2}{\lambda z} + \frac{\pi}{4}\right)}{y} + \frac{\cos\left(\frac{\pi(\theta x - y)^2}{\lambda z} + \frac{\pi}{4}\right)}{\theta x - y}\right)\right.$$

$$\left. + \left(\frac{\lambda z}{4\pi^2}\right)\left(\frac{1}{y^2} + \frac{1}{(\theta x - y)^2} + \frac{2\cos\left(\frac{2\pi xy\theta}{\lambda z}\right)}{y(\theta x - y)}\right)\right]. \tag{28}$$

The diffraction pattern expressed by Eq. (28) is complicated (as one might expect from figures 5.6a, b) containing contributions representing Fresnel fringes near a straightedge, as well as hyperbolic fringes attributable to interference of light from both edges.

As a simple and edifying special case, consider the diffraction pattern along the symmetry axis of the wedge, $y - \theta x/2 = 0$. Neglecting the small, highly

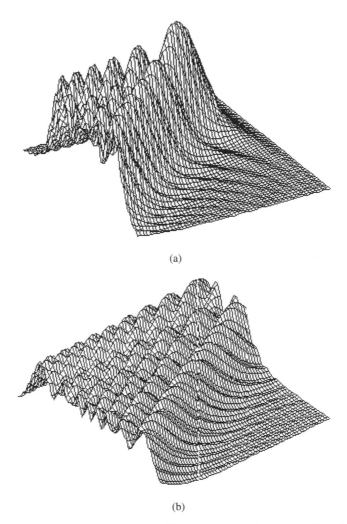

(a)

(b)

Figure 5.6. Surface graphs of near-field diffraction pattern. Scan region (in cm) (a) ($0 \leq x \leq 1.5$; $-0.01 \leq y \leq 0.05$) shows vicinity of vertex with interior fringes bordering edges and exterior hyperbolic fringes; (b) ($1 \leq x \leq 2$; $0 \leq y \leq 0.06$) shows progressive development of interior fringes as wedge widens. Plot parameters: $\lambda = 633$ nm, $\theta = 0.031$ radian, $z = 2$ cm.

oscillatory term in the first bracket and a small term proportional to $1/x^2$ in the second bracket, one can reduce Eq. (28) to the expression

$$I(P) = I_0 \left[1 - \frac{4\sqrt{\lambda z}}{\pi \theta x} \cos \left(\frac{\pi x^2 \theta^2}{4\lambda z} + \frac{\pi}{4} \right) \right] \qquad (29)$$

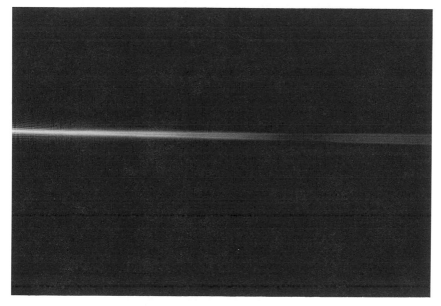

Figure 5.7. Photograph of near-field diffraction pattern showing alternation between bright and dark fringes of the first two maxima and minima along the symmetry axis $(y - x\theta/2 = 0)$. The wedge (of angle 0.0238 radian) widens from left to right. (Two-second exposure without lens on Kodak Tri-X film, asa 400.)

which predicts a slow periodic variation between bright and dark fringes along the optical axis. An example of this behavior is illustrated in the photograph of figure 5.7. The minima (dark fringes) of Eq. (29) occur when the argument of the interference term is an integer multiple of 2π. This occurs at distances x_n from the vertex specified by

$$\theta x_n = \sqrt{8\lambda z \left(n - \frac{1}{8}\right)}, \qquad n = 1, 2, 3, \dots \quad \text{(Minima)}. \qquad (30a)$$

Similarly, there are bright fringes at locations X_n where the value of the argument is π to within an additive multiple of 2π:

$$\theta X_n = \sqrt{8\lambda z \left(n - \frac{5}{8}\right)}, \qquad n = 1, 2, 3, \dots \quad \text{(Maxima)}. \qquad (30b)$$

The predictions of (30a, b) are borne out in figure 5.8, which shows the variation in axial intensity as calculated directly from Eq. (9) without futher approxima-

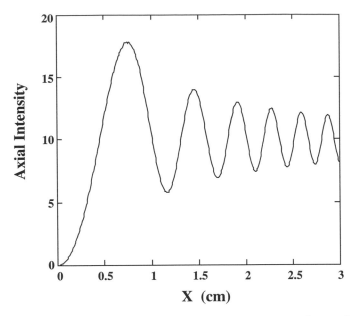

Figure 5.8. Theoretical variation in light intensity along symmetry axis over the range $(0 \leq x \leq 3 \text{ cm})$. Plot parameters: $\lambda = 544$ nm, $\theta = 0.0238$ radian, $z = 2$ cm.

tion. The visibility (or contrast) of the fringes of an interference pattern is defined by the relation

$$V = \frac{I_{max} - I_{min}}{I_{max} + I_{min}} \tag{31}$$

in which I_{max} and I_{min} refer to adjacent maximum and minimum intensities. It then follows from Eqs. (30a, b) that the visibility of the n-th fringe along the symmetry axis is

$$V_n = \frac{\dfrac{2\sqrt{\lambda z}}{\pi \theta}\left(\dfrac{1}{X_n} + \dfrac{1}{x_n}\right)}{1 + \dfrac{2\sqrt{\lambda z}}{\pi \theta}\left(\dfrac{1}{X_n} - \dfrac{1}{x_n}\right)}. \tag{32}$$

With the apparatus at hand it was difficult to determine the absolute light intensity—Newton did not do it either—and therefore to estimate reliably the visibility of the axial fringes. However, for 544 nm light and a wedge with vertex angle 0.0238 radians, measurements of the fringe locations in centimeters ($X_1 = 0.742$, $x_1 = 1.039$, $X_2 = 1.333$, $x_2 = 1.584$) corresponded fairly closely to the maxima and minima predicted from Eqs. (30a, b) ($X_1 = 0.759$, $x_1 = 1.160$; $X_2 = 1.454$, $x_2 = 1.697$); the discrepancies are attributable principally to the uncertainty of visual inspection. Predicted visibilities of $V_1(X_1, x_1) = 54\%$ and $V_2(X_2, x_2) = 34.7\%$ are consistent with the experimental contrast of the CCD camera images observed on a video monitor.

Relations (30a, b) may be understood up to an indeterminate phase on the basis of a simple construction shown in figure 5.9 (such as Fresnel employed in his earlier ray theory of diffraction), where it is supposed—although as yet not justified—that the diffraction pattern is produced by the superposition of a direct plane wave and a wave scattered from each knife edge. Looked at in this way, this near-field diffraction pattern constitutes an in-line hologram of the aperture, such as first proposed by Dennis Gabor for lensless electron microscopy.[8] According to this perspective, maxima occur along the axis when the relative phase between the edge waves and the direct wave is an integer multiple of 2π. The relative phase is due, in part, to the optical path-length difference $\Delta s = \ell - z$ (where ℓ is the distance from the edge to point P) and in part to a phase δ arising from the edge scattering itself. Because of the symmetry of the configuration, the phase of each edge wave relative to the direct wave at P is the same. From the geometry of figure 5.9 it follows that maxima should occur at locations $X_n = \sqrt{8\lambda z(n - \delta/2\pi)}$, which agrees with Eq. (30b) up to the unknown phase δ (which cannot be inferred by kinematics, but must be deduced from a dynamical theory such as that of Fresnel–Kirchhoff).

Figures 5.10a–d show examples of experimental near-field wedge diffraction patterns obained by scanning the CCD camera vertically across the lower knife edge at fixed distances x from the junction. The four selected values of x correspond to the locations of the first two axial maxima and minima (as shown in figure 5.8). The horizontal scale of the figures—which corresponds to the vertical scans in the image plane—is 100 μm/cm. To minimize slight shape distortions arising from vibration of the aperture mount, each image is the average of fifty scans.

The origin of the longitudinal coordinate axis was determined by initially setting the camera to scan in a vertical plane preceding the intersection of the knife edges (where the image is expected to be totally dark) and gradually translating the scan plane toward the junction in steps of 100 μm while observing the resulting images on a video monitor. The origin $x = 0$ was set at the first registration of a clear signal distinguishable from a weak, diffuse background leakage through the junction. Subsequent x coordinates (and corresponding wedge widths $d = x\theta$) were measured with respect to this origin.

Computer simulations of the experimental patterns, based on Eq. (9) with exact numerical evaluation of the Fresnel integrals, are given in figures 5.11a–d. For the small vertex angle of 0.0238 radians, the residual wedge double integral H makes no discernible contribution at the scale of the figures. Calculations were made for a narrow range of x about the nominal experimental value at which each pattern was scanned; the figures shown are those which best matched (by visual inspection) the corresponding recorded image. Given the experimental uncertainties in location of the image plane ($\Delta z \sim 2$ mm as a result of the recessed CCD surface) and in the origin of the horizontal axis

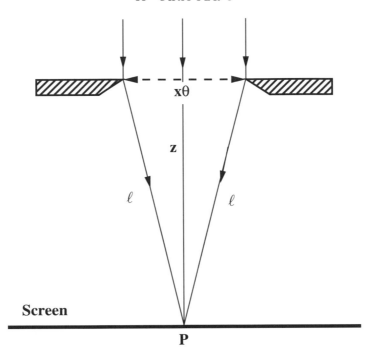

Figure 5.9. Cross section through wedge (at width $d = x\theta$) showing edge waves superposing at point P on the symmetry axis of the diffraction pattern. Maxima and minima can be determined (up to an unknown relative phase) from the optical path difference $2\pi(\ell - z)/\lambda$.

($\Delta x \sim 1.5$ mm as a result of displacement of the optical mount), theoretical and experimental images are in excellent agreement.

In the course of examining many such experimental and theoretical patterns, I could not help wondering: How closely do the diffraction patterns of a narrow wedge, scanned vertically at different distances x from the vertex, match the corresponding patterns made with a true parallel-edge slit of adjustable width $d = x\theta$? This comparison was made in a separate set of experiments, and in all cases (i.e., different values of x) measurement showed that the two patterns were strikingly close with only slight differences (attributable principally to the uncertainty in wedge width in the scan plane) in fringe height or depth, but not fringe number or location. The number and contrast of the fringes were in fact highly sensitive to the width of the slit or wedge, all the more so as the width increased.

Look, for example, at figures 5.12a–b, which show the observed near-field pattern of a parallel-edge slit of width 0.08 ± 0.0015 cm at the image plane $z = 2.0 \pm .04$ cm and the corresponding theoretical pattern, calculated for both slit and wedge apertures of width 0.0813 cm at the image plane $z = 2.00$

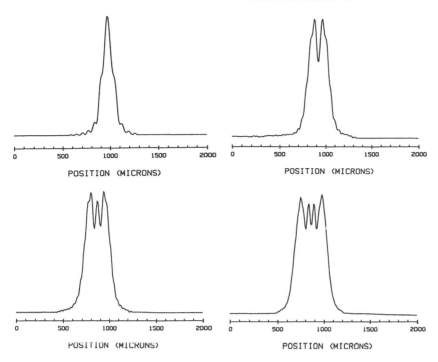

Figure 5.10. Experimental near-field diffraction pattern (average of 50 vertical scans) at distances x (in cm) and widths $d = x\theta$ corresponding to the first two maxima and minima along the symmetry axis. (a) $x = 0.742$, $d = 0.018$; (b) $x = 1.039$, $d = 0.025$; (c) $x = 1.333$, $d = 0.032$; (d) $x = 1.584$, $d = 0.038$. Lateral scale is 100 μm/cm. Experimental parameters: $\theta = 0.0238$ radians, $\lambda = 544$ nm, $z = 2.0$ cm from aperture to image plane.

cm. At the scale of the figure the two superposed theoretical patterns are indistinguishable, and both closely replicate the amplitudes and locations of the fifteen fringes displayed in the experimental pattern. Vary the theoretical values of x and z by as little as $\Delta x \sim 10^{-4}$ cm and $\Delta z \sim 10^{-2}$ cm, and agreement of the experimental and simulated diffraction patterns becomes noticeably poorer. The pattern of fringes is even more sensitive to geometry for the larger slit width $d = 0.10$ cm but is not reproduced here because the larger number of fringes (twenty-three) would have resulted in a figure too compressed after photoreduction.

This sensitivity to geometry can be useful. Under circumstances where scalar diffraction theory is well understood and applicable, observation of a Fresnel pattern can serve to calibrate the measurement of small lengths, a procedure of particular interest in the measurement of subnanometer-sized structures by electron interferometry.

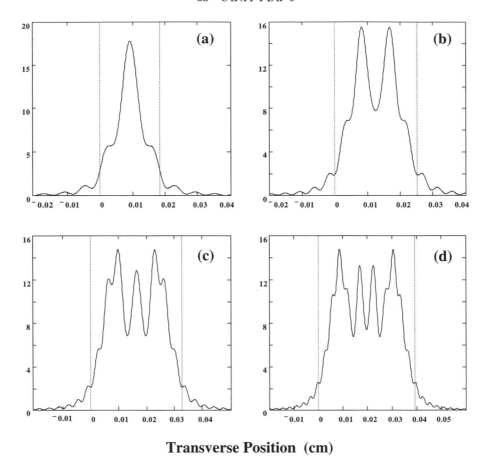

Figure 5.11. Computer simulations of near-field diffraction patterns corresponding to figure 5.10 for x (cm) = (a) 0.77, (b) 1.05, (c) 1.37, (d) 1.64. The vertical intensity scale is arbitrary; the horizontal scale is y in cm. Vertical lines denote the projections of the knife edges. Plot parameters: $\lambda = 544$ nm, $\theta = 0.0238$ radian, $z = 2$ cm.

5.6 WIDE WEDGES

Although the two-knife experiments to this point employed a narrow wedge aperture ($\theta \sim 10^{-2}$ rad $\sim 1°$), let us consider briefly the diffraction of coherent illumination by a wide wedge—by which I still mean $\theta < 1$ rad $\sim 50°$, in which case the approximation $\sin \theta \sim \theta$ remains reasonably valid. It now becomes more convenient to identify the longitudinal (x) axis with the symmetry axis of the wedge, rather than with the lower knife edge, so that vertical scans ($x = $ constant) of the diffraction pattern lead to symmetric images.

The theoretical description of this configuration proceeds just as in sec-

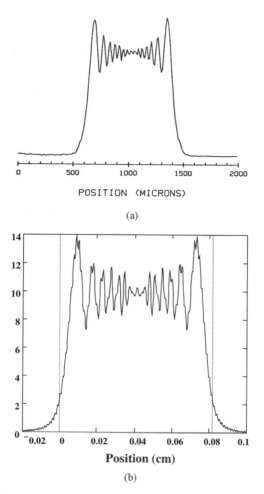

Figure 5.12. (a) Experimental near-field diffraction pattern of a parallel-edge slit of width 0.08 cm (horizontal scale 100 μm/cm); (b) Theoretical pattern for width 0.0813 cm. Parameters: $\lambda = 544$ nm, $\theta = 0.0238$ radian, $z = 2$ cm. At the scale of the figure, (b) is indistinguishable from the diffraction pattern calculated for a vertical section through a wedge of the same theoretical width. The vertical intensity scale is arbitrary; the horizontal scale is y in cm.

tion 5.3, except that the integral over η in Eq. (2) runs from $-\xi \tan(\theta/2)$ to $+\xi \tan(\theta/2)$. In place of Eq. (9), one then obtains

$$\Psi(P) = -i \frac{\Psi_o}{\pi} e^{ikz} \left[\left(\sqrt{\frac{i\pi}{4}} + F(v_0) \right) (F(w_-) - F(w_+)) - G(v_0, w_\pm, \theta) \right] \quad (33)$$

where

$$G(v_0, w_{\pm}, \theta) = \int_{-\infty}^{v_0} \int_{w_+}^{w_+ + v \tan \frac{\theta}{2}} e^{i(u^2 + v^2)} \, du \, dv$$

$$- \int_{-\infty}^{v_0} \int_{w_-}^{w_- - v \tan \frac{\theta}{2}} e^{i(u^2 + v^2)} \, du \, dv \qquad (34)$$

and

$$w_{\pm} = \sqrt{\frac{\pi}{\lambda z}} \left(y \mp x \tan \frac{\theta}{2} \right) \qquad (35)$$

are the dimensionless parameters defining the lower ($w_- = 0$) and upper ($w_+ = 0$) boundaries of the projected wedge. For $\theta < 1$, Eq. (34) is equivalent to

$$G(v_0, w_{\pm}, \theta) = H \left(v_0, w_+, \frac{\theta}{2} \right) - H \left(v_0, w_-, \frac{-\theta}{2} \right) \qquad (36)$$

with the functional form of H defined by Eq. (10) and approximated analytically by the expression (11).

The diffraction pattern of a wide wedge aperture is markedly different from that of a narrow aperture, as shown by the surface graph in figure 5.13 of the far-field ($z = 2$ m) pattern calculated from Eqs. (33)–(36) for a wedge angle of $20°$ (0.349 rad). The length of the displayed pattern is 5 cm. Except for bright fringes close to the shadow boundaries, the image resembles for the most part a uniformly illuminated wedge. The visibility of the hyperbolic fringes in the shadow region is too low to be discernible at the same scale. Such fringes are still marginally present, however, and can be revealed by mathematically "filtering" out a portion of the directly illuminated region to enhance contrast.

The theory outlined above reproduced reasonably well the features of the wide wedge diffraction patterns observed in the laboratory. Figure 5.14, for example, shows a photograph of the diffraction pattern of a $20°$ wedge recorded at $z = 50$ cm from the aperture. An exposure time of 1 second on asa 400 film reveals in detail the straight fringes bordering the edges in the directly illuminated area as well as weak hyperbolic fringes in the vicinity of the vertex. Close scrutiny of the pattern confirms, even more convincingly than does a narrow wedge, Newton's assertion that the straight fringes along the shadow of each knife edge cross one another as they approach the vertex and bend hyperbolically toward the asymptote (along the y-axis).

Further details of wide wedge diffraction may be sought in the cited paper, but the essential results can be summarized as follows. As the angle of the aperture widened, the visibility of the hyperbolic fringes in the shadow region decreased; at $30°$ and $40°$, for example, no such fringes could be seen in the vicinity of the vertex. The differences between the lateral ($x = $ constant) fringe patterns for wedge and parallel-edge geometries became more accentuated—although, surprisingly, the similarity was still striking. These effects were more

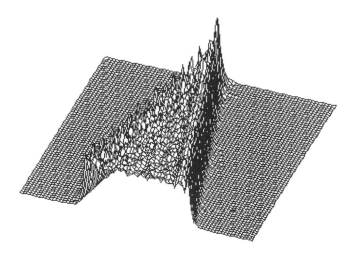

Figure 5.13. Surface graph of the far-field diffraction pattern of a wedge with $\theta = 20°$. Scan region (cm): $0 \le x \le 5$; $|y| - 5\tan(\theta/2) \le 1$. Parameters: $\lambda = 544$ nm, $z = 200$ cm.

Figure 5.14. Photograph of the diffraction pattern of a $20°$ wedge aperture at $z \sim 50$ cm. Recorded on Kodak Tri-X film, asa 400, at 1-second exposure through $f2.8$ lens. Note the Fresnel fringes along the shadow borders and the faint hyperbolic lines close to the vertex.

discernible at greater distances from the vertex where the diffraction pattern was increasingly oscillatory.

Although the question of what is light—wave or particle—has long ago been resolved (it is neither!), Newton's two-knife experiment and the extension to wide wedges nevertheless raise some fundamental questions about the behavior of light that I address in the next section.

5.7 THE SHAPE OF DIFFRACTING WAVES

There are at least two striking features to the diffraction pattern of a long narrow wedge aperture: (1) a family of hyperbolic fringes in the shadow region, and (2) a pattern of straight fringes in the illuminated region remarkably similar to that of a long parallel-edge slit. Although confirmed by experiment and deduced from Fresnel-Kirchhoff scalar diffraction theory, these features are nonetheless puzzling. Is there some simple visualizable way to understand them?

One more-or-less obvious approach to explain the hyperbolic fringes is based on the treatment of Fraunhofer diffraction from a long parallel-edge slit, such as is commonly discussed in elementary physics textbooks like that of Halliday and Resnick.[9] This is the simple heuristic interpretation to which I referred in the introductory remarks of section 5.1. If φ is the angle (small compared to 1 radian), measured with respect to the forward direction (z-axis), at which an image point P on a distant screen is subtended at the center of a slit aperture of width d, then the points of zero light intensity are given by the well-known relation

$$d \sin \varphi = n\lambda \qquad (n = 1, 2, 3, \ldots). \tag{37}$$

Let us assume that Eq. (37) holds as well in each vertical plane ($\xi = x = $ constant) perpendicular to the wedge (as oriented in figure 5.4). The factor $\sin \varphi \sim \tan \varphi$ is effectively y/z. The width of the wedge at a distance ξ from the vertex is $d = \xi\theta$. Equation (37) is then equivalent to

$$\xi y \theta = n\lambda z \qquad (n = 1, 2, 3, \ldots). \tag{38}$$

Note that ξ is an aperture coordinate, whereas y and z are image coordinates. If one can replace ξ by x—that is, if, as assumed, the diffraction pattern is formed projectively in each vertical plane—then Eqs. (37) and (15a) are equivalent. But why should the pattern be formed projectively, for the wedge, unlike the infinitely long slit of fixed width, is laterally asymmetric?

Apart from justification of projective imaging, the foregoing heuristic explanation has some logical flaws. In the elementary textbooks, such as the one just cited, the derivation of Eq. (37) is ordinarily made with *plane* waves. The authors preface their analysis of the single slit with the statement "Although Fraunhofer diffraction is a limiting case of the more general Fresnel diffraction, it is an important limiting case and is easier to handle mathematically. This book deals *only* with Fraunhofer diffraction." They then derive Eq. (37). However, it has been demonstrated in section 5.4 that Fraunhofer diffraction *does not lead* to hyperbolic fringes in the shadow region, but rather to a rectilinear latticework of fringes. Fraunhofer diffraction neglects all wavefront curvature, and one must take account at least of the curvature about the wedge axis (x-axis) to obtain hyperbolic fringes. This feature of the far-field diffraction pattern is intrinsic to Fresnel, not Fraunhofer, diffraction.

In fact, as I indicated in section 5.2, the pattern of hyperbolic fringes is not, strictly speaking, a far-field pattern. One can observe this pattern at a viewing screen placed sufficiently close to the aperture (e.g., $\sim \leq 1$ cm), that the Fraunhofer condition (16) is not even approximately satisfied.

One might try to account for the validity of planar projection and the applicability of Eq. (37)—at least in the case of Newton's experiments with sunlight—by suggesting that the incident radiation is largely incoherent, in which case the wedge aperture acts as an extended incoherent source. The overall image in the x-y plane (parallel to the aperture) would then be an infinite number of partially overlapping vertical single-slit intensity patterns with the slit width varying linearly with position of the pattern along the x-axis. This is not the case, however. The lateral coherence length L_t—in effect the radius of an area (perpendicular to the propagation direction) within which waves from a finite-sized source superpose coherently—can be estimated from the relation $L_t \sim \lambda / 2\beta$, where β is the half-angle of the light source subtended at the diffracting object.[10] Recall that Newton's light source was a small hole 1/42 inch in diameter admitting sunlight onto a wedge some 10 feet distant. Thus, the beam divergence is $\beta \sim 0.03$ cm/300 cm $\sim 10^{-4}$ radians. Assuming a wavelength $\lambda \sim 500$ nm yields a coherence length $L_t \sim 2.5$ cm. Every unobstructed point of the wedge aperture within an illuminated circle of ~ 2.5 cm radius—and not just the points in a narrow vertical strip of approximately constant ξ—coherently emits (according to Fresnel-Kirchhoff) secondary spherical wavelets that superpose at the viewing screen to form the diffraction pattern.

How, then, is the diffraction pattern of the wedge to be interpreted? I believe that a more natural and convincing picture of the process is provided by the idea of boundary waves. I have referred to this model—without explicitly identifying it—several times already: for example, to account for the location of maxima and minima along the symmetry axis, or to explain the formal resemblance of the theory of the far-field shadow pattern (14b) to that of two-slit interference. In contrast to the Huygens-Fresnel picture—in which the diffracted wave arises from a fictitious distribution of secondary emitters over the aperture—one can consider the diffraction pattern to result from a superposition of waves scattered from the edge of the aperture (boundary or edge waves) and the unobstructed portion of the incident wave (the geometric wave); alone, the latter gives rise to a perfect geometric image of the aperture. According to this viewpoint the Fresnel diffraction pattern of the wedge is a kind of hologram.

Scattered from the wedge boundaries are two long cylindrical waves, one centered about each edge, whose axes are inclined to one another at the wedge angle. In the shadow region, these two waves superpose—in the absence of the geometric wave—to give a "two-slit" interference pattern; the natural curvature of the cylindrical waves about the wedge axis results in the hyperbolic pattern of fringes. In the illuminated region, the edge waves superpose with the geometric wave to generate the Fresnel fringes I have previously described. The projective quality of the image in any plane nearly perpendicular (for a narrow wedge) to the cylinder axes is again an expected consequence of slightly inclined cylindrical wavefronts.

Reasonable as the picture of edge waves may be, the mathematical justification is by no means trivial to establish and is beyond the scope of this discussion. The idea of boundary waves was actually first proposed qualitatively by Young, who later rejected it when he learned of Fresnel's integral theory of diffraction. Fresnel—unaware of Young's work, since he could not read English—struggled with the mathematical implementation of superposing edge waves in his incipient ray theory of diffraction before arriving at his definitive theory in terms of Fresnel integrals. It was Arnold Sommerfeld more than a century later who provided the first rigorous formulation of diffraction from a straightedge in terms of cylindrical boundary waves.[11] This approach has been subsequently generalized by others as to be applicable to arbitrary apertures.[12]

The appearance of hyperbolic fringes must undoubtedly have been an embarrassment to Newton. He was never able to deduce—as he had for gravity—the nature of the interaction that caused light to deviate from straight-line motion. Far from the vertex, light passing the edge of a knife was bent—as through the action of a short-range force—into the shadow of that knife. Close to the vertex, however, fringes crossed one another in the bright region between the edges, bending away into the shadow of the opposite blade. How confusing this must have seemed. No wonder Newton wrote, "Are not the Rays of Light in passing by the edges and sides of Bodies, bent several times backwards and forwards, with a motion like that of an Eel?"[13]

Today, virtually all details of the diffraction pattern of a wedge can be predicted quantitatively by the integral theory of Fresnel (and his successors) and the use of computers. And yet, although Newton's dilemma may be a quaint footnote of history, the strikingly beautiful pattern of light that first caught his eye is still not without its mystery and subtlety to those who also discover it.

NOTES

1. Sir Isaac Newton, *Opticks* (New York: Dover, 1952), p. 327.

2. M. P. Silverman and W. Strange, "The Newton Two-Knife Experiment: Intricacies of Wedge Diffraction," *Am. J. Phys.* 64 (1996): 773–87.

3. Newton, *Opticks*. Quotations are from pages 317–18 and 325–35.

4. J. C. H. Spence, W. Qian, and M. P. Silverman, "Electron Source Brightness

and Degeneracy from Fresnel Fringes in Field Emission Point Projection Microscopy," *J. Vac. Sci. Technol. A* 12 (1994): 542–47.

5. See, for example, T. W. Mayes and B. F. Melton, "Fraunhofer Diffraction of Visible Light by a Narrow Slit," *Am. J. Phys.* 62 (1994): 397–403; C. A. Bennett, "A Computer-Assisted Experiment in Single-Slit Diffraction and Spatial Filtering," *Am. J. Phys.* 58 (1990): 75–78; J. T. Wesley and A. F. Behof, "Optical Diffraction Pattern Measurements Using a Self-Scanning Photodiode Array Interfaced to a Microcomputer," *Am. J. Phys.* 55 (1987): 835–44.

6. $\Psi(P)$ and ψ_o differ dimensionally by a factor of length since $\Psi(P)$ corresponds (in the absence of diffraction) to a spherical wave of the general form $(\psi_o/r)e^{ikr}$, and not just to the amplitude ψ_o.

7. M. P. Silverman, *And Yet It Moves: Strange Systems and Subtle Questions in Physics* (New York: Cambridge University Press, 1993), pp. 99–101.

8. D. Gabor, "Microscopy by Reconstructed Wavefronts," *Proc. Roy. Soc. A* 197 (1949): 454–87.

9. D. Halliday and R. Resnick, *Physics* (New York: Wiley, 1978), pp. 1025–27.

10. M. P. Silverman, *More Than One Mystery: Explorations in Quantum Interference* (New York: Springer, 1995), pp. 4–6.

11. A. Sommerfeld, "Theorie der Beugung," in P. Frank and R. v. Mises, *Die Differentialgleichungen und Integralgleichungen der Mechanik und Physik*, 2d ed. (Braunschweig: Vieweg, 1935), part 2, pp. 843–53; id., *Optics* (New York: Academic Press, 1964), pp. 273–89.

12. See, for example, M. Born and E. Wolf, *Principles of Optics*, 4th ed. (New York: Pergamon, 1970), pp. 449–53; A. Rubinowicz, "The Miyamoto-Wolf Diffraction Wave," *Progress in Optics* 4 (1964): 199–240.

13. *Opticks*, Query 3, p. 339.

CHAPTER 6

Young's Two-Slit Experiment with Electrons

It is not surprising that our language should be
incapable of describing the processes occurring
within the atoms, for . . . it was invented to describe
the experiences of daily life, and these consist only of
processes involving exceedingly large numbers of
atoms. . . . Fortunately, mathematics is not subject to
this limitation.
(W. Heisenberg, 1930)[1]

THE INTERFERENCE of two superposed beams leading to bright and dark fringes is considered one of the hallmarks of wave behavior. Thus, on the basis of Young's two-pinhole experiment with light, and the comprehensive experiments of an analogous nature performed independently by Fresnel, one often says without further reflection that light "is" a wave. But "is" and "behaves like" are not the same. A coherently split beam of electrons exhibits interference, too. Examined closely, however, the buildup of the electron interference pattern one electron at a time raises a profound conceptual problem for anyone who tries to imagine what each electron is doing.

6.1 PREDICTIONS

How does one distinguish between waves and particles? The answer is ordinarily obvious. A grain of sand is a material particle. The ripples crossing a pond are waves (made of particles). A particle—to describe it somewhat tautologically—has discretely measurable properties: a certain mass, a certain size, a certain charge, and so on. Waves, by contrast, are diffuse; they transport energy and momentum through matter (or, like light, through vacuum). Waves move particles; moving particles produce waves.

But the answer is not always obvious. The nature of light, for example, was long in dispute. Not until the 1830s, after the fundamental researches of Fresnel on diffraction and polarization, was the conception of light as a transverse wave broadly adopted. (I will discuss seminal advances relating to polarization in chapter 8.) After the 1850s, when Maxwell identified light specifically as an electromagnetic wave and predicted its speed in terms of electric and magnetic constants, there could be no further doubt.

The reader knows, of course, that the neat division of nature into waves and particles did not survive the conceptual upheavals of twentieth-century physics that ultimately led to the quantum theory of matter and radiation. Indeed, the troubling issue of what light is began again on the very threshold of the new century. In a desperate attempt to reconcile the statistical underpinnings of thermodynamics with the observed spectral distribution of thermal (or black-body) radiation, Max Planck introduced around 1900 the concept of a quantum of radiation.

An archly conservative scientist, Planck had no intention to show—nor did he claim to have shown—that light possessed particle attributes. He simply assumed that matter radiated energy in countable bundles, and that the energy of each bundle was proportional to the frequency. In this way Planck was able to avoid the "ultraviolet catastrophe"—the unrestricted growth in radiated energy with increasing frequency predicted by the then-known laws of physics. By focusing attention on the entropy of light, Planck succeeded in deriving a formula for the frequency dependence of the intensity of thermal radiation in close agreement with experiment. He had hoped to salvage this formula somehow even after allowing the proportionality constant (Planck's constant h) to approach zero, but in this he continually failed. No h, no diminution in intensity at high frequency. Still, for many years thereafter, Planck associated the discreteness exclusively with the material oscillators (the radiators), and not with the radiation field itself.

Whereas the mature and cautious Planck hesitated to draw from his research an iconoclastic conclusion, the brash young Einstein showed no timidity. In 1905, twenty-four years of age and employed in a Swiss patent office, Einstein applied Planck's quanta to the photoelectric effect to conclude that "the energy of light spreading out from a point is not continuously distributed over an ever increasing space, but consists of a finite number of energy quanta ... which move without dividing and which can be produced and absorbed only as complete units."[2] (The word "photon," which depicts this intrinsic discreteness of light, was coined some twenty years later by the American chemist Gilbert Lewis.)

Can light be like a particle and yet produce interference? The glimmer of an answer—in the form of an exquisitely simple experiment—was actually provided in 1909 by G. I. Taylor of Trinity College, Cambridge.[3] Acting on the suggestion of J. J. Thomson that "if the intensity of light in a diffraction pattern were so greatly reduced that only a few of these indivisible units of energy should occur on a Huygens zone at once, the ordinary phenomena of diffraction would be modified," Taylor carried out the following test. He projected onto a photographic plate the image of a sharp needle made by the direct light of a gas flame, and noted the time required to achieve "a certain standard of blackness" when the plate was fully developed. Then he made exposures of increasingly long durations whereby the light from the flame was attenuated by increasingly opaque screens. In his longest exposure he photographed the diffraction pattern of the needle over a period of 2000 hours, or about 3 months.

Having ascertained that a standard candle placed 2 meters from the same kind of plate brought it up to his standard of blackness in 10 seconds, Taylor calculated that the light falling upon the longest exposed plate corresponded to a standard candle burning at a distance slightly exceeding a mile. Yet, contrary to Thomson's speculation, Taylor found that "In no case was there any diminution in the sharpness of the pattern. . . ." This remarkable experiment was reported to the Cambridge Philosophical Society by Thomson himself, who concluded that the energy of 1.6×10^{-16} ergs in each cubic centimeter of the radiation which exposed the plate set an "upper limit to the energy contained in one of the indivisible units [of light]."

Since a single visible photon (say, $\lambda = 500$ nm) transports an energy of approximately 4×10^{-12} ergs, Taylor's most attentuated light, if his reckoning is correct, contained on average $\sim 4 \times 10^{-5}$ photons/cm^3.

Was it indeed the case that light illuminated the needle and fell upon the photographic plate one photon at a time? And if this actually were the case, did single photons (instead of superposing wavefronts) produce the interference fringes of the Fresnel diffraction pattern of the needle?

It is difficult to say. Taylor, to judge from his experimental description, never actually detected one photon at a time, but rather examined each fully exposed plate. What does it mean to detect a photon? A photon is pure energy; what ultimately concludes an act of detection, however, is always matter: a grain of silver on a photographic plate; a current of electrons out of a photomultiplier tube. A photon cannot be caught and examined like a charged particle in an ion trap; it either moves at the speed of light, or, once stopped, it vanishes. Also, according to the rigid statistical dichotomy by which all quantum particles are classified, photons are bosons, which signifies that there is no upper limit to the number of such particles that can occupy the same quantum state. The implication—not just for Taylor's experiment, but for any experiment relying on the attentuation of thermal light—is the following: Although the experimenter may reduce the light intensity to an extent such that the mean number of photons reaching a detector within a certain time interval is arbitrarily small, fluctuations about that mean always occur; the probability for more than one photon arriving at a time is never rigorously zero.

The preceding remarks are not objections as much as reminders that things are rarely as simple as they may seem, that Nature is ever elusive when it comes to the nature of light. The notion of a photon has long been a slippery one, and different physicists who use the term do not always mean the same thing (although they may not realize this).[4] Indeed, the argument has been made by reputable physicists that photons are not even needed to account for the very phenomena (e.g., the photoelectric and Compton effects) for which they were invented.[5] Einstein, whose great works revolutionized our concepts of space and time, and whose early researches ultimately led to the field of quantum optics, claimed to have devoted more thought to the enigma of light than to any other problem.

Today, some ninety years after Einstein's radical proposal, the quantum theory of light is a well-developed and mature branch of physics; it is an integral part of quantum electrodynamics, the predictions of which have been experimentally verified to a spectacular degree of precision. As one whose own researches at one time involved high-precision tests of quantum electrodynamics[6]—and this by counting photons—I have no doubt, despite any terminological confusion, that photons exist. All the same, as a matter of personal taste, I find that the intriguing implications of interference are revealed more starkly by the undulatory nature of matter than by the corpuscular nature of light.

Of the various elementary particles of matter, none is more readily available and subject to precise control than the electron. The reader is perhaps by now accustomed to the fact that particle physicists are always predicting new particles, many of dubious existence. Nevertheless, notable successes like the Ω^- meson or the W vector bosons—drawn, as it might seem, out of the thin air of mathematical symmetry—give us pause to reflect on the power of mathematical reasoning to see deeply into the fabric of nature. It is perhaps not widely realized that the electron was the first elementary particle to be predicted theoretically before being experimentally discovered.

The idea that there is a fundamental electric charge of which all electrically charged bodies are composed was advanced by G. Johnstone Stoney in a paper (1874) before the British Association for the Advancement of Science, based on the laws of electrolysis established by Faraday. Stoney later ($\sim$ 1891) named this unit of electricity the "electron." The Dutch theorist Hendrick A. Lorentz, however, is usually credited with giving life to the electron as a fruitful scientific hypothesis. Successfully applied in 1896 to the interpretation of magnetically induced spectral line broadenings observed in metal vapors by his Dutch colleague Pieter Zeeman, Lorentz's electron theory led to the sign and charge-to-mass ratio of the supposed particle. The electron thus became recognized as a fundamental constituent of all atoms. Shortly afterward (1897), J. J. Thomson gave physical evidence of the existence of free electrons by demonstrating that cathode rays emitted from different metals into various gases all had the same charge-to-mass ratio as that deduced by Lorentz and Zeeman.

That the electron was a particle could hardly be doubted for, in contrast to the quantum of light, it had both mass and charge. Nevertheless, convinced that the wave-particle duality, proposed by Einstein specifically for light, was actually general and extended to all of the physical world, Louis de Broglie put forth in 1924 the proposition (elaborated in his doctoral thesis) that there is a wave associated with the motion of any particle. Without being too specific about the nature of the wave, de Broglie produced a relation linking momentum (p) and wavelength (λ) that, together with Einstein's relation for energy (E) and frequency (ν), succinctly recapitulates the wave-particle duality:

$$E = h\nu, \tag{1}$$

$$p = \frac{h}{\lambda}. \tag{2}$$

The examining committee of the University of Paris, before whom de Broglie defended his thesis, were duly impressed with the originality of the work, but not particularly convinced of the reality of these hypothetical matter waves. Asked by Jean Perrin, famed for his experiments on Brownian motion,[7] how one might observe such waves, de Broglie pointed to the possibility of electron diffraction, the regular three-dimensional array of atoms in a crystal serving as a grating. Within a few years, electron diffraction became a reality—and the experiments of Clinton Davisson and Lester Germer, on the one hand, and those of George P. Thomson (son of J. J. Thomson) on the other, now constitute a traditional part of every physicist's course of "modern physics." It is another entertaining irony of history that Thomson *fils* proved to be wavelike what Thomson *père* discovered as a particle.

Against the evidence of electron waves deduced from their passage through a crystal, one could—if he wished to play devil's advocate—raise an objection similar to that brought over a century earlier against Young's two-slit demonstration of the wave nature of light. How does one know that it is not the influence of the crystal, rather than the intrinsic nature of the electrons, that gives rise to the diffracted beams? After all, could one not attribute the diffraction peaks to specular reflection of electrons, as particles, from different atomic planes within the crystal? The objection, to be sure, is not especially convincing—for it suggests the improbable scenario of a handful of birdshot, thrown against stacked layers of cannonballs, rebounding coherently into favored directions rather than dispersing randomly. Nevertheless, how much more satisfying would be an experiment that—like Fresnel's mirror—elicited the interference of electrons without their passage through matter.

In 1986, as the first western scientist invited to the then newly established Hitachi Advanced Research Laboratory (ARL) in Tokyo (since relocated to Hatoyama), I proposed such an experiment employing the state-of-the-art field-emission electron microscope that Hitachi manufactures. The idea, an electron counterpart to G. I. Taylor's light experiment, was quite simple: Attenuate the electron beam to an extent that one electron at a time moves from the source, around one side or the other of a beam-splitting device, to a detector—and record the process on video.

The experiment was eventually done,[8] and when months later the time came for me to return to the United States, my Japanese hosts gave me a copy of this video as a gift. It has long remained a precious reminder to me not only of two most enjoyable stays (I returned to Hitachi again in 1989), but also of the profound strangeness of the quantum world. In the years since, I have been asked to give many lectures on quantum mechanics, and I try, whenever possible, to show this video, for it captures vividly what Feynman has called the "heart of quantum mechanics." No audience—whether of laymen, students, or professional physicists—has ever failed to be intellectually shaken by it, however often they may have read of electron interference in books.

I will discuss the electron two-slit experiment and its outcome shortly. First, recall the quotation from Heisenberg that opened this chapter; it reminds us that the terms *wave* and *particle* were created to describe the experiences of daily life, the purview of classical physics. Within that framework, imagine again the typical two-slit configuration of figure 4.2 and consider the distinctive implications of labeling an electron or photon or any other "-on" as a wave or particle.

If the interfering "-ons" are waves, then an incident wavefront simultaneously illuminating both apertures makes each a source of secondary wavelets that propagate coherently to the viewing screen. Let $\psi_1(\mathbf{x})$ and $\psi_2(\mathbf{x})$ be the amplitudes of the waves from each aperture arriving at a point $\mathbf{x}$ on the screen; then the total amplitude at $\mathbf{x}$ is $\psi(\mathbf{x}) = \psi_1(\mathbf{x}) + \psi_2(\mathbf{x})$, and the resulting intensity is

$$I(\mathbf{x}) = |\psi(\mathbf{x})|^2 = I_1(\mathbf{x}) + I_2(\mathbf{x}) + 2\operatorname{Re}(\psi_1(\mathbf{x})\psi_2(\mathbf{x})^*). \qquad (3)$$

In the second equality the terms $I_1(\mathbf{x}) = |\psi_1(\mathbf{x})|^2$ and $I_2(\mathbf{x}) = |\psi_2(\mathbf{x})|^2$ are the intensities at $\mathbf{x}$ attributable to each aperture alone. The third term is the interference term, which cannot vanish everywhere as long as I_1 and I_2 are nonvanishing. One therefore can make the following

Prediction 1. *Attenuate the beam so as to make the illumination falling upon both slits arbitrarily weak; the interference fringes, however faint, will nevertheless persist.*

If, by contrast, the "-ons" are particles, then each "-on" must pass through one aperture or the other; a particle cannot split itself, traverse both apertures, and then recombine. The fact that interference occurs must therefore signify some kind of cooperative interaction between two or more correlated "-ons" at the diffracting apertures or en route to the viewing screen. This leads, therefore, to the following

Prediction 2. *Attenuate the beam so that one "-on" at a time passes through the apparatus; when there is no interaction between "-ons," the interference pattern must disappear completely.*

Let us now examine the electron two-slit experiment.

6.2 MAKING AND BREAKING ELECTRONS

Although the details of particle production, focusing, and detection differ substantially, electron and optical microscopes operate on the same fundamental principles. A schematic diagram of the 150 kV high-voltage field-emission microscope employed in the Hitachi experiment is illustrated in figure 6.1. Electrons, drawn from a fine tungsten tip by a strong electric field, pass through an array of lenses (condenser, objective, and intermediate) and a device—the electrostatic biprism—that splits the beam into two coherent components. The two components recombine to form an image that is then magnified by another lens (projector).

Although the anatomy of the microscope is not the center of interest in this chapter, it is nevertheless useful, in anticipation of chapter 9, to call attention briefly to the structural complexity of the instrument. This is a large instrument with a microscope barrel over a meter high and a wide console at which the operator sits; the power supplies to produce the extraction potential and the accelerating potential occupy a fair fraction of the working space within the room. The many lenses in the diagram are magnetic lenses that shape the electron trajectories through the microscope by means of the Lorentz force. Like glass lenses, electron lenses give rise to aberrations that pose difficulties for the interpretation of high-resolution images of material structures. One can well imagine a need for a low-voltage, tabletop electron microscope that can yield high-resolution images with *no* lenses at all. Fresnel diffraction makes this possible, as I will discuss later.

The electron microscope is perhaps the most important legacy of the wave-particle duality. All the same, it should be noted that in using the instrument specifically as a microscope (rather than as an interferometer), the wave nature of the electron enters fundamentally at only two points: (1) the image at the "bottom," where, as in an optical microscope, the wavelength sets a lower limit on attainable resolution, and (2) the source at the "top," where quantum tunneling gives rise to electron emission. Otherwise, for most of its passage through the microscope, the electron beam ordinarily can be treated as rays subject to classical forces; the de Broglie wavelength plays no seminal role.

The production of electrons by the process of field emission has been known and understood for a long time. Indeed, the emission of electrons from a cold metal upon application of a strong electric field—first clarified by Fowler and Nordheim ($\sim$ 1928) shortly after the creation of quantum mechanics[9]—was one of the first confirmations of quantum mechanical tunneling. The process is illustrated in figure 6.2. Within the metal tip, conduction electrons (charge $-e$) occupy energy levels up to the Fermi level designated E_F. Although free to wander inside the metal with little gain or loss of energy, electrons are prevented from escaping the metal surface by a contact potential difference

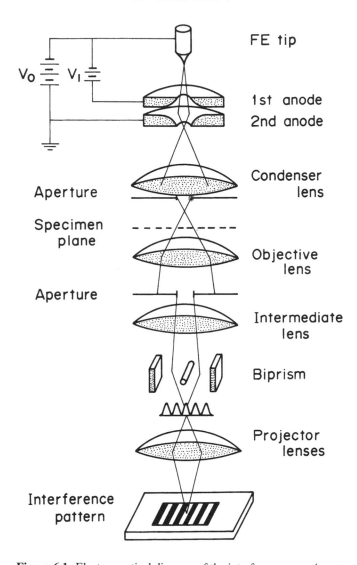

Figure 6.1. Electron optical diagram of the interference experiment.

or work function ϕ. Upon application of a constant external electric field E, the original potential energy function is modified to first approximation by the addition of a term $U_1(z) = -eEz$ that varies linearly with the field and increases in magnitude with distance z from the metal surface. Thus, the height of the original potential barrier is reduced to zero in a distance on the order of ϕ/E. A strong field thereby facilitates electron escape.

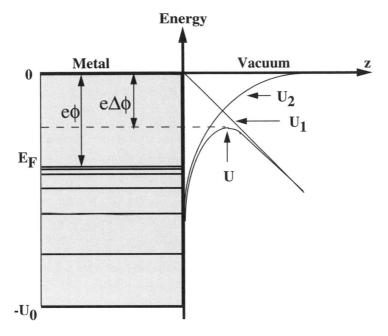

Figure 6.2. Electron field-emission as a metal-vacuum tunneling process.

The situation, however, is actually more complicated in that a charge $-e$, removed to a distance z above the surface, is subject to an image force contributing an additional potential $U_2(z) = -e^2/4z$ decreasing in magnitude with z. (The zero of electron potential energy is at a point infinitely far from the surface in a field-free vacuum.) The total potential energy outside the surface

$$U(z) = U_1(z) + U_2(z) = -eEz - \frac{e^2}{4z} \qquad (4a)$$

produces a barrier of maximum height

$$U_m = -\sqrt{e^3 E} \qquad (4b)$$

at the distance

$$z_m = \sqrt{\frac{e}{4E}} \qquad (4c)$$

from the surface. As figure 6.2 shows, $\Delta\phi = |U_m|/e$ is the effective decrease in the work function. The larger the applied field E, the lower the energy required for electrons to tunnel into the vacuum.

Applying Fermi-Dirac statistics to this simple classical electrostatic model—at which point another facet of the quantum nature of the electron is expressly recognized—Fowler and Nordheim derived an equation of the form

$$\frac{I}{V_1^2} = a \exp\left(-b\frac{\phi^{3/2}}{V_1}\right), \tag{5}$$

predicting the total emission current I in terms of the extraction potential V_1 applied between the tip and a remote plane. For a given metal, the constants a and b depend on E_F, ϕ, a field contraction factor which takes account of the shape of the tip, and an image-correction term attributable to additional effects of the image potential. A key feature of the Fowler-Nordheim theory of field emission is that a plot of $\ln(I/V_1^2)$ vs. $1/V_1$ is linear for a given tip geometry.

To extract electrons from a single crystal of tungsten, which is the electron source of preference in high-voltage field-emission microscopes, requires an electrostatic field of approximately 4×10^7 V/cm in the region of the tip. The sharper the tip—that is, the smaller the radius of curvature—the lower the potential required to attain such a high electric field. A standard field-emission tip may typically have a diameter of 5–10 nm and generate a current of 50–100 μA under an extraction potential of a few kilovolts.

In the Hitachi experiment electrons were extracted under a potential of $V_1 \sim$ 5–10 kV and subsequently accelerated to the anode at a potential $V_0 = 50$ kV. At this potential the speed of the electrons was approximately one-half that of light. For a relativistic particle of rest mass m, the exact expressions for kinetic energy and momentum are, respectively,

$$K = mc^2(\gamma - 1), \tag{6a}$$

$$p = mv\gamma \tag{6b}$$

in which the factor

$$\gamma = \frac{1}{\sqrt{1 - \dfrac{v^2}{c^2}}} \tag{6c}$$

is always ≥ 1; the larger γ, the more relativistic the particle motion.

It is not difficult to show by means of energy conservation ($K = eV_0$) that the de Broglie wavelength ($\lambda = h/p$) of an accelerated electron takes the form

$$\lambda = \frac{h}{\sqrt{2meV_r}} \tag{7}$$

with

$$V_r = V_0 + \left(\frac{e}{2mc^2}\right)V_0^2 \tag{8}$$

as the relativistically corrected effective acceleration potential. Actually, despite the fact that $c/2$ is an enormous speed by ordinary terrestrial standards,

50 kV electrons are not all that relativistic; the factor γ is only ~ 1.1, and one could get by reasonably well with the simpler nonrelativistic relation

$$\lambda = \frac{h}{\sqrt{2meV_0}} \sim \frac{1.23 \text{ nm}}{\sqrt{V_0 \text{ (volts)}}} \tag{9}$$

which neglects the second term in Eq. (8), in essence the ratio of electrical potential energy to the rest-mass energy. From Eq. (9) it follows that a 50 kV electron has a wavelength $\lambda \sim 5.4 \times 10^{-3}$ nm, approximately one-tenth the Bohr radius of a ground-state hydrogen atom.

6.3 COHERENCE AND INTERFERENCE

After collimation by the chain of lenses illustrated in figure 6.1, electrons enter the electrostatic biprism, which is the key component that transforms the microscope into an interferometer. Invented by Möllenstedt and Dücker in 1956, the biprism consists of two parallel plates at ground potential with a fine wire filament of radius $r = 0.5 \mu$m between them at a positive potential of approximately $V_f = 10$ V. In the region between the plates, a distance $2\ell = 10$ mm apart, the potential can be described by a function $V(x, z)$ where the z-axis runs parallel to the barrel of the microscope, and the lateral x-axis is normal to the plates, passing through the center of each.

Although matter-wave interference is a quantum phenomenon, the effect of the biprism can be understood by a simple nonrelativistic classical argument applied to the particle picture of electrons. Directed initially along the z-axis with momentum

$$p_z = \hbar k_z = mv \qquad (\hbar = h/2\pi), \tag{10}$$

each electron acquires a transverse momentum from the electrostatic attraction of the filament. The resulting impulse—which may be positive or negative depending upon which side of the filament the electron passes—is given by

$$\Delta p_x = \int_{-\infty}^{+\infty} F_x \, dt = \int_{-\infty}^{+\infty} F_x \frac{dz}{v} = \hbar k_x \tag{11}$$

where the lateral force F_x is determined from the gradient of the potential

$$F_x = -e \frac{\partial V(x, z)}{\partial x}. \tag{12}$$

We must return, however, to the wave picture to determine what happens next.[10] After diffraction around the filament with resulting deflection toward the z-axis, the waveform can be expressed as a linear superposition of two components

$$\psi(x, z) = e^{ik_z z}(e^{ik_x x} + e^{-ik_x x}) \tag{13}$$

in which the x component of the propagation vector follows from Eqs. (11) and (12):

$$k_x = -\left(\frac{me}{\hbar^2 k_z}\right) \int_{-\infty}^{+\infty} \left(\frac{\partial V(x, z')}{\partial x}\right)_{x=r} dz'. \tag{14}$$

The value of k_x can be estimated by modeling the biprism geometry as an infinitely long straight wire between infinitely extended ground plates. The potential distribution in the vicinity of the filament then becomes

$$V(x, z) = V_f \left[\frac{\ln(\sqrt{x^2 + z^2}/\ell)}{\ln(r/\ell)}\right], \tag{15}$$

and it follows that

$$k_x = \frac{\pi e V_f}{\hbar v \ln(\ell/r)}. \tag{16}$$

Upon arrival at the viewing screen, the electrons, according to Eq. (13), should give rise to a transverse intensity pattern

$$I(x) = |\psi(x, z)|^2 = 4\cos^2(k_x x) \tag{17}$$

with fringe spacing $d = \pi/k_x \sim 90$ nm. In the actual experiment the waveform incident upon the biprism was more like a spherical wave than a plane wave, with the consequence that a somewhat larger fringe spacing was to be expected.

It is instructive to keep in mind that this (effectively) two-slit interference experiment demands much more of the participating waves than does electron diffraction through a crystal. Just as an optical grating will disperse any kind of light, however incoherent, into its spectral colors, so too does the crystal diffract any beam of electrons, however incoherent. We have seen—to play devil's advocate again—how phenomena arising from incoherent waves are more-or-less interpretable in terms of classical particles.

Whereas the only undulatory property that plays a role in the diffraction of incoherent electrons by a crystal is the wavelength, it is not the wavelength per se that matters in the interaction of the field-emitted electrons with the biprism. This is a conceptually significant point. Elementary treatments of physical optics often give the impression that diffraction and interference are noticeable only if diffracting objects and apertures are comparable in size to the wavelength; otherwise the rays, like particles, propagate rectilinearly and can be adequately described by geometric optics. If this were truly the case, then the de Broglie wavelength of a 50 kV beam of electrons would be hopelessly small for the wavelike nature of electrons to be made apparent under the present circumstances. It is not the case, however. What is relevant here is not the wavelength (the unit of periodicity of a hypothetically infinitely long wave train), but the transverse coherence width L_t—that is, the lateral extent of an electron wave packet.

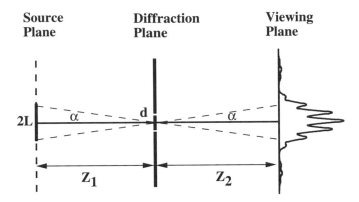

Figure 6.3. Reduced contrast of interference pattern produced by an extended source.

An electron source, however sharply tipped, is an extended object. Particles emitted from different points of the source give rise to correspondingly displaced diffraction-interference patterns at the distant viewing screen. If the source is sufficiently extended that the crests and troughs of different patterns overlap randomly, then the net intensity at the screen will exhibit no interference fringes. Thus, there is a characteristic size

$$L_t = \frac{\lambda}{2\alpha} \tag{18}$$

representing the width of the region in the diffracting plane (perpendicular to the beam) within which arriving wave fronts are coherent.[11] This size depends not only on the wavelength λ, but also on the angular width 2α of the source as viewed from the diffraction plane. (2α is also the angle of beam divergence at the viewing screen.) If diffracting objects or apertures are much larger than L_t, then they scatter radiation incoherently, and the intensities, not amplitudes, of the waves add at the viewing screen.

Consider, for example, the interference pattern produced by a line source of length $2L$ normal to the beam illuminating two pinholes separated by a space d and located a distance Z_1 from the source. Figure 6.3 illustrates the geometry. The pattern on a viewing screen a distance Z_2 from the source is given (up to a normalization factor) by the expression

$$I(x) \sim 1 + \left(\frac{\sin \xi}{\xi}\right) \cos \left(\frac{2\pi d x}{\lambda Z_2}\right) \tag{19}$$

where x is the lateral coordinate at the viewing screen and

$$\xi = \frac{2\pi d L}{\lambda Z_1} \tag{20}$$

is the argument of the bracketed function designated by A. A. Michelson as the "visibility." For a point source, $L = 0$, and the visibility is unity—that is, the contrast in the fringes is 100%. For $\xi = \pi$, however, the visibility vanishes, and the pattern displays no interference. Thus, the lateral coherence length can be defined in this case as the slit separation at the threshold value $\xi = \pi$ or

$$d = \frac{\lambda}{2L/Z_1} \tag{21a}$$

which is equivalent to Eq. (18) (since L/Z_1 corresponds to the angle α in the small-angle approximation $L \ll Z_1$).

Recast in the form

$$s = \frac{\lambda}{(d/Z_1)} \sim \frac{\lambda}{\theta_c}, \tag{21b}$$

the expression relates the coherence angle $\theta_c = d/Z_1$—that is, the angle subtended at the source by a coherently illuminated region of size L_t—to the source size $s = 2L$. The coherence angle, which will be referred to again in the next chapter, is an experimentally useful quantity deducible from diffraction patterns. The form and interpretation of relation (21b) follow as well from the Heisenberg uncertainty principle in which s is a measure of the uncertainty Δx in the lateral position of an electron at time of emission, and $\Delta p_x (t = 0) \sim (h/\lambda)\theta_c$ is the corresponding measure (in a small-angle approximation) of uncertainty in the initial transverse momentum. Thus, $\Delta x \Delta p_x \sim h$.

The divergence angle of the beam employed in the electron two-slit experiment was approximately 4×10^{-8} radians, which led to $L_t \sim 140\ \mu$m, well in excess of the 1.0 μm width of the biprism filament, yet considerably smaller than the 10 mm separation between ground plates.

As discussed in the previous section, the field-emitted current is governed by the Fowler-Nordheim equation. For the choice of extraction potential, the total current emitted into the microscope was limited to 1 μA, only a small fraction ($\sim 10^{-4}$) of which passed through the anodes. An iris placed above the intermediate lens further limited the transmitted current, and by adjusting the focal length of this lens one could restrict the current through the biprism to about 1000 electrons per second. *Note that at a speed of $c/2$, the mean distance between electrons spaced in time by a millisecond is 150,000 m.* As the distance between the field-emission tip and the detecting surface is only ~ 1.5 m, there is little question but that one electron at a time passes through the microscope. When that electron is detected, the succeeding one has not even been extracted from the tip.

The arrival of electrons at the "bottom" of the microscope was registered by a commercially available position-sensitive electron counting system based on a microchannel plate (MCP). A microchannel plate is a honeycomb-like structure—in this case 12 mm in diameter—serving as an electron multiplier. When a photon or charged particle is incident at the input of one of the channels, secondary electrons are generated and accelerated down the channel toward the output end. Each collision of secondary electrons with the channel wall generates additional electrons. In this way a pulse of up to 10^4 electrons can be produced at the output of a straight-channel MCP, and as many as 10^6 electrons for a curved channel. In the Hitachi experiment, an incoming 50 kV electron first struck a fluorescent screen that emitted about 500 photons. The photons then initiated electron emission from the surface of a photocathode. Finally, these electrons impinged upon the upper plate of the MCP to produce a larger pulse of electrons. At the end of this chain of events, a position-sensitive device took note of the location of each excited channel of the MCP and sent the signal to be stored electronically and displayed accumulatively on a TV monitor.

Half a world away from Japan, at another place and another time—I slip a small black cartridge into a video player, and perhaps a hundred or more pairs of eyes, few of which have ever before seen what they are about to see, stare curiously at the monitor as I turn the switch to "on."

6.4 THE VIDEO—AND WHAT IT MEANS

Unlike commercial videos, this one has no suspenseful music—indeed, it has no sound at all. The original experiment lasted about a half hour, but the film runs for only a few minutes. It starts with a slow scan of the camera over the exterior length of the microscope, from the source to the detector. One sees the shiny steel barrel within which electrons travel in ultrahigh vacuum. One sees the binocular optical microscope with which electron images can be viewed. Mr. Matsuda sits at the console. His image suddenly vanishes, and all that remains is an empty field of sky blue.

A small round white spot appears. Then another. It looks like a smooth blue wall hit by randomly thrown snowballs. The snowball splotches accumulate. There is one at the upper right, lower left, middle . . . here, there, everywhere. The snowballs hit steadily. One hundred splotches appear. 200, 300, 400. . . The camera speeds up perceptibly. The spots appear a little faster.

I stop the video—and ask the viewers what they see. In particular, do they see an interference pattern? No, there is only a blue wall with random white spots (figure 6.4). I recall Prediction 1. The beam has been attenuated so that one "-on" at a time hits the detector, and there are *no* interference fringes. If an electron were a wave, it would surely have been diffracted by the biprism filament, for the lateral coherence width is 140 times the diameter of the filament. But there are no fringes. The electron is not a wave.

I restart the video. White spots accumulate more quickly now: 1000, 1500, 2000, 2500, 3000.

I stop the video and ask the audience to squint. Amid all the snowball splotches, there is something barely perceptible forming on the screen. What had hitherto been random spots is organizing into a pattern. It is not quite visible, but one senses it is there. Slightly denser white smudges line the wall in certain places; elsewhere the snowballs seem somehow to miss with greater frequency.

I restart the video and let it run to completion. 10,000, 20,000, 30,000 spots appear. There can be no doubt. The blue wall powdered by snowballs gives way to a white picket fence, the edges of the pickets growing sharper by the second. At the end of the film—the accumulation of over 20 minutes of counting electrons—the pattern comprises more than 1,000,000 electrons, some 240,000 electrons per picket. So sharp are the fringes that one might have thought they were made by a laser beam. But they were made by electrons—one at a time. I recall to the audience Prediction 2. Two "-ons" cannot interact with one another if one "-on" is still within the metal cathode when the other "-on" passes the biprism and arrives at the detector. Yet there are fringes. The electron is not a particle.

So, what is the electron? Photons behave in the same way (although Taylor could not observe them as closely). What, then, is the photon?

I can't answer such questions. They involve semantics, not physics. For want of better terminology, I still think of any "-on" individually as a particle. Patterns indicative of wavelike behavior emerge only from "-ons" in aggregate (or, in some cases, by repetitive observations of a single "-on"). But single observations of a single "-on" never show wavelike behavior. Classical light waves comprise enormous numbers of photons—for example, $\sim 3 \times 10^{18}$ photons of red (633 nm) light leave a 1 Watt helium-neon laser each second (if by photon one means a bundle of energy $h\nu$). No wonder that it is the wavelike, and not the discrete, attributes of light that are ordinarily more obvious. Electrons, as Fermi particles, cannot form classical waves, for there can never be more than two electrons (of opposite spin orientations) in the same spatial quantum state of a particular system. Little wonder that it is the discrete attributes of electrons that are the more often encountered.

The discrete and the undulatory are not confounded; they do not lead to inconsistent experimental results. Quantum mechanics does not permit that. If an experiment allows one to determine the path an "-on" has taken—for example, which side of the biprism each electron has passed—there will never be an interference pattern.

A simple example can help show why. Consider the standard Young's two-pinhole geometry with holes separated a distance d. The holes are illuminated by plane waves of wavelength λ. At the viewing screen, a sufficiently long

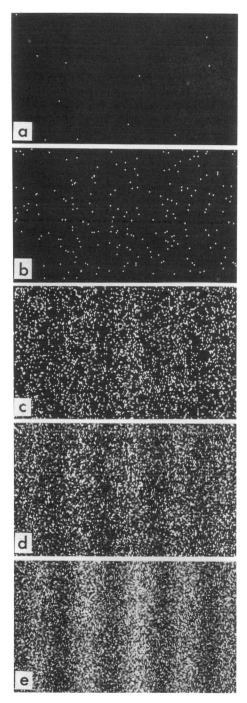

Figure 6.4. Buildup of electron interference pattern one electron at a time at a rate of 1000 electrons/second. The approximate number of recorded electrons in each frame is (a) 10, (b) 100, (c) 3000, (d) 20,000, and (e) 70,000.

distance away such that the angle of diffraction θ is $\ll 1$, an interference pattern forms. The angular location of the first fringe maximum is given by the well-known formula

$$d \sin \theta_1 \sim d\theta_1 = \lambda \tag{22a}$$

or

$$\theta_1 = \frac{\lambda}{d}. \tag{22b}$$

But suppose the incident waves are really composed of "-ons," a kind of particle. Each "-on" is initially incident upon the diffraction screen with a momentum $p_z = h/\lambda$. It then passes through one or the other pinhole and acquires, by diffraction, a transverse component to its momentum, Δp_x. If we are to tell through which hole the "-on" has gone, we must know its transverse coordinate x to within an uncertainty smaller than the separation between the pinholes: $\Delta x < d$. Therefore, by the uncertainty principle—$\Delta x \Delta p_x > h$—the uncertainty in transverse momentum is $\Delta p_x > h/d$. The "-ons" then illuminate the viewing screen within a range of angles α where

$$\alpha = \frac{\Delta p_x}{p_z} > \frac{h/d}{h/\lambda} = \frac{\lambda}{d} \tag{23a}$$

or, from Eq. (22b),

$$\alpha > \theta_1. \tag{23b}$$

The "-ons" arrive at the screen throughout the region where a minimum once neatly separated adjacent maxima. The fringes are wiped out.

Young's two-slit interference experiment does more than confirm that light and electrons both have wavelike properties. It gives dramatic force to British biologist John B. Haldane's insightful remark that the world is not only queerer than we imagine, but queerer than we can imagine.

NOTES

1. W. Heisenberg, *The Physical Principles of the Quantum Theory* (New York: Dover, 1930), p. 11.

2. A. Einstein, "Über einen die Erzeugung und Verwandlung des Lichtes betreffenden heuristischen Gesichtspunkt" (On a Heuristic Point of View concerning the Production and Transformation of Light), *Annalen der Physik* 17 (1905): 132–48.

3. G. I. Taylor, "Interference Fringes with Feeble Light," *Proc. Cambridge Phil. Soc.* 15 (1909): 114–15.

4. R. Kidd, J. Ardini, and A. Anton, "Evolution of the Modern Photon," *Am. J. Phys.* 57 (1989): 27–35.

5. W. E. Lamb and M. O. Scully, "The Photoelectric Effect without Photons," in *Polarisation, Matière, et Rayonnement* (Paris: Presses Universitaires de France, 1969), pp. 363–69. For a discussion of the Compton effect without photons, see J. N. Dodd,

"The Scattering of Electromagnetic Radiation by a Free Electron," *J. Phy. B* 8 (1975): 157–64.

6. An account of this work will be published in a forthcoming book, M. P. Silverman, *Probing the Atom* (Princeton: Princeton University Press).

7. It is worth keeping in mind that, although the electron was established as a constituent of atoms in the 1890s, general acceptance of the atom itself as a physically real object and not just a mathematical hypothesis followed as a consequence of Perrin's investigations around 1910.

8. A. Tonomura, J. Endo, T. Matsuda, and T. Kawasaki, "Demonstration of Single-Electron Buildup of an Interference Pattern," *Am. J. Phys.* 57 (1989): 117–20.

9. R. H. Fowler and L. W. Nordheim, "Electron Emission in Intense Electric Fields," *Proc. Roy. Soc. London A* 119 (1928): 1355–63.

10. It should be stressed that quantum mechanics furnishes a complete and consistent formalism for treating all aspects of the problem including propagation, diffraction, superposition, and interference. It is not necessary to resort to a heuristic classical argument based on force to arrive at the lateral component of the wave vector, Eq. (14). One could have arrived at the same result by solving the Schrödinger equation.

11. M. P. Silverman, *More Than One Mystery: Explorations in Quantum Interference* (New York: Springer-Verlag, 1995), pp. 4–6.

CHAPTER 7

Pursuing the Invisible: Imaging without Lenses

> If a lens is to produce a truthful image of an
> illuminated object, it must have an aperture sufficient
> to transmit the whole of the diffraction pattern
> produced by the object; if but part of this diffraction
> pattern is transmitted, the image will not truthfully
> represent the object, but will correspond to another
> (virtual) object whose whole diffraction pattern is
> identical with that portion which passes through the
> lens.
>
> *(A. B. Porter, 1906)*[1]

ONE OF THE most consequential legacies of the wave-particle duality is the electron microscope. Used as a microscope, it furnishes high-resolution images of structures that cannot be seen with a light microscope. Used as an interferometer, it reveals the relative phase between the components of a coherently split wave that have taken different paths through space or matter before recombining. The two functions are not unrelated. Indeed, all imaging—whether by light or electrons—is an interferometric process. The striking nature of this process is seen perhaps most dramatically in the production of images magnified nearly a million times or more without the use of a single lens or mirror.

7.1 IMAGES AND INTERFERENCE

As a young child reading with rapt enjoyment George Gamow's book *One Two Three ... Infinity*[2]—which played no small role in helping direct and solidify my burgeoning interests in science and mathematics—I came upon an extraordinary photograph. It is a bit chancy to reconstruct in later life the influential events of one's formative years, but of this incident there can be no doubt. I have retained that photograph in my mind for over four decades. It is the image of an organic molecule—hexamethylbenzene to be precise, although the nuances of chemical nomenclature concerned me little at the time—magnified 175,000,000 times!

Curled up on the floor in a corner of my room, staring intently at the page before me, I saw six fuzzy black elongated smudges—each about the size of a U.S. penny—situated at the vertices of a slightly elongated hexagon, and budding radially outward from each was another smudge of the same kind. Twelve smudges in all. The whole was an aesthetically symmetrical pattern that, according to how (and how long) I looked at it, could sometimes take on the appearance of a hexagon inside a larger hexagon, or two W-shaped crowns base to base, or pairs of chromosomes lined up in a peculiar way, or an odd configuration of sunspots, and so on. There is no end to a child's imagination.

The photograph, furnished by a Dr. M. L. Huggins of the Eastman Kodak Laboratory, was made with X-rays, and each fuzzy smudge was the image of an individual carbon atom. The sensation that I was actually staring at single atoms, at the ultimate building blocks of one of the elements, left an indelible impression upon me. More than anything else I had ever read prior to that moment—of chemical combinations, drawn-out oil slicks, Brownian motion, or atomic beams—that X-ray photograph showed me atoms existed. In vain I scrutinized each atomic smudge for some telltale presence of a nucleus and orbiting electrons.

The image, as Gamow was careful to point out—although his explanation escaped me at the time—was in fact not a direct photograph, but a composite of a large number of separate patterns, each pattern characterized by alternating light and dark bands of a particular spatial frequency and geometrical orientation. At the high frequencies of X-rays, the refractive index of nearly all materials is virtually unity. No lenses exist that can refract the rays, as glass refracts visible light, and thereby magnify the image. It was only many years later, as a physicist, that I came to understand how that extraordinary photograph was made. The technique, devised by W. L. Bragg in the 1930s, exploited Ernst Abbe's theory of the microscope to reconstruct an image out of a diffraction pattern. Today Abbe's conception of imaging is a well-understood application of the principles of Fourier optics, but at its inception the idea was a controversial one—and in a roundabout way its influence on one small boy who could marvel at the unseen world was decisive.

X-rays incident on a crystalline sample are scattered by the electrons of the atoms. The overall superposition of many such scatterings results in the diffraction of the radiation, which, unlike refraction, causes substantial angular deviations from the forward direction. A single-crystal X-ray diffraction pattern, as the reader has likely seen before, is a pattern of spots that manifests the symmetry of the crystal and in which are encoded the details of its structure. A simple (although not altogether complete) explanation of the pattern, first given by W. L. Bragg, is to regard the spots as produced by waves specularly reflected from different sets of parallel planes of atoms within the crystal. Although reflection from any plane of atoms is specular, only for certain values of the incident angle θ will reflections from all parallel planes superpose constructively (figure 7.1). With θ the glancing angle between the direct X-ray beam and the plane (not the normal to the plane as in Snel's law) and d the spacing

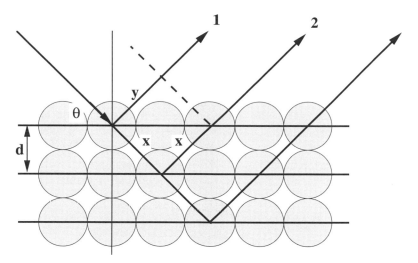

Figure 7.1. Bragg reflection of X-rays from a crystal lattice. Bragg's law follows when the optical pathlength difference $(2x - y)$ between adjacent rays (e.g., 1 and 2) is an integer number of wavelengths.

between atomic planes, the condition of a reflection maximum is expressed by Bragg's law

$$2d \sin \theta = n\lambda \tag{1}$$

where λ is the wavelength of the radiation and n is the diffraction order (the number of wavelengths by which the optical pathlengths from adjacent planes differ). Each plane reflects no more than about one-thousandth the incident radiation; had the first atomic plane reflected 100% of the incident radiation, there would be no interference from underlying planes, and radiation would be reflected at any angle irrespective of wavelength (as governed by the Fresnel relations).

An obvious point, but one worth noting nonetheless: Bragg's law reveals immediately why it is with X-rays (or electron waves), and not with visible light, that the ordering of atoms in matter can be probed. Since the sine of an angle does not exceed unity, the wavelength of the illuminating radiation cannot exceed twice the spacing between atomic planes, a distance scale on the order of angstroms (1 Å = 0.1 nm). Waves in the visible portion of the electromagnetic spectrum have wavelengths on the order of several thousand angstroms.

The effectiveness with which a particular atom (e.g., the j-th) scatters the incident radiation is denoted by a complex scattering amplitude f_j (the atomic form factor), which depends on the number and distribution of electrons in the atom and on the angle of scattering. Radiation scattered from atoms occupying different sites in a unit cell of the material will differ in phase at the detecting surface (e.g., photographic film) because of their differences in optical path length. With respect to the origin ($x = y = z = 0$) of the unit cell, the relative phase of a wave scattered from an atom at a displaced position $\mathbf{r}_j = (x_j, y_j, z_j)$ is

$$\phi_j = 2\pi \left(\frac{hx_j}{a} + \frac{ky_j}{b} + \frac{lz_j}{c} \right) \tag{2}$$

where (a, b, c) are the lengths of the unit cell and (h, k, l) are the so-called Miller indices. The Miller indices designate sets of planes for which the Bragg condition is fulfilled.[3] In other words, planes of atoms separated by a/h along x, b/k along y, or c/l along z scatter X-rays that differ in phase by $360°$. The net effect of scattering from all the atoms in a unit cell is expressed by the structure factor $F(h, k, l)$ of the crystal

$$F(h, k, l) = \sum_j f_j e^{2\pi i (hx_j/a + ky_j/b + lz_j/c)}. \tag{3}$$

The intensity of radiation reflected from a particular set of planes—and that produces a particular spot in the diffraction pattern—is proportional to $|F(h, k, l)|^2$.

Practically speaking, the fundamental problem of diffraction, used as a probe of crystal structure, is to infer from the pattern of spots the three-dimensional distribution of electron density $\rho(\mathbf{r})$. The atomic form factor f_j and the electron density contributed by the j-th atom are connected by a Fourier transform relation. From the inverse transformation one can express the total electron density $\rho(\mathbf{r})$ as a Fourier series in which the structure factors $F(h, k, l)$ are the expansion coefficients

$$\rho(\mathbf{r}) = \frac{1}{V} \sum_{h,k,l} F(h, k, l) e^{-2\pi i (hx/a + ky/b + lz/c)} \tag{4}$$

and V is the volume of the unit cell. If the crystal structure—in effect $\rho(\mathbf{r})$—were known, one could calculate $F(h, k, l)$ and predict the diffraction pattern. It is the "inverse problem," however—to deduce the structure from the diffraction pattern—that physicists and chemists are often faced with, and this is in general a difficult problem to solve. The principal impediment is the loss of phase information; to determine $\rho(\mathbf{r})$ requires knowledge of the complex structure factors, whereas the diffraction pattern furnishes only their absolute magnitudes.

In practice one has approached the inverse problem in an iterative way by assuming a trial structure and calculating the ensuing diffraction intensities. If the hypothetical structure is approximately accurate, the strongest observed intensities will be in reasonable accord with the calculated values, whereupon one then adopts for these reflections the experimentally obtained magnitudes of $F(h, k, l)$ and the calculated phases. Calculation of $\rho(\mathbf{r})$ should then further refine the positions of the atoms, from which new structure constants can be inferred that thereby permit the determination of additional phases. High-speed digital computers greatly facilitate this work, but it is still a difficult and tedious procedure to unravel the structure of large and complex molecules.[4]

The diffractive theory of image formation provides an alternative methodology. From the perspective of optical imaging—rather than the physics of materials—relation (4) has an interesting and suggestive structure. Replace Bragg-diffracted X-rays by Fraunhofer-diffracted light waves, replace the electron density by an optical image—and Eq. (4) is the discrete analog of the fundamental expression that shows how a lens works.

That image formation is at root an interferometric process was probably first recognized by the nineteenth-century microscopist Ernst Abbe, of the University of Jena. Commissioned by microscope manufacturer Carl Zeiss to solve the frustrating problem of why the smallest—and therefore best-crafted—lenses produced images with poorer resolution than larger, less accurately shaped lenses, Abbe discovered ($\sim$1873) in effect the basic principles of Fourier optics. Albert B. Porter's remarks, cited at the beginning of this chapter, capture in their brevity the kernel of Abbe's discovery. Upon illumination, a transparent object (e.g., the specimen in a microscope) diffracts the transmitted light, giving rise to a spectrum of spatial frequencies, that is, rays at different angles of inclination to the optic axis. Those rays propagating closest to the optic axis of the system (low spatial frequencies) contribute principally to the overall illumination of the image; the rays diffracted at large angles (high spatial frequencies) contribute most to sharpness of detail and therefore to image resolution. Small lenses, Abbe found, simply did not capture and refract the high-spatial-frequency rays from a specimen.

The theory of optical imaging as the processing of spatial frequencies has a direct counterpart in the electronic processing of time-varying signals. Here the dc (zero frequency) component of the signal sets the overall electrical bias, whereas the detailed shape of a signal is determined by the higher temporal frequencies. However, in contrast to the sequential computation required by a digital computer to determine the Fourier transform of an electronic signal, the two-dimensional Fourier transform of a transparent object may be generated nearly instantaneously in the back focal plane of an objective lens.

A lens, therefore, is not just the familiar piece of glass used for centuries to focus light. By generating in its back focal plane the Fraunhofer diffraction pattern of the object under scrutiny, it is also an optical computer of Fourier transforms. We may see in a relatively simple way how this works by restricting our attention to a thin double-convex lens (of refractive index n) in the paraxial approximation—that is, for light rays incident on the lens at shallow angles with respect to the optic axis. Both surfaces of the lens are sections of a sphere with radius R. The effect of the lens on a monochromatic incident wave $\psi(\xi, \eta)$— where the origin of the coordinate system is coincident with the center of the lens—is to multiply $\psi(\xi, \eta)$ by a phase factor $t(\xi, \eta)$ of the form (up to an unimportant constant phase)[5]

$$t(\xi, \eta) \approx \exp\left(-i\frac{k}{2f}(\xi^2 + \eta^2)\right) \tag{5}$$

where $k = 2\pi/\lambda$ and the focal length f is defined by the same relation as in geometrical optics

$$\frac{1}{f} = \frac{2(n-1)}{R}. \tag{6}$$

Since f is positive for the lens we are considering, Eq. (5) represents a spherical wavefront converging toward a point a distance f from the lens. Within the thin-lens approximation, once $t(\xi, \eta)$ has been derived, the lens is regarded as a planar object, and no distinction need be made between the locations of its surfaces and center. The lens, then, converts a plane wavefront into a spherical wavefront with radius f.

Suppose next that the lens is to be placed against a thin transparent object. In the thin-lens aproximation it does not matter whether the lens is immediately before or behind the object; the principal effect is the same: to multiply the wave $\psi(\xi, \eta)$ transmitted by the object by the phase factor $t(\xi, \eta)$. Recall (from chapter 5) that in the absence of the lens the wave at some point $P(x, y)$ in the image plane, a distance z from the object plane, takes the approximate form

$$\Psi(x, y) \propto \frac{1}{\lambda z} \iint \psi(\xi, \eta) e^{ikr} \, d\xi \, d\eta \tag{7}$$

where r is the distance between $P(x, y)$ and a point $Q(\xi, \eta)$ in the object plane, and the "standard" assumptions have been made: Obliquity factor is unity, and variation of r in the denominator is negligible. Since the object and image are assumed separated by a distance large compared with their transverse dimensions, one can approximate r by expansion in a truncated binomial series

$$r \approx z + \frac{\xi^2 + \eta^2}{2z} - \frac{\xi x + \eta y}{z}. \tag{8}$$

The term quadratic in ξ and η on the right-hand side expresses the wavefront curvature engendered by Fresnel diffraction. Upon replacing $\psi(\xi, \eta)$ by $t(\xi, \eta)\psi(\xi, \eta)$ and locating the image in the focal plane of the lens ($z = f$), one sees that the lens eliminates the curvature of the diffracted wavefront. The image is then represented by a wave

$$\Psi(x, y) \propto \frac{1}{\lambda f} \iint \psi(\xi, \eta) e^{-ik\left(\frac{x}{f}\xi + \frac{y}{f}\eta\right)} \, d\xi \, d\eta \tag{9}$$

which is precisely the Fraunhofer diffraction pattern, or spatial Fourier transform, of the wave diffracted by the object. The lens, in effect, brings the

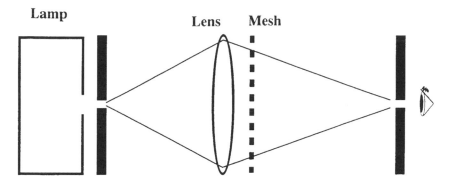

Figure 7.2. Configuration of Porter's spatial filtering experiment. The diffraction pattern of the wire-mesh object is projected onto the viewing screen and transformed into a real-space image by the lens of the viewer's eye.

Fraunhofer pattern in from infinity and locates it at a distance one focal length away.

Abbe and others had devised a number of ingenious experiments to illustrate this idea, but Porter may have been the first to make explicit its mathematical foundation. In Porter's prescient remark—that the object seen through a lens is *not* the object actually present, but the fictitious one whose diffraction pattern the lens passes—is the basis for all optical information processing; by having access to the diffraction pattern of an object, one can modify it and thereby reconstruct the image of an object one wants rather than the object one sees. This procedure is known as spatial filtering, and Porter—decades before the advent of the laser—described one of the most striking and easily reproducible examples based on Abbe's experiments.

Working at a time when a number of his colleagues still believed that microscopic and macroscopic vision were essentially different—and that diffraction played no part in the latter—Porter performed the following experiment to demonstrate "that the images of periodic structures formed by the naked eye itself are due to diffracted light." The configuration of the experiment is sketched in figure 7.2. A piece of wire mesh (with about 30 wires/cm) was coherently illuminated by light from an arc lamp issuing from a pinhole. A double convex lens placed immediately in front of the mesh projected its Fraunhofer diffraction pattern onto a flat cardboard screen located in the focal plane some 25 or 30 cm distant. The basic features of this pattern are quite simple. There is a sequence of vertical spectra ("spots") produced by diffraction from the horizonal wires, and conversely a sequence of horizontal spectra produced by diffraction from the vertical wires, with peak intensities diminishing with diffraction order.

Porter then pierced holes in the cardboard screen so as to allow selected portions of the diffraction pattern to be viewed directly with one eye. When only the central diffraction spot (spatial frequency zero) was passed, the wire gauze, Porter noted, "is quite invisible." No structural details could be seen at all. Piercing a small horizontal slit to permit the central spot and the first lateral spot on each side to pass, Porter saw an object comprised exclusively of vertical wires. Similarly, a vertical slit which admitted the central spot and the first two flanking maxima showed an object with horizontal wires only. And if a slit were pierced diagonally, again allowing the central spot and first two flanking spots to pass, the viewed object appeared to consist of diagonal wires oriented orthogonal to the slit. Finally, if instead of a slit, Porter pierced three separate pinholes so as to pass the central spot and two horizontal second-order spectra, the perceived object consisted of vertical wires as before, but with half the spacing. It is only when the entire diffraction pattern enters the eye of the viewer that the resulting image appears exactly like the original rectilinear wire mesh—although in practice a reasonable replication of the mesh is achievable with only the central spot and first few orders of the horizontal and vertical spectra.

I have often performed with my students a variation of the preceding experiment that (1) replaces the arc lamp with a more convenient helium-neon laser outfitted with pinhole spatial filter—as in the Newton two-knife experiment—to rid the beam of noise, and (2) explicitly employs two lenses (as shown in figure 7.3), preferably identical large-diameter lenses with long focal lengths. The first lens, as already noted, generates the Fraunhofer diffraction pattern of the object; the second lens (the lens of the eye in Porter's configuration), separated from the first by the sum of the two focal lengths, generates in its back focal plane the Fourier transform of the diffraction pattern—or, in other words, reconstructs the image of the original object. Halfway between the two lenses (if one uses lenses of the same focal length), coincident with their common focal plane, is placed a sheet of transparent plastic (such as used with overhead projectors) to receive the diffraction pattern of the copper mesh or whatever object is employed.

The larger the lenses, the larger and more accessible to manipulation is the diffraction pattern on the transparent screen. Now, instead of cutting pinholes into opaque cardboard, one merely puts thin strips of opaque tape across the diffraction orders to be blocked; the effect can be seen immediately on a white viewing screen in the back focal plane of the second lens. The various manipulations reported by Porter—passing the central spot and sundry horizontal, vertical, or diagonal spectra—are easily effected with this experimental arrangement. One particularly interesting experiment is to block only the central spot and admit all the rest. In the reconstructed image the mean level of illumination is now zero. Points on the image that had hitherto been bright now appear dark; points that had previously been dark (light amplitude = 0) are now bright (light amplitude is negative which implies a 180° phase change, but yields a nonvanishing intensity). Thus, blocking the 0-th-order spot generates

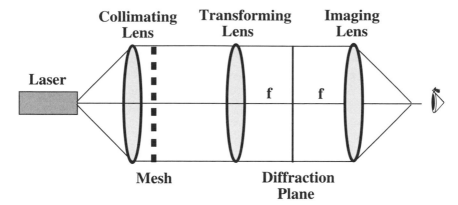

Figure 7.3. Modern variation of Porter's experiment. Laser light, filtered through a pinhole and then collimated, illuminates a wire-mesh object. The transforming lens generates the Fourier transform (Fraunhofer diffraction pattern) of the mesh on a transparent plastic sheet in the back focal plane. This pattern can be modified by taping over selected diffraction orders, and the imaging lens reconstructs the corresponding real-space image.

the photographic negative of the object. This method of dark-field imaging is often employed in microscopy to emphasize certain features of the specimen.

That an experimenter can modify the spatial frequency content of a diffraction pattern to craft an image of his choosing has a cautionary downside, for such modification can also occur unwittingly in the ordinary application of an optical instrument with lenses, like a microscope. Porter gives (see note 1) a striking example of such an occurrence in the unfaithful images of a double grating observed through a microscope with light of different colors—and one would do well even today to take note of his conclusion:

> In the experiments which have been described, the microscope was used under what would be considered normal working conditions, with central illumination, and circular diaphragms centred on the optic axis. Nevertheless, when certain relations existed between the aperture of the lens and the coarseness of structure of the object, images were formed which were utterly false in their smaller details, and other images were profoundly modified by the presence of structure lying entirely beyond the focal plane. It therefore seems that a working knowledge of ... the laws of diffraction might well form a part of the equipment of anyone whose uses the microscope and attempts to interpret its indications.

The photograph of the single hexamethylbenzene molecule that had intrigued me as a child is another example of spatial filtering, or rather the opposite process: optical Fourier synthesis. I do not know with certainty the exact details of its preparation, but it is likely that it was prepared from a mask in which circular holes were punched or drilled at the locations of the X-ray diffraction spots, the mask then being illuminated with visible light. In order that the intensity of the diffracted light rays be proportional to the intensity of the corresponding X-rays, the area of each hole was made proportional to the magnitude $|F(h, k, l)|$ of the associated structure factor. Replication of the correct relative phases, in general difficult to do, is simplified if a molecule (like hexamethylbenzene) is centrosymmetric. In that case relative phases are either 0 (the Fourier coefficient is positive) or π (the coefficient is negative) and can be set by placing an appropriately oriented piece of mica, a birefringent material, across each hole. The entire procedure is a very tedious job even for a symmetric molecule, as described by colleagues of Bragg who continued work on the "X-ray microscope":

> About eighty brass cylinders were prepared, each about 1 cm long, with outer diameter 4 mm. and inner diameter 2 mm. One end was counterbored to a diameter of 3 mm., and a depth of about 2 mm. A mica disc was placed in the end of each cylinder, and held firmly in place with a piece of wire spring.
>
> Reciprocal-lattice sections [i.e., diffraction spots] were represented by drilling holes of 2 mm. diameter at appropriate points in an opaque plate about 1 cm. thick, counterboring these holes to a depth of about 7 mm., and a diameter of 4 mm., and placing a brass cylinder in each hole. The required intensities and phase relations could then be introduced by appropriate rotations of the cylinders.[6]

Thus, to make the sort of photograph shown in Gamow's book may well have taken several days at least. For this reason alone, and the fact that high-speed digital computers can now replace all this manual labor with a rapid simulation of any object—even nonexistent ones—from a theoretical diffraction pattern, Bragg's technique is rarely used today for "seeing" atoms.[7] Also, to one whose vision of microscopy was molded by the small light microscope with triple-objective turret that he played with as a child, the so-called X-ray microscope turned out to be a bit of a swindle. Nevertheless, it has the distinction, broadly speaking, of being one of the first lensless imaging methods with atomic-scale resolution.

With wavelengths considerably shorter than X-rays, it is the high-voltage electron microscope, such as the field-emission instrument introduced in the previous chapter, that affords a view of the structure of matter with greatest magnification and resolution. Unfortunately, the device is limited by its lenses. Whether based on electrostatic or (as is more commonly the case) on magnetic fields, electron optical lenses, like their light optical counterparts, encounter difficulties not only of the kind signaled by Abbe and Porter, but also for more conventional reasons attributable to shape (e.g., geometrical aberrations) and dispersion (e.g., chromatic aberration). Electron lenses must have cylindrical symmetry. Since the motion of an electron in a magnetic field is intrinsically three-dimensional, the theoretical analysis of magnetic lenses is fairly complex, and lens design remains to a large extent empirical. To minimize aberrations only very small angular apertures can be used, with the consequence that achievable resolution may be well below what is theoretically possible.

It was precisely to circumvent such difficulties in electron optics that Gabor developed the technique of holography,[8] although the lack of a coherent electron source at the time made it impossible to implement his idea. The invention of the laser (which has since made optical holography commonplace in both science and art) and the highly coherent field-emission electron source brought electron holography to fruition as a practical tool, although the microscopes used for that purpose still require electron lenses. Like the X-ray "image" of the hexamethylbenzene molecule, present-day electron holography is effectively a two-step process requiring, first, the creation of an electron diffraction pattern (the hologram) and, second, the reconstruction of the visible image through illumination, usually with the light of a helium-neon laser.

There are, besides the field-emission microscope (FEM), more recent lensless electron-imaging processes capable of rendering atomic-scale resolution. These instruments, such as the scanning tunneling microscope (STM), atomic force microscope (AFM), and an "alphabet soup" of other microscopes, provide spectacular images of atoms on surfaces.

It may at first seem surprising that the STM (and related technologies) can achieve atomic-scale resolution, because the de Broglie wavelength of the typically 1 eV electrons employed is ~12 Å—about 200 times larger than the electron wavelength in a high-voltage FEM and about 4–5 times the typical interatomic distances in solids. Standard optics texts teach that one cannot resolve structure smaller than the wavelength of the illumination; that was ostensibly the reason for constructing electron microscopes in the first place. This limitation applies, however, only to far-field viewing. With the emission tip just fractions of a nanometer from the specimen, the STM operates in the so-called near-field regime in which the image is not diffraction limited. There is no electron beam to speak of, and the images of near-field viewing need not resemble those of far-field viewing which correspond to what one ordinarily means by "seeing" in the traditional sense of Abbe's theory.

One of the most recent advances in electron imaging is a sort of hybrid between the STM and the FEM: the low-voltage field-emission point-projection microscope (PPM).[9] Employing what is perhaps the oldest known method of creating images, this tool holds out the promise of fulfilling many a scientist's dream: to illuminate a sample and simply project upon a distant screen a ready-made, blown-up picture of its atomic structure.

7.2 CASTING SHADOWS

To someone not familiar with the sources and characteristics of particle beams, the question "What is the brightest beam known to science?" may well conjure up a picture of the Sun or some exotic astrophysical object pouring light and matter into interstellar space, or perhaps an immense high-power laser used in nuclear fusion, or the high-energy beam of a gargantuan particle accelerator like the (now defunct) superconducting supercollider. But these images would be incorrect. The brightest particle source currently available, developed at the IBM Research Laboratory in Zurich,[10] is produced by the field emission of electrons from a minute tungsten tip of basically atomic size.

Designated a nanotip for its nanometer-scale emission region, or a teton tip for its nipple-like shape, this new electron source is brighter than a synchrotron and—per unit of energy—more than a billion times brighter than the Sun. Brightness, as I will discuss in more detail shortly, is related to virtually all the principal attributes of a beam—intensity, divergence, coherence, degeneracy— and is perhaps the single most significant determinant of the overall performance of microscopes and interferometers.

For one thing, the size of a particle source (in either light or electron optics) is ultimately related to the resolution of the images it produces. An ultrasharp metal tip of a fraction of a nanometer effective radius of curvature—often referred to, somewhat misleadingly, as a "point" source, although it terminates in a single atom or small cluster of atoms—can in principle give rise to images of the surface features of material samples with atomic-scale resolution.

Besides being small—or, rather, partly as a consequence of this very small radius of curvature—the application of a low extraction potential (for example, less than 100 V) produces an intense electric field in the vicinity of the tip, a field sufficiently high for electrons to be drawn off by field emission. In the previous chapter it was noted that extraction potentials of several thousand volts are required for conventional field-emission sources—and the electrons are subsequently accelerated to energies of over a 100,000 eV. By contrast, the "gentle" low-energy electrons from a nanotip can be a notable advantage in the investigation of fragile specimens such as those of biological origin. For example, the range of electrons in organic films reaches a minimum of about 0.6 nm at 100 eV, but increases dramatically below this value. There is hope, therefore, that nanotip microscopes can provide useful high-resolution images of organic materials with much less damage than synchrotron or high-energy electron beams. The damage these cause has been estimated to be comparable to that arising in close proximity to a nuclear explosion.[11]

The greatest consequence of source size, however, concerns the property of coherence, that is, the capacity of superposed waves to interfere. It is this property that has made the field-emission electron microscope one of the most important tools of modern physics to exploit the quantum behavior of electrons. In discussing imaging and interferometry with electrons it is necessary, if we are to avoid cumbersome locutions, to switch freely between the classical imagery of waves and particles.

Electrons, as particles, leave the nanotip source along electric field lines and can project shadowlike images of an opaque object onto a distant viewing

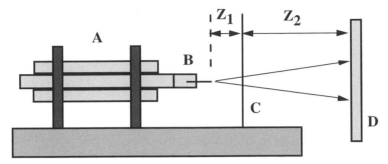

Figure 7.4. Geometric configuration of a point-projection electron microscope with inchworm positioner (A), nanotip (B), specimen stage (C), and microchannel plate detector (D). Z_1 is the source-object distance; Z_2 is the object-image distance.

screen. Like the shadow images on a far wall created by holding one's hands up to a bare incandescent light bulb, the projected electron images can exhibit enormous magnifications dependent only on geometry, that is, on the ratio of screen-to-sample and source-to-sample distances. For example, if the nanotip is brought to within 10 nm of a specimen, the magnification of the resulting shadow image viewed on a screen some 10 cm distant would be on the order of 10^7. Molecules of 10 Å (1 nm) size would be enlarged to the macroscopic scale of 1 cm. The hexamethylbenzene molecule shown in Gamow's book might actually be photographed some day directly in a single step.

From the perspective of waves, however, the wavefront emitted by a true point source is everywhere spatially coherent—and it is therefore no surprise that the electron "waves" from a real nanotip likewise display a large transverse coherence, theoretically estimated to be on the order of centimeters. (The transverse coherence of the beam produced by a standard field-emission microscope such as employed in the Young's two-slit experiment is a few hundred microns.) One manifestation of this wavelike behavior is the appearance (under sufficiently high magnification) of Fresnel fringes in the electron shadow images of objects sufficiently small as not to block the electron beam entirely. These fringes, which constitute a form of holographic recording since they are formed by the superposition of scattered and reference waves, provide pertinent information about the characteristics of the source.

A schematic diagram of an electron PPM is shown in figure 7.4. Note the simplicity of the device compared with the high-voltage electron microscope (figure 6.1); there are no lenses at all, and the entire instrument is sufficiently compact to fit on an optical table. The emission tip, attached to a piezoelectric positioner, is brought to within the desired range of a specimen; emitted electrons are received at a detector (a microchannel plate) a macroscopic distance away.

If the specimen is sufficiently large to intercept the entire electron beam, and sufficiently thin to permit electron transmission, the resulting pattern is not in the conventional sense a shadow image, but a diffraction pattern characteristic of the lattice structure.[12] There is a particularly striking feature associated with the images of a *periodic* object formed by projection from an ideal point source. In this special case, sharply magnified reproductions of the object produced occur in a denumerable number of focal planes. This process—which is technically not holography as there is neither a distinct reference wave nor a hologram to be reconstructed—was serendipitously discovered by British photography pioneer H. F. Talbot around 1836 and partially explained by Lord Rayleigh nearly 50 years later. I will return to the curious phenomenon of self-focused lattice images in a later section.

Actually, projective imaging of electrons is not new but was demonstrated as long ago as 1939 by two scientists, G. A. Morton and E. G. Ramberg of the RCA Electronic Research Laboratory.[13] The first electron microscope with electron optical lenses had only just been built a little earlier, around 1931, by Max Knoll and Ernst Ruska of the Technische Hochschule in Berlin, and the field-emission microscope appeared four years later. With their lens-free instrument Morton and Ramberg were able to magnify the mesh of an electroplated copper screen approximately 8000 times, but the mechanical steadiness of the instrument was inadequate to permit photography; thus, the images they published of the holes in the mesh showed a maximum magnification of only 3000. Although the idea of lensless electron microscopy was "in the air," the PPM did not catch on and was not pursued further until much later when the needs of an entirely different mode of viewing—scanning tunneling microscopy—prompted technological developments that made a high-resolution point-projection instrument possible.

Like the FEM—in which electrons tunnel from a metal emission tip into a vacuum through a barrier lowered by an externally applied potential—the STM also operates on a principle of quantum mechanical tunneling. In the latter case, however, a bias voltage is applied between a conducting sample (e.g., a metal or semiconductor) and a sharp metal probe that is brought to within a few Angstroms of the material surface under investigation. Electrons tunnel between the surface and the probe with a probability that diminishes exponentially with distance. In operation, the tip is scanned over a minute portion of the surface; the tunneling current is kept constant by an electronic feedback system, and the compensating variations in probe-surface separation map out the topography of the surface.

It is the capacity to fabricate an ultrasharp tip and position it with great precision at extremely small distances from a sample that makes both the STM and PPM possible. Thus, the development of a new microscope led to the resurrection of an old one.

7.3 BRIGHTNESS, COHERENCE, AND DEGENERACY

Although the concept of brightness can be more rigorously defined, it is sufficient for the present discussion to define the mean axial brightness B of an electron source as the current I emitted through cross-sectional area A into a solid angle Ω about the forward direction

$$B = \frac{\partial^2 I}{\partial A \, \partial \Omega}. \tag{10}$$

Divided by the relativistically corrected acceleration potential (Eq. 6.8), the resulting "reduced" brightness is an invariant along the axis of a symmetric electron-optical system. Thus, for a nonrelativistic electron source, B effectively scales with applied potential.

There is a theoretical maximum brightness that any electron source of given energy and energy dispersion can have. This limitation, which does not apply to light, is imposed by quantum mechanics on fermionic—that is, spin-$\frac{1}{2}$—particles like electrons, neutrons, and protons. To understand this, let us return to the particle picture and try to imagine an electron beam as a sort of fluid in which the state of motion of each constituent particle is completely specified by coordinate (q) and linear momentum (p) variables. Specification of q and p for each degree of freedom of each particle in the system uniquely determines the microscopic state of the system as a point in a multidimensional space, or phase space. In order to enumerate in some way the possible states of the system, one can arbitrarily choose fixed intervals, δq and δp, into which the full range of coordinate and momentum variables are to be subdivided. Thus, the entire phase space can be partitioned into small cells of equal size and volume $(\delta q \delta p)^{Nf} = h^{Nf}$, where f is the number of degrees of freedom (1 for pointlike particles moving along an axis, 2 for particles moving in a plane, etc.), N is the number of particles in the sytem, and h is an arbirary constant with dimension of angular momentum (coordinate $\times$ linear momentum). The state of the system can then be localized to a particular cell of phase space by specifying each coordinate to lie between certain values of q and $q + \delta q$ and likewise each momentum to lie between p and $p + \delta p$.

Were electrons classical particles, one could specify the system ever more precisely by choosing the basic volume h as small as desired. However, quantum theory, as is well known, limits this value to (approximately) Planck's constant, below which further precision is physically meaningless; the precision with which momentum and coordinate variables can be simultaneously determined is restricted by the Heisenberg uncertainty principle whereby $\delta q \delta p \geq h/2\pi \equiv \hbar$.

Beam degeneracy refers to the mean number of particles per cell of phase space. In this regard, however, quantum physics imposes another restriction beyond the simultaneous precision of canonically conjugate variables (variables that satisfy a commutation relation of the form $[q, p] \equiv qp - pq = i\hbar$). No two identical fermions characterized by the same quantum numbers can occupy the same cell of phase space. The maximum phase space density of electrons is therefore two electrons with opposite spin components per cell corresponding to a "degeneracy parameter" δ of unity. Whereas the degeneracy of a light source like a laser can be enormous (e.g., $\delta > 10^{12}$), in practice the degeneracy of a real electron source is orders of magnitude smaller than one.

Nevertheless, all things being equal, the brighter a source, the more degenerate it is. If Δn is the number of particles received in a time interval Δt within a solid angle $\Delta \Omega$ through a detecting surface ΔA, then (from the definition of brightness)

$$\Delta n = \frac{B}{e} \Delta A \Delta t \Delta \Omega. \tag{11a}$$

(All quantities designated by Δ are assumed small—in principle infinitesimally small.) Following the preceding discussion of phase space and degeneracy, one can write an alternative, but equivalent, expression for Δn by considering the phase space volume element $\Delta V \Delta \mathbf{p}$ of a flux of particles with momentum vector $\mathbf{p}$ and speed v:

$$\Delta n = (\text{number of particles/cell of phase space}) \times (\text{number of occupied cells})$$
$$= 2\delta \left(\frac{\Delta \mathbf{p} \Delta V}{h^3} \right) = \frac{2\delta}{h^3} \left[(p^2 \Delta p \Delta \Omega)(v \Delta t \Delta A) \right] \tag{11b}$$

where δ is the degeneracy parameter and the factor 2 refers to the two spin states of the electron. Comparing the two preceding relations for the case $\delta = 1$ leads to the maximum theoretical brightness

$$B_{\text{max}} = \frac{2evp^2 \Delta p}{h^3} = \frac{2ep^2 \Delta E}{h^3} = \frac{2e \Delta E}{h\lambda^2} \xrightarrow{NR} \frac{4me E \Delta E}{h^3}. \tag{12}$$

The second equality in Eq. (12) follows from the relativistically exact relations connecting energy, momentum, and speed (Eqs. 6.6a, b), whereas the third equality, also relativistically exact, results from substitution of the de Broglie wavelength $\lambda = h/p$. For the case of nonrelativistic electrons with *kinetic* energy $E = p^2/2m$, the foregoing expressions reduce to the fourth (approximate) relation, which is ordinarily the easiest to apply since it depends only on the (kinetic) energy and energy width of the beam.

From relations (11b) and (12) one sees that the degeneracy parameter is related to the actual and maximum beam brightness by the simple expression

$$\delta = \frac{B}{B_{\text{max}}}. \tag{13}$$

It is useful to relate explicitly the "corpuscular" degeneracy parameter and the "undulatory" coherence parameters by noting that the volume of a cell in phase space takes the form

$$(p^2 \Delta p \Delta \Omega)(L_c A_c) = h^3. \tag{14}$$

Here, $A_c \sim L_t^2$ is the lateral coherence area in which the transverse coherence length L_t has already been discussed in chapter 6 (see Eq. 6.18). The longitudinal coherence length L_c depends on the coherence time T_c

$$L_c = vT_c \qquad (15)$$

which is related to the energy uncertainty of the beam in the following way:

$$T_c = \frac{h}{\Delta E}. \qquad (16)$$

L_c, loosely speaking, is the length of the electron wave packet emitted by the source. If an electron wavepacket were coherently split (e.g., by amplitude division in a Michelson or Mach-Zehnder interferometer) and subsequently recombined, no interference fringes would be seen if the difference in optical path lengths traveled by the two components greatly exceeded L_c. In effect, the two components would not overlap at the viewing screen or detector.

From the definition of brightness and the various coherence parameters, one may express the degeneracy parameter as

$$\delta = \frac{j}{e} A_c T_c. \qquad (17)$$

In other words, the degeneracy of a beam is the mean number of particles passing through a coherence area (normal to the beam) in an interval of a coherence time. The beam degeneracy—and therefore brightness—fundamentally determines the magnitude of novel quantum interference effects involving correlated particles.[14] I will return to the significance of the nanotip electron source for fundamental quantum physics at the end of this chapter.

Although brightness is one of the most critical parameters for gauging the performance of an electron optical instrument, it is also one of the most difficult to measure. The determination of brightness can be made by two general approaches rather different in principle. On the one hand, by the so-called minimum probe size method, one can try to assess directly the experimental quantities—total current, focused probe radius (and hence beam cross-sectional area), and beam divergence—that appear in the defining Eq. (10). This is the standard approach implemented on lens-focused beams. On the other hand, one can deduce brightness through knowledge of the coherence parameters of the beam. Theoretical considerations of spatial coherence in standard field-emission microscopes have shown that the two methods give essentially the same results when the source is incoherent—as would be the case for independent random emissions from widely separated and uncorrelated emission sites on a broad tip—but not for the opposite case when the illumination radius is comparable to, or smaller than, L_t. In this latter case, which applies to the nanotip source, the intensity method gives underestimated values. The reason for this is that the true size of the source (the Gaussian source radius) cannot be inferred in the coherent case by direct measurement of a focused beam intensity; the Gaussian image of the source is hidden in the coherent pattern of the condenser lens aperture.

Nevertheless, in recent experiments applying the standard method to a lensless point-projection configuration,[15] the average brightness of an ultrasharp tungsten single-crystal tip of approximately 1 nm radius—that is, terminating in a single atom—was estimated to be 3.3×10^8 Amperes per cm^2 per steradian (A/cm^2-sr) at an energy of 470 eV. To put this number into perspective, note that this is the same order of magnitude brightness as produced by a 100 kV field-emission microscope with standard-size tip. Since brightness scales with potential, the nanotip source would have an equivalent brightness of 7.7×10^{10} A/cm^2-sr at an energy of 100 keV—that is, two orders of magnitude greater than a conventional field emitter.

For purposes of comparison with other sources, it is useful to consider the actual particle flux per unit energy. The Sun, for example, emits roughly 2×10^{21} photons/sec-cm^2-sr per eV (deduced from a solar constant of 135 mW/cm^2 measured over the bandwidth 330–1100 nm with mean wavelength of ~600 nm.)[16] Advances in "wiggler" and "undulator" technology make possible synchroton sources producing 10^{25}–10^{26} photons/sec-cm^2-sr per eV (assuming a bandwidth of 10 keV).[17] A nanotip with the brightness cited above and a typical energy dispersion of approximately $\Delta E = 0.2$ eV corresponds to 2.4×10^{30} electrons/sec-cm^2-sr per eV—at least four orders of magnitude greater than a synchrotron and a billion times the reduced brightness of the Sun.

7.4 THE WORLD'S SIMPLEST ELECTRON INTERFEROMETER

Although microscopes are ordinarily employed to magnify small objects for viewing, they can also be used to provide information about the source of illumination; indeed, they can serve both functions at the same time. One important example of this is provided by the field-ion (not to be confused with the field-emission) microscope; here the source and specimen are the same object. In the field-ion microscope (FIM) a large positive potential is applied to a sharp tip made of a refractory metal (e.g., tungsten) located opposite to a grounded fluorescent screen. The tip is in a vacuum chamber filled with a gas such as helium. Because of the strong and highly nonuniform electric field at the tip, helium atoms are polarized, attracted to the tip, subsequently ionized, and then repelled and accelerated to the screen where they produce a magnified image of the surface features of the metal. Since the field is most intense at protuberant atoms on the surface, and ion trajectories are highly sensitive to variations in field strength, the resulting field-ion images can show surface features with atomic-scale resolution.

The FEM and FIM are actually quite similar in operation. Indeed, by reversing the potential of the tungsten tip from negative to positive and admitting the imaging gas, one can convert a FEM into a FIM. (Converting one mode of operation into another is actually a useful procedure employed during the fabrication of PPM emission tips to ensure that the tip terminates in a single atom or small cluster of atoms.) The imaging gas atoms arriving at the detector appear to come from a point at the center of a spherical tip. Thus, they form a projective image of the surface analogous to that produced by the electron point-projection microscope.

This interpretation helps us understand why atoms were first observed by field-ion microscopy rather than by electron microscopy. Mechanical vibration of the source with respect to the sample is magnified in any electron projection imaging system, resulting in blurring of the image. It was precisely this impediment that frustrated the efforts of Morton and Ramberg to record higher magnifications. But for the FIM, the virtual source (inside the tip) cannot move with respect to the sample (the surface of the tip), and any movement of both together is not magnified. It is not surprising, therefore, that it was many years later ($\sim$1958) that electron microscopy attained the stability required for atomic resolution.

The use of a nanotip source in a field-emission PPM allows one to determine other properties of the source besides the arrangement of surface atoms. There is much interest, for example, in being able to ascertain the size, brightness, and degeneracy of these new sources.[18] One way to do this is by using the electron microscope "out of focus" as an electron interferometer. In fact, in the experimental configuration illustrated in figure 7.4 the lensless PPM constitutes what may well be the world's simplest electron interferometer.

Because electrons do not penetrate far into matter even when accelerated to energies on the order of 100 keV, transmission electron microscopy is largely limited to thin samples, in contrast to light microscopy in which both optically thick and thin samples can be studied. Thick samples are amplitude objects; they can be observed as a result of the varied light transmission through different parts of the sample. For optimal clarity, the image is recorded at exact Gaussian focus when the object and image fall on conjugate planes within the microscope, that is, at locations connected by application of the "lens equation." For thin lenses one has the familiar relation

$$\frac{1}{\text{object distance}} + \frac{1}{\text{image distance}} = \frac{1}{\text{focal length}}, \tag{18}$$

but, in the case of thick lenses, which ordinarily characterizes electron microscopes, more general relations are needed.[19] In contrast to amplitude objects, thin specimens are often largely transparent phase objects. Viewed by unpolarized light at the Gaussian focal plane under conditions of either coherent or incoherent illumination, a phase object is essentially indistinguishable from the background since it affects the phase, and not the intensity, of the transmitted light.

One way to enhance the contrast of a transparent phase object is to illuminate it with a source of coherent radiation and record the image slightly out of focus. This basic feature can be demonstrated readily by a simple experiment employing a helium-neon laser as the source and a transparent microscope slide with small "dip," etched out by a drop of hydrofluoric acid, as the specimen. At Gaussian focus, the dip is practically invisible, but under conditions of under- or overfocus it stands out strongly. The observed phase contrast is essentially an interference effect deriving from refraction of waves at the dip and superposition with undiffracted incident waves. The same procedure has its parallel in point-projection electron microscopy where Fresnel edge effects are examples of phase contrast.

Although the PPM was first demonstrated over a half century ago, it is only recently that instruments of sufficient resolving power and coherence have been developed that can produce electron Fresnel fringes and atomic resolution. In the instrument of figure 7.4, constructed at the National Center for High-Resolution Electron Microscopy at Arizona State University, the tip-to-specimen separation Z_1 is precisely controlled by an ingenious positioning device known as an "inchworm." The inchworm, which can translate the tip in steps of ~ 2 nm, comprises a shaft (that contains the tip and does the actual "walking") held by three sequential piezoelectric elements serving as clamps. Suppose the three clamps are "open." When a potential is applied to one of the two outer elements, the diameter of the element shrinks, and the clamp firmly grasps the shaft. The central element, which makes no direct contact with the shaft, is then excited by an applied voltage, whereupon its length is extended and it pushes the first clamp (with the grasped shaft) forward. The shaft is then grasped by a third clamp; the first clamp is opened, and the potential is removed from the middle element, which recovers its original length. The inchworm is then ready for another step forward or backward depending on the sign of the applied voltages. In this way the distance Z_1 and hence the magnification of the microscope

$$M = \frac{Z_1 + Z_2}{Z_1} \xrightarrow{Z_2 \gg Z_1} \frac{Z_2}{Z_1} \tag{19}$$

can be changed continuously. (The resolution of the microscope, however, is approximately equal to the size of the emitting area on the source.) Piezoelectric devices are also employed to move the tip up or down.

The projected electron images are detected by a single-stage microchannel plate and phosphor screen and recorded using a CCD (charged-coupled device) camera system with 384×532 pixels. The microchannel plate, described in chapter 6, is in effect an electron multiplier capable of generating $\sim 10^4$ output electrons for each electron incident at one of the channels. The plate used with the PPM had an active surface area of 12.5 cm^2 and a resolution of approximately 10 mm.

Figures 7.5 and 7.6 show examples of PPM images of a thin perforated carbon film produced by a beam of 90 eV electrons ($\lambda = 0.13$ nm).[18] Under low magnification (figure 7.5) one sees the sharp shadow image of the copper microscope grid covered with film; each square of the grid is 150 μm across. Under high

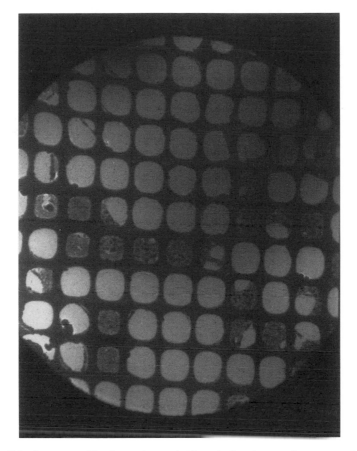

Figure 7.5. Low-magnification point-projection shadow image of a copper electron microscope grid covered with a thin carbon film containing holes; each square is 150 microns across.

magnification (figure 7.6), the edge of a single hole occupies a large fraction of the viewing window, and Fresnel fringes are clearly visible. In this config-uration, the microscope is functioning like an interferometer. The visibility of Fresnel fringes testifies to the high transverse coherence of the electrons emitted from the nanotip and to the mechanical stability of the instrument, which was achieved in part by "floating" the entire vacuum-sealed microscope on rubber tires.

Under the condition, applicable in this experiment, that the electrons do not undergo multiple scattering within the foil sample, the fringes may be analyzed exactly as if they were produced by light—viz., by the Fresnel-Kirchhoff scalar theory. Figure 7.7 illustrates the pertinent geometry for both the PPM, in which the sample is illuminated with spherical waves from a proximate source

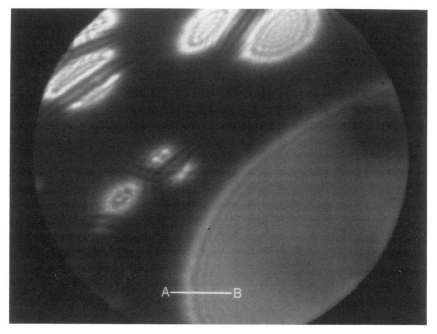

Figure 7.6. High-magnification PPM image of perforated carbon film recorded on a channel plate at 90 eV and beam current less than 1 nm. The tip-to-specimen and specimen-to-screen distances are, respectively, $Z_1 \sim 461$ nm and $Z_2 \sim 14$ cm with resulting magnification $M \sim 3 \times 10^5$. Fresnel fringes, extending over a transverse coherence length $L_t \sim 26$ nm in the specimen plane, were scanned along line AB.

at distance Z_1, and a conventional transmission electron microscope (TEM), in which the sample is illuminated with plane waves from a remote source. There is a sort of reciprocity relation between the two configurations that, once recognized, greatly facilitates the analytical treatment of PPM images. The pattern of fringes produced by the TEM at a distance (the objective lens defocus) $\Delta f = Z_1$ beyond the object is the same as that of the PPM although the conditions of illumination are quite different.

Application of geometric optics to the ray diagram of figure 7.7a, assuming $Z_2 \gg Z_1$ and angles $\alpha \sim \beta \ll 1$ radian, leads to the expression

$$\frac{1}{2}Z_1\alpha_n^2 = n\lambda + \Lambda_0 \tag{20}$$

for the angular location α_n of the n-th-order fringe (maximum). Here Λ_0 results from the phase change introduced by scattering at the edge and depends (for illumination by electrons) on the mean inner potential V_0 of the object and on its thickness. As discussed previously (chapter 5) in the context of the Newton

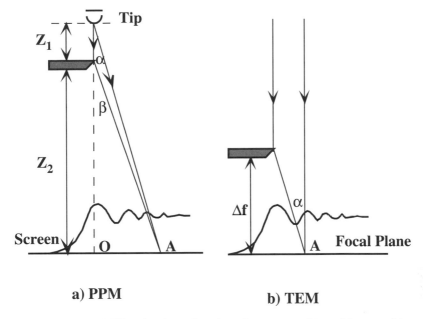

a) PPM   **b) TEM**

Figure 7.7. Fresnel diffraction from the edge of an opaque object. The same fringe pattern is produced by (a) spherical wave illumination in a PPM and (b) plane-wave illumination in a conventional TEM with $\Delta f = Z_1$.

two-knife experiment, the Fresnel fringe pattern for an opaque object may be regarded as the result of interference between an unobstructed incident wave cut off sharply at the geometrical shadow and a cylindrical edge wave diverging from the boundary of the object. Early electron microscopists found good agreement with scalar optical theory in assigning $\Lambda_0/\lambda = \frac{3}{8}$ for the edge wave scattered outside the edge shadow and $-\frac{1}{8}$ for the edge wave scattered inside the edge shadow.[20]

Applying Fresnel-Kirchhoff scalar diffraction theory to a straightedge, one can show that the lateral position X_n and width δX_n of the n-th fringe at the detector are, respectively,

$$X_n = M\sqrt{2Z_1(n\lambda + \Delta_0)}, \tag{21}$$

$$\delta X_n = \frac{M^2 Z_1 \lambda}{X_n} \tag{22}$$

where M is the magnification (Eq. [19]). From a profile of the image density along the line A-B in figure 7.6, the first six Fresnel fringes (with image-object distance $Z_2 = 14$ cm) were measured on the channel plate screen and found to be 4.02, 5.36, 6.26, 6.89, 7.43, and 7.88 mm relative to the edge of the shadow image. These fringes were recorded with a CCD camera capable of resolving the $10~\mu$m fibers of the channel plate, from which a position uncertainty of 0.01 mm may be assumed. The set of fringe locations $\{X_n\}$ together with Z_2 led to a tip-to-specimen distance $Z_1 \sim 461.2$ nm and associated magnification $M \sim 304,000$ by means of Eqs. (21), (22), and (19). One can also deduce the phase shift Λ_0/λ independently of M and Z_1 by applying Eq. (21) to the ratio $\zeta = (X_n/X_{n+1})^2$ of the positions of succesive fringes to obtain

$$\frac{\Lambda_0}{\lambda} = \frac{[(n+1)\zeta - n]}{1 - \zeta}. \tag{23}$$

As just illustrated, Fresnel diffraction affords perhaps the simplest way to determine a very small tip-to-specimen distance (and therefore the magnification), which, by other methods, have required either cumbersome calibration of a piezoelectric tip-deflecting device or manipulation of images of standard objects of known size.

The pattern of Fresnel fringes also leads to the effective size of the source r_e, an experimental quantity of particular significance, for it is a measure of the attainable instrumental resolution. Since interference fringes are produced only if the superposing beams are coherent, the angle α_N corresponding to the highest-order observable Fresnel fringe is correspondingly the coherence angle of the source as expressed by Eq. (6.21b). To a good approximation, the width of the entire band of Fresnel fringes can be equated with the transverse coherence width L_t. The effective source size, defined as the radius of the emission region of an equivalent cylindrically symmetric source, is then usually given by

$$r_e = \frac{\lambda}{\pi \theta_c} \approx \frac{\lambda}{(\pi L_t/Z_1)} \tag{24}$$

which corresponds to Eq. (6.21b) to within a factor of π. The preceding source-size relation is derivable rigorously from application of the Van Cittert–Zernike theorem, one of the most important theorems in modern optics.[21] This theorem relates by a Fourier transformation the coherence of waves at an observation plane to the intensity distribution across the source.

Applied to the measurements on figure 7.6, the foregoing theory yields a lateral coherence $L_t \sim 26$ nm and source size $r_e \sim 0.73$ nm—considerably smaller than the ~ 100 nm size of a standard field-emission tip. Nevertheless, the emission region in this experiment is about twice the interatomic spacing, and it is therefore quite likely that the nanotip in this case did not terminate in a single atom. Because the tip was not observed by means of field-ion microscopy during its preparation, it is not possible to know its end geometry exactly. Measurement of the source brightness, which turned out to be approximately that of a standard 100 kV FEM, also suggests that the nanotip was not as sharp as it could be. To achieve such perfection, the tungsten tip must be prepared according to a tairly elaborate procedure involving chemical etching, electropolishing, annealing, resistance heating (to remove a tungsten oxide layer), ion sputtering (for tip sharpening), and high-field evaporation of atoms (for tip shaping). The tip in the experiment described here was simply formed electrochemically with a potassium hydroxide solution and then cleaned briefly in vacuum at room temperature by field evaporation.

As a practical guide, it is worth noting that r_e is approximately equal to the width of the finest resolvable Fresnel fringe. This can be seen from the relations

$$\frac{\delta X_N}{M} = \frac{\lambda}{\left(\frac{X_N}{M}\right)/Z_1} = \frac{\lambda}{L_t/Z_1} \approx r_e \tag{25}$$

where the first equality is a rearrangement of Eq. (22) evaluated for the outermost fringe N, the second equality recognizes that X_N/M is the transverse coherence length at the specimen plane, and the third (approximate) relation for source size follows from Eq. (24).

The content of Eq. (25) has a close counterpart in the theory of the Fresnel zone plate used in light optics. Indeed, we may think of Fresnel fringes as a one-dimensional Fresnel zone plate. A zone plate is essentially a screen inscribed with concentric circles, the radii of which are very nearly proportional (to within negligible terms containing the square of the wavelength) to the natural numbers, that alters the light in either amplitude or phase coming from every *other* annular region. Each annulus, or zone, has the same area (to within the preceding approximation). Were all zones equally transmissive, light from adjacent zones would reach a certain point on the optic axis of the plate nearly 180° out of phase. However, alternate zones are made opaque, and light received from the remaining open zones arrive at this point in phase, thereby producing intense illumination such as at the focal point of a converging lens.[22] The width of the outermost annulus in a zone plate equals the size of the focused image (or source size).

A zone plate is therefore a flat multiple-focal point lens. It may also be regarded as a hologram of a point scatterer—and, as illustrated by the superposition of "reference" and scattered waves in figure 7.7, Fresnel edge fringes similarly form a hologram of an edge.

Theoretical studies of the lenslike behavior of a nanotip have shown that its spherical and chromatic aberration coefficients are negligibly small. Nevertheless, chromatic aberration does play a role in limiting the resolution of interference fringes. In the Fresnel diffraction pattern of figure 7.6, the dispersion in electron energy ΔE leads to an uncertainty in fringe location ΔX_n (not to be confused with δX_n of Eq. [22]) expressible as

$$\Delta X_n = \frac{dX_n}{d\lambda}\frac{d\lambda}{dE}\Delta E = -\left(\frac{nM^2 Z_1 \lambda}{2X_n}\right)\left(\frac{\Delta E}{E}\right). \tag{26}$$

In order that the fringes not be washed out by chromatic broadening, it is necessary that

$$\delta X_n \geq |\Delta X_n|. \tag{27}$$

By comparison with Eq. (22), it then follows that the highest resolvable fringe number must satisfy the inequality

$$N \leq \frac{2E}{\Delta E}. \tag{28}$$

In the experiment that produced figure 7.6, in which $E = 90$ eV and $\Delta E \sim 0.2$ eV, the highest resolvable fringe number is estimated at $N \sim 900$. Far fewer fringes are actually observable in the figure, so clearly chromatic aberration is not the principal limit of resolution here. The limiting factors in this case are incoherent instabilities such as mechanical vibration of the tip and fluctuating fields in the vicinity of the tip. Such effects lead to an underestimate of the coherence width and therefore to an overestimate of the source size.

7.5 SELF-IMAGING OF A PERIODIC OBJECT

"Although so much has been explained in optical science by the aid of the undulatory hypothesis," wrote H. F. Talbot, Fellow of the Royal Society, in 1836—not that many years after Fresnel's pioneering investigations into diffraction—"yet when any *well-marked phenomena* occur which present unexpected peculiarities, it may be of importance to describe them."[23] So began Talbot's account of his chance observation of a most curious phenomenon encountered while attempting to observe the diffraction of light more closely with a magnifying lens. Like Newton, he found it "requisite to have a dark chamber and a radiant point of intense solar light," and therewith proceeded to investigate the pattern of lines transmitted by one of Joseph Fraunhofer's high-quality gratings, fabricated by covering a plate of glass with gold leaf into which several hundred parallel lines were cut. Talbot's description of his experiment is crystal clear (see note 23), and one could do no better than reproduce it:

> About ten or twenty feet from the radiant point, I placed in the path of the ray an equidistant grating ... with its lines vertical. I then viewed the light which had passed through this grating with a lens of considerable magnifying power. The appearance was very curious, being a regular alternation of numerous lines or bands of red and green colour, having their direction parallel to the lines of the grating. On removing the lens a little further from the grating, the bands gradually changed their colours, and became alternately blue and yellow. When the lens was a little more removed, the bands again became red and green. And this change continued to take place for an indefinite number of times, as the distance between the lens and grating increased.
>
> It was very curious to observe that though the grating was greatly out of the focus of the lens, yet the appearance of the bands was perfectly distinct and well-defined.
>
> This however only happens when the radiant point has a *very small* apparent diameter, in which case the distance of the lens may be increased even to one or

two feet from the grating without much impairing the beauty and distinctness of the coloured bands.

Mystifying in its simplicity—for, after all, others had illuminated gratings before him—Talbot's experiment did not find an explanation until Rayleigh encountered it almost a half century later.[24] Apart from the striking aesthetic appearance of alternating bands of complementary colors, what is truly remarkable here—to extract the essence from Talbot's account—is the appearance of a sharply focused image of the grating at seemingly any distance from the lens, but only when the grating was illuminated with a point source of light. Ordinarily, one illuminates a grating with collimated light—a circumstance that probably explains why this phenomenon was not reported earlier.

The curious effect observed by Talbot is actually a special case, restricted to nearly sinusoidal objects, of a more general phenomenon associated with the point-projection imaging of any transparent periodic object. Projective self-imaging, like spatial filtering, is another example of the principles of Fourier optics as embodied in Fresnel's theory of diffraction. Indeed, it is instructive to contrast the two. The optical microscope and many spatial filtering experiments are usually devised (by means of lenses) so that a specimen is illuminated by plane-wave radiation. Masking or phase shifting selected components of the spatial frequency spectrum in the diffraction pattern is frequently used to suppress periodic detail, such as scan lines,[25] from a composite optical image. In an ideal point projection geometry, however, no lenses are employed, and the sample is illuminated with spherical waves. Astonishingly, a set of focal planes results, within which appears a magnified image of periodic structure, all aperiodic structure being largely suppressed.

Although made independently, the discoveries of spatial filtering and Fourier self-imaging have an interesting historical point in common: in a manner of speaking both evolved from focusing problems. Recall that Abbe was called upon to explain why large lenses gave higher resolution than better crafted small ones. Similarly, Fox-Talbot noted difficulties in obtaining a unique focus condition for the images of one of the best spectroscopic gratings provided him by Fraunhofer in Germany.

Consider again the configuration of figure 7.4 with tip-to-specimen distance Z_1 and specimen-to-detector distance Z_2. If the specimen is sufficiently thin so that multiple scattering does not occur, then the optical effect of a periodic object is simply to multiply an incident wave by a transmission function of the form

$$\Psi(X, Y) = \sum_{hk} F_{hk} \exp\left[-2\pi i \left(\frac{hX}{a} + \frac{kY}{b}\right)\right]. \tag{29}$$

The coordinates (X, Y) locate a point in the exit plane of a flat object with respective spatial periodicities (a, b). F_{hk} (with integer h, k), the analog of the structure factor in X-ray diffraction (Eq. [4]), is the amplitude of the transmitted wave corresponding to an incident plane wave with respective horizontal and vertical spatial frequencies $2\pi h/a$, $2\pi k/b$. When a radiant point is the light source, the wave that propagates from the source through the object to the detector with surface coordinates (x, y) takes the form

$$\Psi(x, y) = \{U(X, Y; Z_1)\Psi(X, Y)\} * U(x, y; Z_2) \tag{30}$$

in which

$$U(x, y; z) = \exp\left[\frac{i\pi}{\lambda z}(x^2 + y^2)\right] \tag{31}$$

is called the Fresnel propagator, and the asterisk connotes the convolution integral defined by

$$f * g = \iint f(x, y)g(x_0 - x, y_0 - y)\,dx\,dy. \tag{32}$$

Note that the lens transmission function $t(\xi, \eta)$, discussed earlier (Eq. [5]), is a special case of the Fresnel propagator for light converging to a point at a distance of one focal length from the lens.

Substitution of expressions (29) and (31) into Eq. (30) and evaluation of the resulting integral leads (to within a constant factor K dependent upon the specified geometry and wavelength) to the wave at the detector

$$\Psi(x, y) = K \sum_{hk} F_{hk} \exp(-i\theta_{hk}) \exp\left[-2\pi i\left(\frac{hx}{Ma} + \frac{ky}{Mb}\right)\right]. \tag{33}$$

The magnification M is again given by Eq. (19), and the phase θ_{hk} takes the form

$$\theta_{hk} = \frac{\pi\lambda Z_1 Z_2}{Z_1 + Z_2}\left(\frac{h^2}{a^2} + \frac{k^2}{b^2}\right). \tag{34}$$

Comparison of Eqs. (29) and (33) shows that, exclusive of the unimportant factor K, the image waveform replicates the object waveform when the phase factor in Eq. (33) is unity—that is, when θ_{hk} is an integer multiple of 2π. Since a global phase factor of $e^{i\pi} = -1$ does not affect the resulting *intensity*, the recorded object and image are actually identical when θ_{hk} is an integer multiple of π. From the above expression for θ_{hk}, it follows that this condition is met whenever Z_1 and Z_2 obey the relations

$$\frac{1}{Z_1} + \frac{1}{Z_2} = \frac{\lambda}{2na^2} \tag{35}$$

and

$$\frac{a}{b} = \sqrt{\frac{m}{n}} \tag{36}$$

where m and n assume integer or half-integer values. For any periodic pattern, there is an infinite number of possible choices for the lattice constants a and b—that is, for the shape of the unit cell; relation (36) restricts the allowed choices for self-imaging.

Eq. (35) is analogous to the familiar thin-lens formula of Eq. (18), except that there are now multiple focal lengths inversely proportional to the wavelength

$$f_n = \frac{2na^2}{\lambda} \tag{37}$$

with $n = 1/2, 1, 3/2$, and so on. The physical conditions underlying expression (37) are of course quite different, as there is no lens in the system at all.

One result that follows readily from the preceding considerations is that a purely sinusoidal object is focused at *all* distances from the object plane, for Eq. (29) then involves no summation over Fourier components, and the phase in Eq. (33) is a global, rather than relative, phase. This phase vanishes from the expression for intensity, whereupon object and image distances are not restricted by Eq. (35). In the case investigated by Talbot, the grating was not a pure sinusoid, although it undoubtedly possessed a single dominant spatial frequency. However, as a consequence of the broad spectrum of wavelengths contained in the white-light illumination, Talbot saw sharply defined lines of the grating over an apparently continuous range of distances from his lens.

Equation (37) bears a superficial resemblance to the focal relation for a Fresnel zone plate, but there are significant differences since the zone plate is not a periodic object; the width of each annular zone narrows with increasing radius. The distances from the zone plate of axial points at which pronounced transmission peaks occur are given by the expression

$$f_n = \frac{R_N^2}{N\lambda n} \qquad \text{[Zone Plate]} \tag{38}$$

where R_N is the radius of the outermost zone N, and $n = 1, 3, 5, 7, \ldots$ is the order of the focal point. Note that the order number enters the denominator in Eq. (38), rather than the numerator as in Eq. (37). Thus, the larger the order, the closer the focal plane occurs to the zone plate—in contrast to point-projective imaging of lattices in which (for fixed Z_1) larger values of n lead to more distant focal planes and to more highly magnified images.

Abbe's theory of image formation affords a heuristic interpretation of self-focused lattice images. In an actual experiment the illumination does not, of course, issue from a mathematical point, but rather as a conical beam of angular width (full apex angle 2α) dependent on the source size. Upon traversing a periodic object with lattice spacing d, the beam is split by Bragg diffraction into multiple cones (as shown in figure 7.8), whose axes differ in orientation by twice the Bragg angle θ_B defined by

$$d \sin\theta_B = \lambda. \tag{39}$$

(There is no factor 2, as in Eq. [1], since the specimen surface is now illumi-

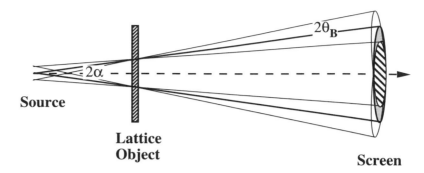

Figure 7.8. Formation of self-focused lattice images as a result of Bragg diffraction and interference of coherent cones of radiation from a point source. Unaberrated images are produced along the optic axis if the Bragg angle θ_B is smaller than the half-apex angle α of the cone.

nated at normal incidence.) These diffraction cones define an array of virtual point sources (in a plane perpendicular to the optic axis of the system) that are necessarily coherent with one another since they are all virtual images of the same real source. The resulting pattern in the image plane consists of the superposition of all sets of interference fringes produced by the array of point sources. If $\alpha < \theta_B$, the cones do not overlap, and formation of self-focused lattice images does not occur. For $\alpha > 2\theta_B$, unaberrated self-focused images are produced on the optic axis as shown in the figure. Similarly, for the range $\theta_B < \alpha < 2\theta_B$, self-focused (but distorted) images are produced off-axis.

The half-angular width α of a beam issuing from an optical pointlike source produced by a pinhole spatial filter of radius R is effectively given by the location of the first diffraction minimum in the corresponding Airy disc

$$R \sin \alpha = \lambda. \tag{40}$$

In the small-angle approximation (applicable to the experiments I will discuss), it follows from Eqs. (39) and (40) that point-projective lattice images should occur along the optic axis when the relation

$$\frac{\alpha}{2\theta_B} = \frac{d}{2R} > 1 \tag{41}$$

is satisfied.

The essential features of Fourier self-imaging are easily demonstrable in a simple tabletop optical experiment (with the geometry of figure 7.4; see note 9) employing a helium-neon laser ($\lambda = 633$ nm) with pinhole spatial filter ($R = 12.5 \ \mu$m) to serve as the point source and a Ronchi grating (figure 7.9) with frequency 100 lines/cm to serve as a one-dimensional periodic object (lattice constant $d = 0.01$ cm). Images of the grating were recorded at different

Figure 7.9. Ronchi grating with lattice constant $a = 100\ \mu$m.

TABLE 7.1

Optical Point-Projection Self-Images:
Focal Planes and Magnifications*

n	$Z_2^{(theory)}$	$Z_2^{(expt)}$	$M_n^{(theory)}$	$M_n^{(expt)}$
0.5	1.84	—	1.17	—
1.0	4.43	—	1.40	—
1.5	8.33	—	1.76	—
2.0	14.85	14.60	2.35	2.44
2.5	28.02	27.80	3.55	3.64
3.0	68.54	69.10	7.23	7.35

*Z_2 expressed in cm. Source-to-sample separation fixed at $Z_1 = 11.0$ cm.

distances Z_2 with a CCD digital camera (480 vertical pixels × 592 horizontal pixels). Each image was projected onto a video monitor for clear viewing as well as scanned (with resolution 10 μm/pixel) into a computer that generated the Fourier transform or spatial frequency spectrum. Of particular interest are the location and magnification of the images and the fidelity of the images to the original object.

With the source-object distance Z_1 held fixed at 11.0 cm, the focal planes were quickly found by manually adjusting Z_2 with a micrometer and viewing the grating on the monitor. The table below compares focal plane location predicted by Eq. (3.5) and magnification predicted by Eq. (19) with the observed values. Note that for an object of one-dimensional periodicity, Eq. (36) is not relevant. Also, the exact expression for M must be used, since the ratio Z_2/Z_1 in the optical experiment is not huge compared to unity as is the case for the electron PPM.

The uncertainty in predicted values of Z_2 is approximately 0.01 cm, as limited by the precision with which Z_1 was determined. Direct measurements of

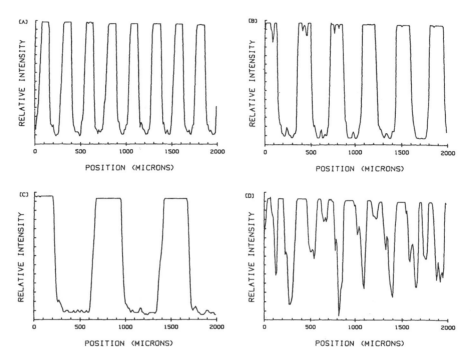

Figure 7.10. Optical PPM images (for fixed $Z_1 = 11.0$ cm) of a Ronchi grating with lattice spacing 0.01 cm recorded at distances Z_2 corresponding to (a) $n = 2$ focal plane, (b) $n = 2.5$ focal plane, (c) $n = 3$ focal plane, (d) midway between focal planes $n = 2.5$ and 3.

Z_2 could in principle be made with a precision of 0.05 cm if limited only by the calibration of the optical bench, but the primary uncertainty derived from the somewhat subjective location of the plane of optical focusing. Nevertheless, agreement between predicted and measured values is reasonably good. The gaps in the table for focal planes $n = 0.5$, 1, and 1.5 occur because the corresponding object-image separations were shorter than the distance of the (recessed) CCD surface from the front end of the camera.

Figure 7.10 shows the images of the grating recorded by the CCD camera for the three sequential focal planes $n = 2$, 2.5, and 3, and for a nonfocal plane midway between planes $n = 2.5$ and 3. Each image comprises 480 points, of which only 200 are displayed in the figure (to avoid severe image compression). As expected, the square-wave pattern of the grating is nicely reproduced at the theoretically predicted focal planes, whereas the image recorded between focal planes does not resemble the original periodic object at all. Higher values of n correspond to larger distances Z_2 and greater magnifications M, and hence to fewer periods observable within the fixed scan range of 2000 μm.

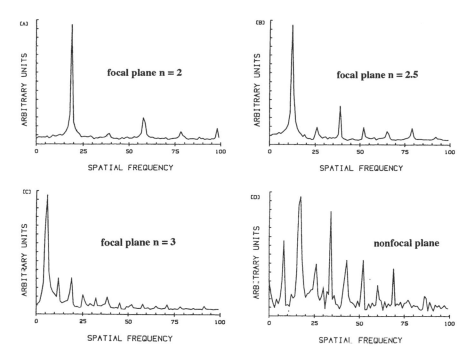

Figure 7.11. Spatial frequency spectra (i.e., Fourier transformations) of the images of figure 7.10a–d.

The corresponding spatial frequency spectra (i.e., Fourier transformation of the images of figure 7.10), displayed in figure 7.11, provide a sensitive measure of image quality. More precisely, what is shown is the distribution of magnitudes $f_k = \sqrt{c_k^2 + s_k^2}$ where c_k and s_k are, respectively, the amplitudes of cosine and sine components at the spatial frequency k. The term "spatial frequency" here refers to the number of fringes in a full vertical scan of 480 pixels. Each faithful self-image, as the figure confirms, is characterized by a single principal Fourier amplitude f_k. Thus, f_6 is the principal amplitude for $n = 3$, and f_{12} is the principal amplitude for $n = 2.5$, since the real-space images in a full vertical scan of 480 pixels contained 6 and 12 fringes, respectively. (For clarity, only a truncated portion of the full recorded pattern is displayed.) By contrast, the Fourier transform of an image located midway between the two focal planes reveals a complex spatial frequency spectrum corresponding to the disordered real-space image.

The advantages of projective imaging from a pointlike source for examining periodic lattices directly—as an alternative to the tedious procedure of deducing such structure computationally from a diffraction pattern—were recognized in the 1950s by J. M. Cowley and A. F. Moodie, who were apparently unaware of the previous studies of Talbot and Rayleigh.[26] The electron microscopes of the day being inadequate to the task, the two researchers demonstrated the principle optically by making experimental images of various gratings with light from a mercury arc lamp. However, without the incentives of coherent electromagnetic waves or electrons, the subject again largely passed into oblivion until nanotip electron sources once more revived interest in Talbot's discovery.

The potential utility of projective self-imaging lies not only in its capacity to generate without lenses magnified images of a lattice, but also in providing a method for selectively retrieving periodic detail in a complex multiperiodic structure possibly containing random noise as well. This is an area in which investigations have only just begun.[27]

Consider, for example, two distinct one-dimensional lattices $L1$ and $L2$ with spectrum of spatial frequencies $\{F_k\}$. If the lattices were superposed to create a single biperiodic object $L1 + L2$, would it be possible, by locating the detector in selected planes, to create images of $L1$ and $L2$ individually? Computer simulations show that the broader the Fourier spectrum, the more distorted are the non-self-focused images, and consequently the more specific is point-projective imaging for lattice recognition and retrieval. Optical experiments again employing laser illumination have confirmed the foregoing ideas. The periodic object, a Ronchi-like grating with lattice constants $a_1 = 89.5$ μm and $a_2 = 100$ μm, was designed on a computer, laser-printed, and subsequently made into a 35 mm slide transparency. Although some of the resulting images contained noise, there was no difficulty in recognizing clearly each component lattice. In particular, the spatial frequency spectrum of each self-focused lattice was dominated by the same principal peak characterizing the spectrum of the corresponding monoperiodic object.

The retrieval of periodic detail from what in outward appearance is a random object constitutes an especially interesting example of this procedure. Figure 7.12 shows a "random-line" grating, fabricated by using a random-number generator to determine the parameters of lines within a specified field; the lines were laid out with black tape on white cardboard, and the ensuing pattern was photographed onto a 35 mm slide transparency. As a totally aperiodic object, the random-line grating was found to be out of focus at all distances when viewed by projection from a point source. The Fourier spectrum of a typical vertical scan is also random, although the composite spectrum of many such scans yielded a characteristic distribution with maximum in the vicinity of the origin and falling monotonically in amplitude with increasing spatial frequency.

A planar quasi-periodic object was then made from a 35 mm slide transparency comprising the random-line grating and a Ronchi grating with lattice constant 0.01 cm (as in figure 7.9). Although direct visual inspection showed an essentially aperiodic object, positioning the viewing screen of the PPM in one of the focal planes of the grating revealed the principal structural features

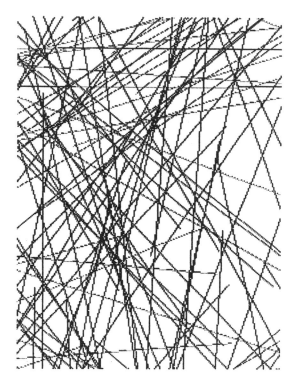

Figure 7.12. Random-line grating generated by computer.

of the underlying lattice (figure 7.13). Despite some residual noise, comparison of the retrieved image and its transform with the image and transform of the original lattice viewed alone shows that the desired periodic structure has been recovered with little loss of significant information.

Traditionally, periodicity in lattice-like objects was sought by inspection of their diffraction patterns rather than through direct real-space images. Point-projection microscopy with a nanotip electron source, however, has the potential to take a priori advantage of the periodicity inherent in an object to achieve higher resolving powers than in a conventional system while employing relatively simple and less expensive apparatus.

7.6 IN-LINE ELECTRON HOLOGRAPHY: SHARPENING THE IMAGE

The success of modern optical holography has obscured somewhat the historical origins of the field. It is often forgotten that Gabor invented holography for electrons (not light) using the in-line geometry (rather than the off-axis geom-

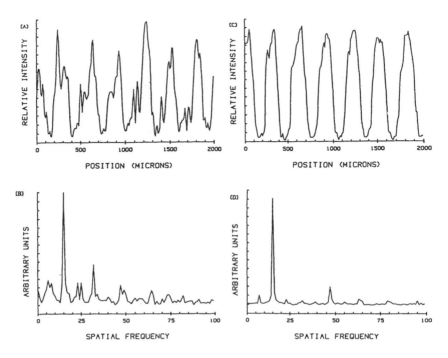

Figure 7.13. Projective images of a quasi-periodic object comprising a random-line grating and periodic grating with lattice constant 0.01 cm: (a) $n = 1$ focal plane of the grating; (b) associated spatial frequency spectrum. For comparison are shown the corresponding (c) real-space image and (d) spatial frequency spectrum of the original lattice imaged separately.

etry normally used in light optics) for the purpose of correcting electron lens aberrations (and not for producing a three-dimensional virtual reality). The low-voltage point-projection configuration shown in figure 7.4 corresponds exactly to the in-line geometry envisaged by Gabor. However, since nanotips had yet to be developed, the first experimental tests of Gabor's ideas in the early 1950s used a demagnified thermal electron source. Lack of source brightness and the severity of lens aberrations prevented the success of this approach, but subsequent attempts with in-line geometry and kilovolt field-emission sources eventually proved successful.

Since point-projective imaging is an example of in-line holography, the question naturally arises as to whether a shadow image, such as shown in figure 7.6, can be reconstructed as an electron hologram. One purpose for doing so is to remove the Fresnel fringes that accompany highly magnified images—in effect, to bring the image "back into focus" and thereby attain subnanometer resolution of structural detail. It was pointed out earlier—in the specific case of Fresnel edge diffraction—that spherical-wave illumination of an object in a PPM with object-tip distance Z_1 is equivalent to plane-wave illumination in a TEM with defocus $\Delta f = Z_1$ (as long as the transmission object is weakly scattering). The reasoning is quite similar to that leading to Eq. (30), except that in this case there is no requirement of periodicity. In the simple case of a one-dimensional planar object with transmission function $F(X)$ illuminated by a spherical wave emitted from a radiant point upstream a distance Z_1, the resulting amplitude at a distance Z_2 downstream from the object, and displaced by x from the optic axis, takes the form

$$\Psi(x) = F\left(\frac{x}{M}\right) * U\left(\frac{x}{M}; Z_1\right) \tag{42}$$

where U is the Fresnel propagator defined in Eq. (31) and $M = Z_2/Z_1$ is the magnification. Equation (42) represents an ideal magnified image out of focus by Z_1 and, although obtained for a PPM configuration, is the identical expression for an out-of-focus TEM image formed by plane-wave illumination.

To focus the PPM image would require setting $Z_1 = 0$, in which case M would become infinite. Clearly, this adjustment cannot be made mechanically, for electron field emission will not occur if the source (cathode) is in direct contact with the specimen (anode). To sharpen a shadow image, therefore, one must resort to analytical means, that is, to holographic reconstruction.

An extensive discussion of electron holography would take us too far afield, but the basic reconstructive algorithm can be succinctly expressed in a qualitative way for a simple masklike object, that is, an opaque object with holes. Treating the PPM image as an out-of-focus TEM image, one first takes the Fourier transform of the recorded intensity, multiplies it by the complex conjugate of the Fresnel propagator (for the appropriate distance Z_1), and finally computes the inverse Fourier transform. Needless to say, these various transformations are executed electronically on a computer. The conjugate Fresnel propagator represents a spherical wave propagating backward a distance Z_1 to the PPM point source—that is, to the exact focal plane of the corresponding TEM—thereby compensating for the defocusing error.

The mathematical expression that embodies the preceding reconstruction is actually a superposition of three different components signifying: (1) a perfect reconstruction of the original object, (2) an image of the "complementary" object (i.e., an object opaque where the original object has holes, and vice versa) out of focus by twice the distance Z_1, and (3) a hologram of the complementary object. As Gabor himself realized, an exact reconstruction of a masklike object is actually not possible, in part because a reference wave does not exist deep within the shadow region, and in part because of the "twin image" problem characteristic of in-line geometry.

The twin image refers to the virtual image from the conjugate waveform ψ_0^* in the overall intensity resulting from superposition of a reference wave ψ_r and a (much weaker) object wave ψ_0

$$I = |\psi_r + \psi_o|^2 \approx I_r + \psi_r^* \psi_o + \psi_r \psi_0^*$$
$$= \text{(Background)} + \text{(image)} + \text{(twin image)}. \qquad (43)$$

In in-line holography both images fall within the line of view. It was for this reason that Gabor's in-line geometry was largely abandoned and off-axis holography was adopted in light optics. The twin image is not too distracting, however, if the hologram is recorded in the far field of the object, that is, under the condition characteristic of Fraunhofer diffraction: $Z_2 > D^2/\lambda$, where D is the size of the object. If the object is small enough, the virtual image is then sufficiently far out of focus to avoid interfering with the real image. Optical holography of macroscopic objects does not ordinarily satisfy the Fraunhofer condition. For example, to record the far-field pattern of 500 nm light scattered from a 1 cm object, the detector would need to be located at some 200 m from the object. By contrast, the far-field condition is achieved anywhere beyond ∼800 nm for a 10 nm object illuminated with 100 eV electrons ($\lambda \sim 0.12$ nm).

A carbon film with holes is a good example of a masklike object for which holographic reconstruction by computer has been successfully implemented (see note 9). Under the reasonable assumption that the film, itself, is completely opaque, the only contrast features are the Fresnel edge fringes, which holographic reconstruction seeks to remove by correcting for the PPM defocus. Figure 7.14 shows an experimental PPM image of a thin carbon film and holographic reconstructions for three assumed focus settings. The reconstruction displaying the sharpest edges and no internal fringes corresponds to a focus correction of 1850 nm, which is taken to be the tip-to-sample distance in the original experiment. If the reconstruction at 1850 nm is compared with the experimental photomicrograph, the single bridging fibers appear with sharp edges, and the background of Fresnel fringes is much wider than in the original. This background is the twin image of the fiber and is wider because it is twice as far out of focus (3700 nm) when the real image is brought into focus. The sharply delineated shape of the fibers retrieved from the holograph with 1850 nm focus correction is also shown in the figure. For masklike objects, holographic focusing of the PPM works reasonably well.

One potential application of low-voltage electron PPM is to determine the shapes of large molecules—for example, those of biological significance—drawn across holes. Progress will depend greatly on the results of more detailed studies of radiation damage at these low energies, a field in which there are presently few data. It is hoped that images may be obainable at very low energies (let us say 10 eV), where penetration increases greatly, radiation damage is expected to be small, and resolution is not limited too severely by the longer electron wavelength (∼0.4 nm). In view of the importance of atomic structure information in molecular biology and the gravity of the radiation damage problem for other techniques (which may be limited to the study of crystallizable

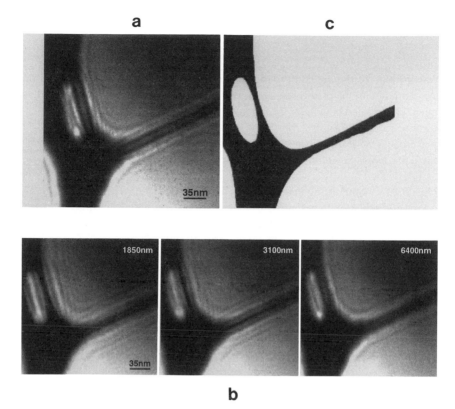

Figure 7.14. Holographic reconstruction of a PPM image with removal of Fresnel fringes. (a) Experimental image of perforated carbon film; (b) reconstructed amplitudes recovered from (a) for different defocus corrections (marked on images); (c) shape of carbon fibers retrieved from the holographic reconstruction at 1850 nm.

molecules), the further development of low-voltage holography with nanotips is likely to proceed apace.

7.7 QUANTUM PHYSICS IN THE NEW MILLENNIUM

Apart from imaging (and interferometry related to imaging), the point-projection microscope has an important—indeed, unique—role to play in the investigation of the fundamental principles of quantum physics. The practical and conceptual importance of a bright electron source that emits coherent and focused beams of electrons can be easily imagined when one recalls how much of modern light optics relies on the availability of optical sources with the same characteristics.

What the laser is to quantum optics, the atomic-sized field-emission electron source may one day become to the quantum optics of matter waves.[28]

Of all the phenomena that fall within the purview of quantum physics, it is the process of matter-wave interference that distinguishes most starkly the strange possibilities of the quantum world from the familiar events of everyday life. As intimated in the previous chapter, the interference of quantum particles has been well studied both theoretically and in the laboratory. Most quantum interference experiments, however, involve the self-interference of single-particle wave packets—for it has not been possible to produce bright beams of correlated particles (except in the case of light). Were such a source available, then, as I described in my earlier book devoted to quantum interference, *More Than One Mystery*, interesting new processes—some analogous to those of quantum optics and others with no optical analogs at all—could be examined.

One set of proposed experiments, for example, is based on the distinctive quantum features of electron intensity correlation interferometry in which the statistical properties of electron beams are examined in ways similar to those pioneered by Hanbury Brown and Twiss for optical fields. At least three different types of experiments can be envisioned: (1) measurement of the variance in electron counts at a single detector, (2) measurement of the conditional probability of electron arrival at a single detector as a function of the time interval between two consecutive detections, and (3) measurement of the correlated count rate of electrons at two detectors as a function of time delay. These experiments probe the quantum correlations of electrons arising from spin and statistics.

A chaotic beam of fermionic particles—a beam in which each particle is in a quantum state of sharp energy, momentum, and spin (analogous to frequency, wave vector, and polarization of thermal light)—should exhibit the phenomenon of "antibunching," that is, a negative cross-correlation, in experiment (3), a degree of second-order coherence less than unity for time intervals shorter than the coherence time T_c in experiment (2), and a variance in (1) smaller than that of Poissonian statistics. Other types of electron beams are predicted to exhibit different correlations. The quantum contribution in each type of experiment is proportional to the degeneracy parameter δ. This is understandable. For specifically quantum mechanical correlations to arise, an electron source must produce pairs of electrons whose wave functions overlap to some degree at the time of emission. The degeneracy parameter characterizes the extent to which these correlated pairs are produced. With a degeneracy at least two orders of magnitude higher than any other source of free particles, the electron nanotip should make these proposed experiments more than just marginally feasible.

Another set of suggested experiments, which probes the long-range correlations of entangled electron states, concerns the nonlocal effects of magnetic fields, as in the once highly controversial Aharonov-Bohm (AB) effect.[29] In the AB effect a single-particle wave packet is coherently split; each component propagates around (but never through) a confined magnetic field, and the two components are recombined at a detector to produce a pattern of interference fringes that depends on the amount of enclosed magnetic flux. One can conceive, however, of coherently emitted pairs of electrons in which the two particles move off in opposite directions and diffract around separate magnetic field regions far downstream from the source. The cross-correlation of electron arrivals at the two distant detectors and the arrival probability of single electrons at an individual detector lead to unusual variations of the (single-particle) AB effect. Such experiments probe the "doubly" nonlocal nature of quantum phenomena embodied in the nonclassical interaction with the magnetic field (AB effect) and the long-range correlations of the electron wave function (as in the Einstein-Podolsky-Rosen paradox).[30]

To produce the requisite pairs of correlated electrons, imagine crafting an electron nanotip with multiple emission sites in which two or more atomic-scale asperities are positioned within a lateral coherence length of one another. It is then conceivable that coherent emission of two (or more) electrons may occur. To my knowledge, no experiment of this kind has yet been done, nor has one determined theoretically whether electrons emitted under such circumstances are entangled in their momenta. Nevertheless, computer modeling of field emission from one or a few atoms on a nanotip indicates that the electron emission pattern is highly sensitive to the tip geometry.

Finally, the nanotip source itself is an object of quantum mechanical interest. Recent experimental work seems to show significant differences between the properties of field-emitted beams from tips of atomic size and standard tips some two or more orders of magnitude broader. Apart from their higher intrinsic brightness and larger transverse coherence under comparable extraction potentials, beams from nanotips exhibit strong quantum mechanical self-focusing effects, current-voltage characteristics different from those described by the Fowler-Nordheim equation discussed in the previous chapter, and multiply peaked emission energy spectra. Thus, the nanotip appears to operate through unusual quantum processes. This is perhaps to be expected, since the source size is comparable to the electron wavelength inside the tip—and physicists have already noted strong quantum effects in the conductance of electrons through narrow constrictions in mesoscopic semiconductor systems.

As a first step in the execution of experiments with beams of correlated free electrons, experiments are now under way to examine the quantum statistics of electron emission from nanotips.[31] There is yet much experimental and theoretical work to be done before the potential of these extraordinary sources can be fully realized. Of this, though, one can be sure: The brightest source of matter waves in science will assuredly stimulate challenging experiments and through-provoking ideas.

NOTES

1. A. B. Porter, "On the Diffraction Theory of Microscopic Vision," *Phil. Mag.* 11 (1906): 154–66.

2. George Gamow, *One Two Three ... Infinity: Facts and Speculations of Science* (New York: Mentor, 1947). Plate 1 shows the photograph of a magnified hexamethylbenzene molecule.

3. The Miller indices (h, k, l) specify an atomic plane by denoting the reciprocal of the intercept of the plane with the axes of the unit cell. Each intercept is expressed in terms of the unit cell length for the corresponding axis. For example, a plane parallel to the z-axis that passes through the points $a/2$ and $b/3$ along the x- and y-axes, respectively, would be labeled $(2, 3, 0)$.

4. The mathematical insights of H. A. Hauptman and J. Karle, who received the 1985 Nobel Prize in Chemistry for their work, have have made it possible in many cases to determine atomic coordinates directly from the intensities of the diffraction spots. The constraint that electron density must be non-negative leads to an infinite set of inequalities among phases and magnitudes of the scattering amplitudes. Further constraints posed by the much greater number of diffraction spots than atoms and by judicious use of symmetry significantly reduce the number of unknown phases to be determined. See, for example, "Nobel Prize in Chemistry to Hauptman and Karle," *Physics Today* 38 (December 1985): 20–21.

5. J. W. Goodman, *Introduction to Fourier Optics* (New York: McGraw-Hill, 1968), ch. 5.

6. A. W. Hansen and H. Lipson, "Optical Methods in X-ray Analysis. III: Fourier Synthesis by Optical Interference," *Acta. Cryst.* 5 (1952), 363–66. I thank Brian J. Thompson for bringing this reference to my attention.

7. Optical Fourier synthesis by illumination of photographically prepared templates has been used to produce magnificent diffraction patterns of use to crystallographers and undoubtedly artists as well. See G. Harburn, C. A. Taylor, and T. R. Welberry, *Atlas of Optical Transforms* (Ithaca, N.Y.: Cornell University Press, 1975).

8. D. Gabor, "Microscopy by Reconsructed Wave-Front," *Proc. Roy. Soc. London A* 197 (1949): 454–61.

9. M. P. Silverman, W. Strange, and J. C. H. Spence, "The Brightest Beam in Science: New Directions in Electron Microscopy and Interferometry," *Am. J. Phys.* 63 (1995): 800–13.

10. H.-W. Fink, "Point Sources for Ions and Electrons," *Physica Scripta* 38, (1988): 260–63.

11. I thank my colleague John C. H. Spence for this eye-opening comparison.

12. Strictly speaking, a shadow is itself part of a diffraction pattern. From the perspective of geometrical optics, light propagates through empty space in straight lines, and a shadow results where no light reaches. According to physical optics, however, a shadow is produced by destructive interference of light diffracted into this region. I will leave to philosophers the semantic question of whether or not light actually "goes" into a region where its net amplitude is zero; the mathematical details of shadow formation are given by A. Sommerfeld, *Optics* (New York: Academic Press, 1964), pp. 210–12.

13. G. A. Morton and E. G. Ramberg, "Point Projector Electron Microscope," *Phys. Rev.* 56 (1939): 705.

14. M. P. Silverman, *More Than One Mystery: Explorations in Quantum Interference* (New York: Springer-Verlag, 1995).

15. W. Qian, M. Scheinfein, and J. C. H. Spence, "Brightness Measurements of Nanometer-Sized Field-Emission-Electron Sources," *J. Appl. Phys.* 73 (1993): 7041–45.

16. J. M. Pasachoff, *Contemporary Astronomy* (Philadelphia: Saunders, 1977), pp. 167–68.

17. J. L. Laclare, "ESFR X-ray Synchrotron Radiation Source," *Rev. Sci. Instr.* 60 (1989): 1399–1402.

18. J. C. H. Spence, W. Qian, and M. P. Silverman, "Electron Source Brightness and Degeneracy from Fresnel Fringes in Field Emission Point Projection Microscopy," *J. Vac. Sci. and Tech. A* 12 (1994): 542–47.

19. See, for example, M. Born and E. Wolf, *Principles of Optics*, 4th ed. (Oxford: Pergamon, 1970), pp. 161–63.

20. J. Hillier and E. G. Ramberg, "The Magnetic Electron Microscope Objective: Contour Phenomena and the Attainment of High Resolving Power," *J. Appl. Phys.* 18 (1947): 48–54. The "phase jump" that occurs at the shadow of the edge itself is not real but rather an artifact of the mathematical approximations employed. Sommerfeld (*Optics*, p. 263) points out that phase, like amplitude, is continuous at the origin "as far as it is permissible to talk of a phase of the complicated oscillations in that vicinity."

21. J. W. Goodman, *Statistical Optics* (New York: Wiley, 1985), pp. 207–22.

22. Lord Rayleigh was pehaps the first person to construct such a plate by blocking out with black ink the odd-numbered zones of a large-scale drawing that was to be photographically reduced directly onto a glass plate. Rayleigh also noted in an article on wave theory in the *Encyclopaedia Britannica* that a fourfold increase in intensity would result at the focus if the alternate zones, instead of being opaque, were made perfectly transparent but with a reversal of phase. Robert Wood constructed such a phase-reversal zone plate by making the zones of a thin film of gelatin on glass. See R. Wood, *Physical Optics* (New York: Macmillan, 1934), p. 38.

23. H. F. Talbot, "Facts Relating to Optical Science," *London and Edinburgh Philosophical Magazine and Journal of Science* 9, no. 56 (1836): 401–7.

24. Lord Rayleigh, "On Copying Diffraction-Gratings, and on Some Phenomena Connected Therewith," *London and Edinburgh Philosophical Magazine and Journal of Science* 5, no. 67 (1880): 196–205.

25. The horizontal scan lines, encountered, for example, in satellite images of the Moon's surface, produce a regular vertical array of diffraction spots that can be masked, thereby resulting in a reconstructed image free of scan lines. The degradation of the image by removal of a small part of the diffraction pattern is hardly perceptible since each point on the original transparency is mapped onto all points in the diffraction plane.

26. J. M. Cowley and A. F. Moodie, "Fourier Images: I: The Point Source," *Proc. Roy. Soc. London B* 70 (1957): 486–96.

27. M. P. Silverman and W. Strange, "Projective Imaging and Interferometry with Light and Electrons," *Bull. Am. Phys. Soc.* 40, no. 2 (April 1995): 944.

28. M. P. Silverman, "Applications of Photon Correlation Techniques to Fermions," in *Photon Correlation Techniques and Applications*, ed. J. B. Abiss and E. E. Smart (Washington, D.C.: Optical Society of America, 1988), pp. 26–34.

29. Y. Aharonov and D. Bohm, "Significance of Electromagnetic Potentials in the Quantum Theory," *Phys. Rev.* 115 (1959): 485.

30. A. Einstein, B. Podolsky, and N. Rosen, "Can Quantum Mechanical Description of Reality Be Considered Complete?" *Phys. Rev.* 47 (1935): 777–80. See also Niel Bohr's reply of the same title, *Phys. Rev.* 48 (1935): 696–702.

31. M. Scheinfein, M. P. Silverman, and J. C. H. Spence (unpublished work in progress).

Part Three

POLARIZATION

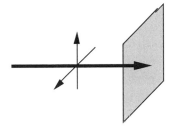

CHAPTER 8

Poles Apart: The Mysteries of Light Polarization

> These observations lead me to conclude that light
> acquires . . . properties that are independent of its
> direction with respect to the reflecting surface and
> that are the same for the south and north sides . . .
> and different for the east and west sides. . . . I will call
> polarization the modification that gives light
> properties relative to these poles.
> (*Etienne Malus, 1811*)[1]

DRAW A SMALL dark spot on a piece of paper and place over it a transparent rhomb of calcite (Iceland spar). With little effort you can find an orientation of the crystal such that the image will appear as two spots. Rotate the crystal; one spot will rotate about the second, which remains fixed. View the crystal through one lens of polarizing sunglasses; rotate the lens slowly in its plane. At a certain orientation one spot will vanish; at 90° to this orientation the other spot will have vanished, and the first will appear at its darkest. These observations reveal something profound about the nature of light, but in 1819, when Fresnel received the Academy prize for his essay on diffraction, no one, not even Fresnel, could interpret their significance.

The odd effects of calcite on light were already known to both Huygens and Newton. Indeed, Huygens had studied calcite in considerable detail. Illuminating one face of the crystal at normal incidence with a narrow light beam, he saw two parallel beams emerge from the opposite face: one (the "ordinary" ray) passing straight through, the other (the "extraordinary" ray) refracted at both surfaces (figure 8.1). The ordinary ray always obeyed Snel's law—that was why he called it ordinary; the refraction of the extraordinary ray, however, depended on its direction of propagation through the crystal and, in marked violation of Snel's law, occurred even for normal incidence.

Passing the two transmitted beams through a second calcite crystal, Huygens discovered that four beams emerged from the latter. However, by rotating the second crystal he could make first one pair and then the other pair vanish. More perplexing still, for selected orientations of the two crystals the ordinary ray in the first became the extraordinary ray of the second (and vice versa). Huygens tried to account for these results by assuming that light was a wave that propagated through matter with a direction-dependent speed. As discussed

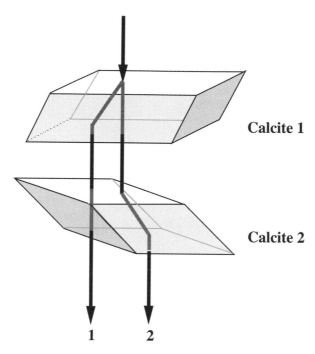

Calcite 1

Calcite 2

1   2

Figure 8.1. Light propagation through calcite. A ray incident upon calcite 1 splits into an extraordinary ray (1) and an ordinary ray (2). A second calcite is so oriented that incident ray 1 becomes the ordinary ray and ray 2 the extraordinary ray. Another orientation can be found for which emerging rays 1 and 2 recombine into a single ray. For a general orientation, four rays will emerge from calcite 2.

in chapter 2, however, he could in no way provide a mechanism for such behavior.

Newton, who revealed much of his thoughts on light through queries in *Opticks*, adopted, in keeping with his corpuscular perspective, a different working hypothesis (Query 26): "Have not the Rays of Light several sides, endued with several original Properties?" He conjectured in fact that light had four sides, two of which resulted in ordinary refraction, and two through which "unusual," that is, extraordinary, refraction occurred. Furthermore, this sidedness was an intrinsic property, and no interaction with material objects, according to Newton, could modify the "original and unchangeable Properties of the Rays."[2] This view of light rays was to dominate scientific thinking for well over 120 years until sometime after the 1830s.

The passage from the obscure notion that light manifests a quality of "sidedness" in matter to the precise realization (by Fresnel) that light is a transversely vibrating wave was long and often stormy. Though many have contributed, the drama unrolled principally in France—the predominant players being Malus, Arago, Biot, and, of course, Fresnel.

Older than Fresnel by thirteen years, Etienne Malus, like Fresnel, was an engineer schooled at the Ecole Polytechnique; like Fresnel, he was a skilled experimenter as well as an adept mathematician; regrettably, like Fresnel, he died young, succumbing (in 1812) at age thirty-seven to the plague, which he contracted while serving in Egypt during the Napoleonic wars. To physicists today, the name Malus probably conjures up the "cosine squared" law for the intensity of light passing through a linear polarizer. The relationship, as one knows from elementary physics texts, is trivially derived by projecting the electric vector of light onto the transmission axis of the polarizer and squaring the result to obtain the light intensity. The analysis was far from trivial, however, around 1807 when Malus was investigating reflection and double refraction of light.

One evening, so the story goes, while passing by the Luxembourg Palace in Paris, the thought occurred to Malus to look through a piece of calcite at the setting sun reflected from a palace window. As expected, he saw two images, but to his amazement he found he could extinguish one or the other image by rotating the crystal to certain orientations. Double refraction, then, was not the only process by which light exhibited its quality of "sidedness." Malus had discovered that light is polarized by reflection. Following up this chance observation, he experimented with reflection of candlelight from water, examining quantitatively the properties of the polarized light. It was Malus, in fact, who coined the term "polarization"—although the connotation of the word differed markedly from its usage today. Like most nineteenth-century physicists, Malus adhered to the corpuscular view of light and regarded polarization as a property of physically real rays.

I have been asked a number of times, when relating Malus's seemingly arbitrary action to a lecture audience or to a physics class, whether this anecdote is apocryphal. Why would someone, presumably walking through beautiful gardens, spontaneously decide to analyze reflected sunlight and just "happen" to have a calcite prism in his pocket? The answer, I can only speculate, is simply out of curiosity—the hallmark of a true scientist. I find nothing unbelievable about the account. For nearly forty years—long before Malus's name meant anything to me—I have, myself, kept two small squares of Polaroid sheet in my pocket for viewing reflected and scattered light whenever the whim strikes me.

Stimulated by Malus's investigations, François Arago, then recently elected (1809) to the Paris Academy at the exceptional age of twenty-three, examined Newton's rings through a calcite crystal, observing that the rings, viewed both by reflection and transmission, were polarized, although in a way at variance with the theory of partial reflection that Malus had developed. The details of this puzzle—one of many that would confront the seemingly fruitful, but at root deeply flawed, ray theory of light—are not significant enough to pursue here. What ensued, however, was to have a profound effect on the evolution of physics—not only by its scientific import but also as an outcome of the bitter personal schisms that developed.

Sometime around 1811 Arago replaced the variable air gap in the Newton's rings apparatus with a sheet of mica. Applying a calcite crystal to view sky-light transmitted through the mica, he saw that the two resulting images showed complementary colors that varied with the Sun's position and vanished altogether on overcast days. Arago reasoned that skylight must be polarized—a significant deduction in its own right—and set about to enhance the newly discovered phenomenon, which he termed chromatic polarization, by using light more strongly polarized by reflection from glass.

Arago's discovery, which we would today recognize as an example of interference colors from a birefringent medium, makes for an impressive demonstration that one could construct at home. Take a glass microscope slide and lay upon it, in pyramid-like fashion, ten or more strips of cellophane tape in layers of diminishing length. (Use the kind of tape that is transparent, not the kind with a translucent or "frosty" appearance.) Place the slide between two linear polarizers (again, the lenses of polarizing sunglasses work fine) and hold the "sandwich" up to an incandescent light source. Bands of bright colors should be seen in what initially was a structure devoid of all color (except perhaps for the neutral gray-green tint of the polarizers). Note that the color of each layer of uniform thickness changes to the complementary color when the relative orientation of the two polarizers is changed from parallel to crossed positions. One can produce a particularly spectacular effect by replacing the taped microscope slide with a crumpled wad of cellophane; the variations in thickness lead to a random profusion of bright colors. To see these demonstrations for the first time is to understand how intrigued Arago must have been nearly two centuries ago.

Arago's experiments on chromatic polarization uncovered another effect, the eventual implications of which were also far-reaching. Replacing the sheet of mica by a thin lamina of quartz, and observing the quartz between two polarizers (a glass reflector and calcite prism), he again noted bright colors. Unlike mica, however, rotating the quartz sheet in its plane had no effect at all upon the colors. The correct explanation of this effect anticipates phenomena (optical activity) to be discussed later in this book. Suffice it to say at present that quartz, although linearly birefringent like mica for arbitrary direction of light propagation, is, in contrast to mica, circularly birefringent for light propagating along the optic axis. Arago was perhaps the first to see a conspicuous effect of circularly polarized light passing through an optically active medium, a medium whose structure exhibits an intrinsic left- or right-handedness.

The discovery of chromatic polarization was of major importance; its various ramifications set much of the agenda for the study of light during the next two decades. Arago had presented a paper before the Academy in 1811, expecting to follow through this initial work with more comprehensive studies. But he did not have the opportunity. In 1812 Jean-Baptiste Biot read two memoirs on chromatic polarization that in effect took over the field Arago opened and made his work of marginal interest. The sense of betrayal Arago must have felt—for Biot and he had recently collaborated on other projects—the reader can well imagine. On the other hand—as any present-day physicist can attest—physics is a competitive activity, made all the more so by the offerings of prizes, fellowships, professorships, and the like.

The rancorous exchanges that followed, with Arago's charges and Biot's rebuttals, need not be pursued further except to mention the most significant outcome of the controversy. Arago, who until that time had thought about light largely from a corpuscular perspective, came to regard this doctrine with repugnance, for it reminded him of Biot, who remained a lifelong partisan of the ray theory. Upon switching allegiance, Arago—in a manner of speaking—subsequently made the most important discovery of his life: Augustin Fresnel.

At a time when he was still unknown to his scientific contemporaries and unaware of their work, Fresnel had convinced himself that Newton was wrong and that light was a wave. In 1814 he sent a paper with his first ideas on the subject to Ampére, who brought it to the attention of Arago. Fresnel subsequently visited Arago in Paris, initiating what was to become a long and cordial collaboration: Fresnel, for the most part, unraveled to the core the manifold implications of wave theory, and Arago in turn encouraged him, made suggestions, and took advantage of membership in the Academy to publicize and defend his work. Had Arago, like most French "opticians," not been receptive to the wave theory, it is unlikely he would have been as receptive to Fresnel.

In partnership with Arago, Fresnel had by 1816 created a wave theory of diffraction of unparalleled predictive power. Not only could the theory provide the locations of interference fringes of any object or aperture, but it could yield their relative intensities as well. As described in chapter 4, Fresnel's lengthy memoir on diffraction was honored by the Academy. However, it said nothing about polarization. Fresnel had indeed thought about the question, but he was at a loss then for a suitable explanation. However, since he was convinced that light was a wave, he was also certain that polarization was somehow intimately connected with the mode of wave vibration.

In an ingenious and revealing experiment, Fresnel illuminated a double-slit apparatus in which he had covered each slit with a thin lamina of gypsum, a bire-fringent crystal. The two laminae were oriented with their axes perpendicular, and Fresnel observed the region where the ordinary beam of one slit overlapped the extraordinary beam of the other. There was *no* interference pattern. He tried another approach at Arago's suggestion, this time coherently superposing light that had been orthogonally polarized by reflection from two stacks of glass plates. Again, no interference pattern. From these results Fresnel concluded that orthogonally polarized light cannot interfere—that polarization must be a consequence of a transverse vibration, that is, a vibration within the wavefront, the vibration axis providing the asymmetry or "sidedness" that Newton had ascribed as an intrinsic and unmodifiable property of rays.

There remained, however, a simple yet vexing question: What is *unpolarized* light? All experiments indicated that unpolarized light was characterized by perfect symmetry about the ray direction. To Fresnel this implied that unpolarized light waves must vibrate longitudinally, that is, perpendicular to the wavefront. How, then, could unpolarized light give rise—for example, by passage through calcite—to polarized light? Did this not imply that all light contained a longitudinal component?

These questions were to perplex Fresnel for some five years until the correct solution at last ($\sim$ 1821) occurred to him. *All* light vibrates transversely. The appearance of symmetry about the ray of unpolarized light is a consequence of a transverse vibration rapidly and randomly varying over time. There is no longitudinal component. To polarize naturally occurring light, therefore, is to select out—ordinarily by transmission or reflection—only those waves vibrating in a particular plane through the ray.

Fresnel's conceptual breakthrough enabled him to conceive of states of polarization which one could not even define within the framework of the extant ray theory. Ray-theory adherents recognized three types of light: polarized, unpolarized, and partially polarized—as determined by the observable effects of passing such light through polarizing crystals. Imagine, however, a light wave in which the wavefronts rotated rapidly clockwise about the direction of propagation. (For visible light the angular frequency of such a rotation is of the order of 10^{15} rad/sec.) Observed through a polarizing crystal, the light would give rise to no perceptible variation in intensity as one turns the polarizer orientation. And yet it is distinguishable from unpolarized light. Circularly polarized light passed through a thin birefringent crystal and observed through a linear polarizer gives rise to interference colors; unpolarized light does not.

The idea of a rotating wavefront is merely one way of conceiving of two mutually orthogonal linear vibrations occurring 90° out of phase. Depending upon whether the relative phase represents a lead or a lag, the resultant wave rotates in a clockwise or counterclockwise sense to an observer facing the source. (Physicists, in contrast to engineers, ordinarily refer to these two states, respectively, as right and left circular polarizations.) Similarly, any state of linear polarization may be decomposed into a superposition of the two states of circular polarization.

To show the existence of circularly polarized light Fresnel constructed a composite prism made up of alternating sections of intrinsically left- and right-handed quartz. Quartz displays circular birefringence; that is, the speed of light propagating along the optic axis depends upon the sense of circular polarization. According to Snel's law, therefore, left- and right-handed light, incident at an interface, should be refracted at different angles. The angular difference is very small, for the difference in refractive indices for left- and right-handed light in quartz is only about 7×10^{-5}—but by putting together oppositely handed quartz segments Fresnel was able to separate the two beams to a measurable extent.

Combining his understanding of light interference and polarization, Fresnel was able to provide a correct interpretation of chromatic polarization, the phenomenon first observed by Arago and subsequently dominated by Biot. Recall that, viewed through a linear polarizer (calcite prism), a slab of birefringent crystal (mica) illuminated with linearly polarized white light appeared colored, the ordinary and extraordinary images showing complementary colors. Let us examine this phenomenon quantitatively, but with modern formalism for handling the superposition of waves. (Fresnel, who worked with real trigonometric functions instead of complex exponentials, would have had a more cumbersome algebraic analysis to perform.)

Orient the birefringent slab, as shown in figure 8.2, in the x-y plane, where for simplicity it is assumed that waves vibrating along the x-axis are the ordinary waves (with refractive index n_o) and those vibrating along the y-axis are the extraordinary waves (with refractive index n_e). The waves propagate in the $+z$ direction, first traversing a linear polarizer with transmission axis at the angle θ_1 to the x-axis. Upon emerging from the crystal of thickness d, the polarization of a spectral component of wavelength λ can be described by the vector

$$\mathbf{E} = \mathbf{x} \cos \theta_1 e^{i \varphi_o} + \mathbf{y} \sin \theta_1 e^{i \varphi_e} \tag{1}$$

in which $\mathbf{x}$, $\mathbf{y}$ represent unit vectors along the principal axes of the crystal, and the changes in phase of the ordinary and extraordinary components of the wave are, respectively,

$$\varphi_o = \left(\frac{2 \pi d}{\lambda} \right) n_o \tag{2a}$$

and

$$\varphi_e = \left(\frac{2 \pi d}{\lambda} \right) n_e. \tag{2b}$$

The total amplitude of the wave has been set to unity since it is not a quantity that matters in the following discussion. Similarly, a common factor $e^{-i \omega t}$ representing the vibration of both components at angular frequency $\omega = 2 \pi c / \lambda$ has been omitted since it, too, does not affect the results.

If one looks at the the transmitted light without a second polarizer, there is no preferential coloration since the intensity

$$I_\lambda = |\mathbf{E}|^2 = 1 \tag{3}$$

of each transmitted spectral component is simply the same as for each spectral component incident on the crystal—and the crystal, we assumed, was illuminated with white light.

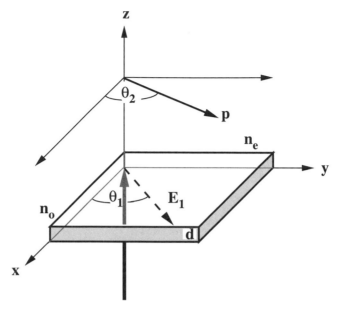

Figure 8.2. Geometry of interference colors. Linearly polarized white light with electric vector $\mathbf{E}_1$ at angle θ_1 to the x-axis is incident upon a transparent slab of birefringent material with refractive indices n_o and n_e and thickness d. The transmitted light traverses a second linear polarizer with pass-axis $\mathbf{p}$ at angle θ_2 to the x-axis. The resulting color of the slab corresponds to the wavelength for which o- and e-waves interfere constructively. Rotating $\mathbf{p}$ by $90°$ generates the complementary color.

When the transmitted light traverses a second linear polarizer with pass-axis (represented by unit vector $\mathbf{p}$) at angle θ_2 to the x-axis, each component of relation (1) is projected onto that axis, and the amplitude of the resulting wave (up to a global phase factor) becomes

$$E_p = \mathbf{E} \cdot \mathbf{p} = \cos\theta_1 \cos\theta_2 e^{i\Delta\varphi} + \sin\theta_1 \sin\theta_2 \quad (4)$$

where the relative phase between the ordinary and extraordinary components is

$$\Delta\varphi = \frac{2\pi d}{\lambda}(n_o - n_e). \quad (5)$$

The intensity of an observed spectral component is now

$$I_\lambda = (\cos\theta_1 \cos\theta_2)^2 + (\sin\theta_1 \sin\theta_2)^2 + 2\cos\theta_1 \sin\theta_1 \cos\theta_2 \sin\theta_2 \cos\Delta\varphi \quad (6)$$

and depends sensitively on the phase difference $\Delta\varphi$.

Suppose, for purposes of illustration, the first and second polarizers are both oriented at $\theta_1 = \theta_2 = 45°$. Then the intensity of the particular spectral component λ would be

$$I_\lambda^{(0)} = \tfrac{1}{2}(1 + \cos \Delta\varphi) \tag{7a}$$

where the superscript label represents the relative orientation $\Delta\theta = (\theta_2 - \theta_1)$ of the two polarizers. If one now rotates the second polarizer by $90°$ so that $\theta_2 = 135°$, the intensity of the same spectral component becomes

$$I_\lambda^{(90)} = \tfrac{1}{2}(1 - \cos \Delta\varphi). \tag{7b}$$

The relative phase that enters the interference term and the direction of polarization are closely coupled—a subtle, but key, point that Fresnel eventually realized.

The implications of the preceding two relations for the production of colors follow immediately. Let us take the wavelength to be $\lambda = 440$ nm, which corresponds to blue, and assume that for the given thickness and optical constants of the birefringent crystal the relative phase is $\Delta\varphi = 4\pi$. Then the intensity of blue light is maximum at the polarizer orientation $\theta_2 = 45°$, but vanishes entirely at the orientation $\theta_2 = 135°$.

Since, for a given material, the product $\lambda(\Delta\varphi)$ is constant (see Eq. [5]), one can expect to find another wavelength λ' with associated phase $\Delta\varphi' = 3\pi$ which suffers destructive interference where the blue is maximally bright and constructive interference where the blue is absent. This wavelength—$\lambda' = (4/3)\lambda = 587$ nm—corresponds to yellow.

When looking at a mica slab through an appropriately oriented calcite prism (equivalent to the second polarizer), the ordinary and extraordinary images correspond to settings of **p** at $90°$ apart—and the two images show up in complementary colors.

Thus did Fresnel make a giant stride toward resolving the fundamental question: What is light? It is a transverse wave. From the perspective of physics, little more of a fundamental nature would be added until Maxwell's formulation of electromagnetism. With regard to personal fulfillment of the man who once wished to acquire a small bit of glory, Fresnel's "cup runneth over." Look at the index of any optics book, and there you will find eponymous laws (Fresnel-Arago laws), processes (Fresnel diffraction), formulas (Fresnel reflectance and transmittance coefficients), experiments (Fresnel double mirror), and devices (Fresnel double prism, Fresnel lens, Fresnel zone plane), to mention but a few. He died in 1827 within about six years of mastering the mystery of polarization. I do not know if he was satisfied. Arago delivered his eulogy before the Paris Academy.

As for Arago, he lived long enough (1786–1853) to see the unrivaled ascendency of the wave theory—and to wreak his revenge on Biot. Armed with Fresnel's theory of chromatic polarization, he delivered before the Academy in 1821 a devastating polemic against Biot's corpuscular theory, effectively recapturing the field he had lost some ten years earlier. Was Biot crushed? I don't know. In contrast to the brief lifetime of a number of key figures in optics (Malus, Fresnel, Maxwell, and others), he succumbed in 1862 at the extraordinary age of eighty-eight—still faithful to the particle theory.

To many scientists the equations and apparatus that form a familiar part of professional life are but useful tools for research. I use these tools myself but have come to see much more in them than mere utility. Behind each is a human story: the commitment to a philosophical ideal, the strenuous investment of time and effort, the pleasure of success, the pain of humiliation. Strident echoes of the passionate debates out of which new physics was forged have long died away; the printed memoirs, yellowed and brittle with age, are buried in dusty volumes no modern scientist will likely look at. Who today smells the acrid odor of gunpowder and hears the cries of the dying when he adjusts a linear polarizer and applies Malus's law? Who picks up a thin sheet of mica and imagines the angry public exchange between two bitter rivals? Who reflects light from a piece of glass or focuses it with a zone plate and sees the triumphant expression of a short-lived, rebellious young man striving for recognition? Not too many, I would suppose. Could it be, I wonder, that is why so few students, and the adults they become, find physics interesting?

NOTES

1. Etienne Malus, "Mémoire sur de nouveaux phénomènes d'optique," *Mém. Inst.* 11 (1811): 105–11.
2. Sir Isaac Newton, *Opticks* (New York: Dover, 1952), p. 358.

The State of Light

Particles with zero rest-mass have only two
directions of polarization. . . . This contrasts with the
$2S + 1$ directions of polarization for particles with
nonzero rest mass and spin S. . . . [Light] is the most
familiar example for this phenomenon. The "spin" of
light is 1, but it has only two directions of
polarization, instead of $2S + 1 = 3$. The number of
polarizations seems to jump discontinuously to two
when the rest-mass decreases [to] zero. . . . Instead of
the question: "Why do particles with zero rest-mass
have only two directions of polarization?," the
slightly different question, "Why do particles with a
finite rest-mass have more than two directions of
polarization?" is proposed.

(Eugene Wigner)[1]

THOUGH THEY differ in fundamental ways, electrons and light can both be
treated as two-state systems under appropriate circumstances. In this regard
quantum physics furnishes a powerfully simplifying formalism for solving
problems involving light polarization. Experimentally, the measurement of
polarization can be performed with an instrument of remarkable versatility—
the photoelastic modulator. Together, the two-state description of light and the
technique of phase modulation greatly facilitate the study of a wide range of
optical materials and processes.

9.1 HIDDEN POLARIZERS

There is no property of light that is not exploited in some way for its information
content. Photography, holography, and microscopy, for example, are concerned
with the spatial distribution of intensity I and wave vector $\mathbf{k}$ that constitutes
images and diffraction patterns. Spectroscopy, which provides penetrating
details of atomic and molecular structure, probes the distribution of wavelength
λ or frequency ν. Mining the riches of light polarization is the province of
ellipsometry and polarimetry.

Whereas the human visual system is sensitive to variations of intensity and color (at least within the narrow spectral window identified as visible light), it cannot, for the most part, unaidedly discern variations in light polarization—as can various insects (most notably bees), birds, and fish. This is in some ways regrettable, for without special appurtenances like polarizing sunglasses, we humans pass obliviously through a world of striking optical phenomena. Over the many years I have taught physics, I have never encountered a single student who had observed that rainbows are polarized. Indeed, precious few ever noticed the polarization of skylight.

Polarized light can be mystifying. I recall walking into my kitchen one bright day many years ago and noticing on the formica tabletop a crumpled cellophane bread wrapper. Illuminated principally by an open window behind it, the wrapper cast upon the smooth surface an impressive array of bright colors.[2] That thin-film interference, responsible for the colors often seen in soap bubbles and oil slicks, played no significant role here could be readily established. The effect disappeared when I put the wrapper on a mat surface such as a sheet of paper, or illuminated it with an incandescent or fluorescent lamp instead of light from the window.

I have often demonstrated for students the delightful profusion of colors to which a birefringent material of variable thickness gives rise when inserted between two Polaroid sheets. These multicolored patches, which vary in hue with relative orientation of the polarizers, are attributable—as discussed in the preceding chapter—to (1) the existence of two orthogonally polarized waves of different phase velocity within the sample, and (2) the wavelength dependence of the retardation of these waves.

Let us recall briefly the role of the two polarizers. Because the incident radiation is ordinarily unpolarized, the first element, the polarizer, thereby transmits waves of well-defined plane of vibration. The birefringent material then splits each monochromatic component into an ordinary wave with electric vector normal to the optic axis and an extraordinary wave polarized parallel to the optic axis. The two components originated from the same wave, so they are correlated in phase but cannot interfere because they are orthogonally polarized. This necessitates the second polarizing element, the analyzer. The analyzer projects the electric vectors of both waves onto a common axis, permitting their linear superposition and interference.

The bread wrapper in my kitchen was not inserted between anything, but rested crumpled by itself on the table. If the profusion of colors I saw were an interference phenomenon of the kind just described, then where were the polarizer and analyzer? Although initially puzzling, a little thought (and experimentation) revealed that the hidden polarizers came from two different polarizing mechanisms: Rayleigh (or molecular) scattering and specular reflection.

Sunlight scattered by air molecules is polarized—in fact, quite strongly. Skylight received at 90° to the incident illumination should theoretically be

100% polarized perpendicular to the plane defined by the incident and scattered wave vectors (figure 9.1a), although in practice it is somewhat less due to various depolarizing mechanisms. Nevertheless, a test with a Polaroid sheet showed that the light entering my window remained very noticeably polarized. That Rayleigh scattering played the role of polarizing element was shown to be plausible by placing the cellophane wrapper on a slab of plate glass (a flat casserole top worked fine) and examining the colors by a window through which sunlight entered directly and by a window on the far side of the house through which entered light scattered at nearly 90°. In the first case the polarization was markedly reduced as were the colors reflected in the glass; in the second case the colors stood out strongly.

Besides molecular scattering, specular reflection at Brewster's angle also polarizes light (figure 9.1b); in this case, the light is 100% polarized perpendicular to the plane of incidence defined by the surface normal and the incident wave vector. That polarization by reflection simulated the analyzing element was verified by examining the colors at different reflection angles. As I viewed the image of the cellophane in the table from directions approaching the normal to the surface, the colors faded away.

One does not have to go far from home to witness the aesthetic properties and enigmatic play of polarized light. I did not realize it at the time, but countless manifestations of light polarization associated with reflection and scattering were to occupy my thoughts for years to come.

9.2 DEFINING THE STATE OF LIGHT

As a quantum physicist whose professional interests in physical optics developed later, I have always been intrigued by the fact that light has two basic polarization states—just like an electron. The elementary excitation of light, the photon, is classified quantum mechanically as a spin-1 massless particle, and one might have expected light to have three states of polarization—as pointed out in the opening quotation by Eugene Wigner, recipient of the Nobel Prize for his profound mathematical investigations of symmetry in physics. In a delightfully paradoxical way, Wigner turns the issue upside down to suggest that the real question is not why photons have two states of polarization, but why massive particles have more. He answered this question by examining the physical consequences of Lorentz invariance, and I recommend to the reader his insightful essay on the subject.

To many of my optical colleagues, thought of the quantum implications of light polarization is but a remotely useful indulgence in academic exercise. Maxwell's equations predict two states of polarization, and why should anyone want to contemplate more? I agree—to a certain extent. Nevertheless, even exclusively within the domain of classical optics, the quantum analog between

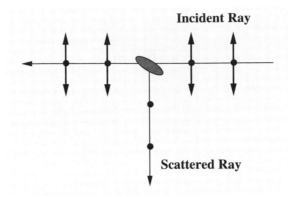

(a) Rayleigh Scattering at 90⁰

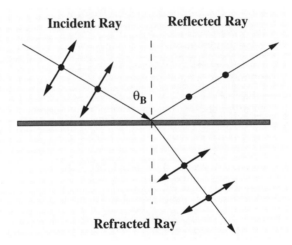

(b) Reflection at Brewster's Angle

Figure 9.1. Complete polarization by (a) Rayleigh scattering at 90° and (b) specular reflection at Brewster's angle θ_B. Double arrows signify polarization in the plane of incidence (π-polarization); black circles signify polarization normal to the plane of incidence (σ-polarization).

electron spin and light polarization is highly useful to remember. It is this analog that furnishes the simplest theoretical framework for treating the passage of light through a system of optical components.

   Perfectly polarized monochromatic light is one of nature's "irreducibles"; frequency, wave vector, and electric field orientation (corresponding to the quantum properties of energy, momentum, and spin) are sharply defined and cannot be decomposed into anything simpler. Such a light field exists, of course, only as an idealization in much the same way as frictionless bearings and point masses exist. The light issuing from any real source has a certain spectral content, beam divergence, and spatial distribution of field vectors. Of these various attributes of light, the concern of this chapter is only with light polarization. In effect, light is reduced to two degrees of freedom—for example, two orthogonal states of linear or circular polarization—and is amenable to treatment with the same mathematical formalism as that used in quantum physics to treat a spin-$\frac{1}{2}$ particle or the "two-level atom."[3]

In a typical experimental problem, incident light of known polarization passes through a suite of optical elements, and the properties of the light emerging from the final element are sought. Although many different kinds of measurements could conceivably be made on the detected beam, there are, in fact, only four empirically accessible independent quantities that embody the full information content of the light. Traditionally, these four quantities are identified with the Stokes parameters (of which I will say more), and the influence of each optical element is correspondingly represented by a four-dimensional matrix known as a Mueller matrix. The net effect of an entire optical system is then obtained by routine multiplication of a chain of such 4×4 matrices.

I must admit that I have never liked or used Mueller matrices. Finding them mystical and incomprehensible at the outset of my optical studies, I eschewed them then and never bothered to master them later. Even to their creator they must have had a certain quality of black magic. "How are Mueller matrices derived?" ask the noted authors of one text on polarized light. They continue, "One might guess that they stem from the electromagnetic theory, but this is not the case. They rest on a phenomenological foundation, i.e., on experiment. Professor Mueller has declared, only partly jokingly, that if his matrices should be found to disagree with the electromagnetic theory, '. . . so much the worse for the electromagnetic theory.'"[4] Statements like that make me uneasy.

Apart from any ambiguities associated with their origin, these matrices each have sixteen elements, many of them redundant. There are matrices, for example, which correspond to no possible optical components. Is it really necessary to treat light—the analog of a two-level system—with four-dimensional matrices? The standard reply appears to be in the affirmative, that anything short of the Mueller calculus is suitable only for perfectly polarized light in the absence of any depolarizing interactions. I am dubious.

One of the more vivid recollections of my undergraduate days as a chemistry student is a homework problem assigned in a class of physical chemistry. We were studying the electronic structure of aromatic hydrocarbons—the beautiful (on paper—*not* in the laboratory) ringlike structures with alternating (Linus Pauling would say "resonating") single and double bonds. The homework problem was to calculate, within the framework of the Debye-Hückel theory, the energy-level structure of the parent molecule, benzene. For what I wish to say here, the reader need know nothing more about Debye-Hückel theory or benzene than that the molecule is a six-carbon ring and that the required calculation led to a six-dimensional matrix for which we students were to deduce the sixth-degree secular equation and find its eigenvalues. It was a slave's work. Sometime after completing it, however, I learned of group theory—the branch of mathematics whose applications to physics were pioneered by Wigner—taught myself its rudiments, and realized to my great chagrin that the six-dimensional benzene problem could have been reduced to two one- and two-dimensional equations and completed in a minuscule fraction of the time previously spent. I vowed never to make that mistake again.

In the course of time, as my interests in quantum physics led me as well into various fields of optics, I worked out for myself a simple calculus to handle the sequential effects on light propagating through a chain of optical elements. I never sought to publish the method nor inquired as to its originality, but simply used it, much like Feynman originally employed his self-styled "funny little diagrams," as a personal shorthand way of arriving quickly at answers. Nevertheless, by using this simple calculus in invited talks or in the course of discussing details of a research problem with colleagues and collaborators, I began to realize how unfamiliar it was.

Looking over my shoulder as I scribbled equations on paper in my office at the Ecole Supérieure de Physique et Chimie Industrielles (ESPCI) in Paris, a colleague might point to a cluster of Greek symbols and ask, "What's that?" to which I would reply, "Those are Pauli spin matrices." "Where do they come from?" he would ask, and I would explain that they originated in the quantum theory of electrons. "What does a Pauli spin matrix have to do with a light wave?" inevitably came the rejoinder, and I would describe anew the universal applicability of these two-dimensional matrices to any system that could be reduced to two degrees of freedom irrespective of their actual physical nature— that is, whether the object under scrutiny was an electron with spin, a molecule with two electronic states, a Josephson junction between two superconducting wires, or—as the case actually was—a light wave with two polarization states. Looking skeptical, my bemused colleague would inquire if we would arrive at the same result using Mueller matrices. In the beginning I often had to answer, "I don't know; I think so, but I never learned to use Mueller matrices. I don't even know where to find them." In time, however, I could say with confidence, "Yes, of course; the two methods, if employed consistently, must give the same result." And, mindful of my painful chemistry lesson, I would add, "But why bother with four-dimensional matrices when two-dimensional matrices suffice?"

The calculational procedure that I transplanted from quantum physics provides many useful relations that tie together different experimentally accessible quantities. It enables one to handle in the most straightforward way (at least to my way of thinking) not only problems in which the optical components (polarizers, retarders, etc.) are homogenous and give rise to matrices with constant elements, but also problems in which the properties of these components may vary spatially and their corresponding matrices must be integrated or exponentiated. (Try evaluating e^M where M is a general four-dimensional matrix. Benzene!) Although I employed bits and pieces of this polarization calculus in various technical papers over the years treating light reflection, refraction, modulation, scattering, and other topics, I have never before, apart from lectures, recorded the work in a comprehensive and logical fashion. As I believe this is a useful activity, I take the opportunity to do it here.

9.3 THE DENSITY MATRIX

All that one can know about the polarization state of light is contained in what quantum mechanicians call the density (or statistical) matrix of the system. For light, or any other physical system with two degrees of freedom, this matrix is two-dimensional with at most four independent complex-valued elements. If, as is customary, the state of light is defined in terms of its electric vector $\mathbf{E}$, then the density matrix ρ can be defined by the relation

$$\rho \equiv \overline{\mathbf{E}\mathbf{E}^\dagger} = \begin{pmatrix} \rho_{11} & \rho_{12} \\ \rho_{21} & \rho_{22} \end{pmatrix} \tag{1}$$

where the overbar signifies an average over a time long compared to an oscillation period, but short in relation to a characteristic measurement time. The dagger signifies the adjoint or Hermitian conjugate; that is, if $\mathbf{E}$ is represented as a column vector, then $\mathbf{E}^\dagger$ is a row vector whose components are the complex conjugate of the components of $\mathbf{E}$.

Consider, for example, a light wave defined with respect to an orthonormal triad of basis vectors $\mathbf{e}_1, \mathbf{e}_2, \mathbf{k}$ (the unit wave vector) with right-handed sense $\mathbf{e}_1 \times \mathbf{e}_2 = \mathbf{k}$,

$$\mathbf{E} = a_1\mathbf{e}_1 + a_2\mathbf{e}_2. \tag{2}$$

The density matrix deduced from Eq. (1) is then

$$\rho = \begin{pmatrix} \overline{|a_1|^2} & \overline{a_1 a_2^*} \\ \overline{a_1^* a_2} & \overline{|a_2|^2} \end{pmatrix}. \tag{3}$$

If the basis vectors are real and the coefficients are $a_1 = a_2 = 1$, the matrix in Eq. (3) takes the form $\begin{pmatrix} 1 & 1 \\ 1 & 1 \end{pmatrix}$, which signifies linearly polarized light. By contrast, if $|a_1| = |a_2| = 1$, but the relative phase of the amplitudes is not well defined (i.e., if it is a fluctuating quantity), then the time-averaging process eliminates the off-diagonal elements, and the form of the matrix $\begin{pmatrix} 1 & 0 \\ 0 & 1 \end{pmatrix}$ signifies unpolarized light. These are but two special cases. The density matrix formalism allows one to handle all forms of light polarization.

It is worth noting, for readers interested in analogs to quantum physics, that the distinction between linearly polarized and unpolarized light, as represented by the preceding two density matrices, is precisely the same distinction in quantum mechanics between a *linear superposition* of states which gives rise to quantum interference, and a *mixture* of states which does not. A mixture is specified only by the populations of the states (i.e., the diagonal elements of the density matrix), whereas in a linear superposition there are coherence terms (the off-diagonal elements). The difference between two such systems—a mixture and a linear superposition—can have dramatic experimental consequences. For example, atoms raised incoherently into a mixture of excited states fluoresce in the expected way by decaying exponentially in time. Radiative decay from a superposition of states, however, can be modulated in time, showing "quantum beats" which reveal the energy intervals (Bohr frequencies) between the excited states.[5]

Besides the explicit representation with elements ρ_{ij} ($i, j = 1, 2$), as shown above, the density matrix can be cast into a particularly useful form by expanding it in terms of the unit two-dimensional matrix $\mathbf{1}$ and the Pauli spin matrices σ_i ($i = 1, 2, 3$) in the following way:

$$\rho = \rho_0 \mathbf{1} + \sum_{i=1}^{3} \rho_i \sigma_i = \rho_0 \mathbf{1} + \boldsymbol{\rho} \cdot \boldsymbol{\sigma} \tag{4}$$

where $\boldsymbol{\rho}$ is a vector with (numerical) elements ρ_1, ρ_2, ρ_3 and $\boldsymbol{\sigma}$ is a vector with (matricial) elements $\sigma_1, \sigma_2, \sigma_3$. When it is clear that an equation involves two-dimensional matrices, one can ordinarily dispense with the symbol $\mathbf{1}$ without incurring any confusion; it must then be remembered, however, that ρ_0 is a scalar coefficient and not a matrix.

The spin matrices are defined by the algebraic relation

$$\sigma_i \sigma_j = \delta_{ij} + i \sum_{k=1}^{3} \varepsilon_{ijk} \sigma_k \tag{5}$$

in which the Kronecker symbol δ_{ij} equals unity if $i = j$, and zero otherwise, and ε_{ijk} is the totally antisymmetric permutation symbol whose properties are succinctly described as follows. If the indices ijk form an even permutation of the numbers 123 (e.g., 231 or 312) the symbol is $+1$; if the indices form an odd permutation of 123 (e.g., 213 or 132), the symbol is -1; finally, if any two indices are equal (e.g., 112 or 333), the symbol is 0. Thus, the content of Eq. (5) is that the product of any two different spin matrices yields the third ($\sigma_1 \sigma_2 = i\sigma_3$, or $\sigma_2 \sigma_1 = -i\sigma_3$), and the square of any spin matrix is the unit matrix ($\sigma_i^2 = \mathbf{1}$). There are many (in principle, infinite) ways to represent the spin matrices explicitly. Often it suffices to know only their multiplication properties. Nevertheless, a widely used representation particularly convenient for physical optics is the following:

$$\sigma_1 = \begin{pmatrix} 0 & 1 \\ 1 & 0 \end{pmatrix}, \qquad \sigma_2 = \begin{pmatrix} 0 & -i \\ +i & 0 \end{pmatrix}, \qquad \sigma_3 = \begin{pmatrix} 1 & 0 \\ 0 & -1 \end{pmatrix} \tag{6}$$

in which case the expansion of the density matrix in Eq. (4) takes the form

$$\rho = \begin{pmatrix} \rho_0 + \rho_3 & \rho_1 - i\rho_2 \\ \rho_1 + i\rho_2 & \rho_0 - \rho_3 \end{pmatrix}. \tag{7}$$

To relate the coefficients in Eq. (4) to the density matrix elements in Eq. (1), one could compare matrices (1) and (7) and solve the resulting set of algebraic equations. It is mathematically simpler and more elegant, however, to utilize the multiplication properties of the spin matrices and the fact that these matrices are traceless. The trace (Tr) of a matrix is the sum of the elements along the principal diagonal (from upper left to lower right) and is one of the invariant properties of a matrix—that is, the trace remains the same regardless of the matrix representation. From Eq. (5) and the relations $\text{Tr}\{\sigma_i\} = 0$, $\text{Tr}\{1\} = 2$, one readily obtains

$$\rho_0 = \frac{1}{2}\text{Tr}\{\rho\} = \frac{\rho_{11} + \rho_{22}}{2}, \tag{8a}$$

$$\rho_1 = \frac{1}{2}\text{Tr}\{\rho\sigma_1\} = \frac{\rho_{12} + \rho_{21}}{2}, \tag{8b}$$

$$\rho_2 = \frac{1}{2}\text{Tr}\{\rho\sigma_2\} = \frac{\rho_{21} - \rho_{12}}{2i}, \tag{8c}$$

$$\rho_3 = \frac{1}{2}\text{Tr}\{\rho\sigma_3\} = \frac{\rho_{11} - \rho_{22}}{2}. \tag{8d}$$

In quantum theory, taking the trace of the product of the density matrix ρ with an arbitrary matrix M is equivalent to calculating the mean or expectation value of M (symbolized by angular brackets: $\langle M \rangle$) for the system described by ρ. Thus, one can write that

$$\rho_0 = \frac{1}{2}\langle 1 \rangle, \qquad \boldsymbol{\rho} = \frac{1}{2}\langle \boldsymbol{\sigma} \rangle; \tag{9}$$

the coefficient ρ_0 is the mean value of the unit matrix, and each component of the vector $\boldsymbol{\rho}$ is the mean value of the corresponding spin matrix.

For a light wave represented by the density matrix (3), the associated coefficients ρ_0 and ρ_i ($i = 1, 2, 3$) are

$$\rho_0 = \frac{|a_1|^2 + |a_2|^2}{2}, \tag{10a}$$

$$\rho_1 = \text{Re}\left(\overline{a_1^* a_2}\right), \tag{10b}$$

$$\rho_2 = \mathrm{Im}\left(\overline{a_1^* a_2}\right), \tag{10c}$$

$$\rho_3 = \frac{\overline{|a_1|^2} - \overline{|a_2|^2}}{2} \tag{10d}$$

which correspond, respectively, to one-half the Stokes's parameters usually designated I, U, V, Q.

$I = 2\rho_0$ is the mean intensity of the wave and is an invariant of the system; the value of I is independent of the choice of basis states and coordinate system. It is worth commenting explicitly on the physical significance of invariance. Although the analysis of an experimental configuration often calls for specifying a particular coordinate system and set of polarization basis vectors, ideally one would like to extract information that is characteristic of the light itself and independent of the location or orientation of the apparatus or any other decision made for mathematical convenience. In this regard, being able to recognize the invariants of the density matrix and relate them to experimental measurements is important.

$Q = 2\rho_3$ is the difference in intensities of the two chosen polarization basis states and therefore a measure of the degree of polarization—for example, the degree of linear polarization in the case where the basis vectors $\mathbf{e}_1$ and $\mathbf{e}_2$ are real-valued. This parameter is not, however, an invariant of the system. If the basis vectors $\mathbf{e}_1$ and $\mathbf{e}_2$ are transformed by a rotation, the value of ρ_3 will change.

As the real and imaginary parts of a bilinear product of amplitudes, the parameters $U = 2\rho_1$ and $V = 2\rho_2$ are clearly sensitive to the relative phase of the component polarization states. In contrast to ρ_0 and ρ_3, the interpretation of ρ_1 or ρ_2 is not readily apparent. It turns out—as I will demonstrate shortly—that ρ_1 is a measure of the degree of linear polarization relative to a new set of basis vectors rotated by $45°$ about $\mathbf{k}$, and ρ_2 is a measure of the degree of circular polarization. Like ρ_3, ρ_1 depends on the orientation of the basis set, but ρ_2 is a rotational invariant of the system—a surprising statement, perhaps, in view of the ostensibly unrevealing form of Eq. (10c). In fact, ρ_2, like I, is an intrinsic property of the light. Why, one might wonder, should the imaginary part of a function have special properties that either the real part or the entire function does not have? We will consider these points soon.

From the formalism established above, one can extract fairly easily another invariant quantity less obvious than that of total light intensity I. To do so, consider the diagonalized form of the density matrix with eigenvalues ρ_+ and ρ_-. Although there is a well-known standard procedure for diagonalizing a matrix, the eigenvalues of a two-dimensional matrix can be obtained almost by inspection simply by noting that the trace and determinant are invariant properties of a matrix. Starting with ρ expressed in Eq. (7), one then can write

$$\mathrm{Tr}\{\rho\} = 2\rho_0 = \rho_+ + \rho_- \tag{11}$$

and

$$\det\{\rho\} = \rho_0^2 - |\rho|^2 = \rho_+\rho_- \tag{12}$$

in which

$$|\rho| = \sqrt{\rho_1^2 + \rho_3^2 + \rho_3^2} \tag{13}$$

is the magnitude of the "density vector" in the expansion (4). From Eqs. (11) and (12) the eigenvalues can be extracted immediately:

$$\rho_\pm = \rho_0 \pm |\rho|. \tag{14}$$

The eigenvalues of ρ signify the intrinsic mix of orthogonal polarization states. It is reasonable, therefore, to define the difference of these eigenvalues— normalized by the light intensity (to within a numerical factor)—as the "degree of polarization" τ:

$$\tau \equiv \frac{\rho_+ - \rho_-}{\rho_+ + \rho_-} = \frac{|\rho|}{\rho_0}. \tag{15a}$$

That τ is an invariant property of the light (and not dependent on the orientation of the measuring apparatus, etc.) is readily demonstrable by rewriting Eq. (15a) in terms of the invariant trace and determinant of the density matrix

$$\tau = \sqrt{1 - \frac{4\det\{\rho\}}{(\mathrm{Tr}\{\rho\})^2}}. \tag{15b}$$

Consider again the two special cases of light introduced previously,

$$\rho_p = \begin{pmatrix} 1 & 1 \\ 1 & 1 \end{pmatrix}, \qquad \rho_u = \begin{pmatrix} 1 & 0 \\ 0 & 1 \end{pmatrix},$$

both with unit amplitudes in the initially chosen (linear) basis set. It follows from Eq. (15b) that $\tau = 1$ for ρ_p, signifying 100% polarized light, and $\tau = 0$ for ρ_u, indicative of totally unpolarized light.

The degree of polarization defined in Eqs. (15a,b) is actually too comprehensive, for it comprises two independent invariant quantities, each of which is of experimental interest in itself. Starting with the second equality of Eq. (15a), one can express τ in the alternative form

$$\tau = \sqrt{\tau_L^2 + \tau_C^2} \tag{15c}$$

in which

$$\tau_L = \sqrt{\frac{\rho_1^2 + \rho_3^2}{\rho_0^2}} \tag{16}$$

is the "degree of linear polarization" and

$$\tau_C \equiv \frac{|a_L|^2 - |a_R|^2}{|a_L|^2 + |a_R|^2} = \frac{\rho_2}{\rho_0} \tag{17}$$

is the "degree of circular polarization," defined in terms of amplitudes a_L, a_R of left and right circular polarization (LCP, RCP).

I will justify these terms soon, as well as explain how one can measure these functions experimentally by using one of the most versatile devices in optics, the photoelastic modulator.

9.4 TRANSFORMATIONS

Having represented a light wave $\mathbf{E}$ in terms of a particular basis set, such as in Eq. (2), suppose it is necessary to rotate that set by an angle θ about the beam direction $\mathbf{k}$. Specifically, let us choose the original basis $(\mathbf{x}, \mathbf{y})$ along the x- and y-axes of our coordinate system with light propagating in the $\mathbf{k} = \mathbf{z}$ direction. The rotation is effected in a positive or right-hand sense, that is, from x toward y. Mathematically, a rotational transformation, represented by a matrix $R(\theta)$, affects both basis vectors and the projection of amplitudes along the associated axes, but cannot, of course, cause any physical change to the light wave.

To express this invariance of $\mathbf{E}$ under rotation, as well as to avoid confusion later over which matrix—R or R^{-1}—rotates basis vectors or amplitudes, write Eq. (2) explicitly in the form of an invariant scalar product of a row vector of unit vectors and a column vector of amplitudes:

$$\mathbf{E} = (\mathbf{x} \quad \mathbf{y})\begin{pmatrix} a_1 \\ a_2 \end{pmatrix} = (\mathbf{x} \quad \mathbf{y})RR^{-1}\begin{pmatrix} a_1 \\ a_2 \end{pmatrix} = (\mathbf{x}' \quad \mathbf{y}')\begin{pmatrix} a_1' \\ a_2' \end{pmatrix}. \tag{18}$$

Insertion of the unit matrix $\mathbf{1} = RR^{-1}$ cannot change the form of $\mathbf{E}$, but shows that the basis vectors and amplitudes transform according to the rules

$$(\mathbf{x}' \quad \mathbf{y}') = (\mathbf{x} \quad \mathbf{y})R(\theta), \tag{19}$$

$$\begin{pmatrix} a_1' \\ a_2' \end{pmatrix} = R^{-1}(\theta)\begin{pmatrix} a_1 \\ a_2 \end{pmatrix}. \tag{20}$$

For a positive rotation, $R(\theta)$ takes the form

$$R(\theta) = \begin{pmatrix} \cos\theta & -\sin\theta \\ \sin\theta & \cos\theta \end{pmatrix}. \tag{21a}$$

Rotation is a length-preserving transformation, which, in the present case, means that the intensity $|a_1|^2 + |a_2|^2$, or ρ_0, is invariant. R is an orthogonal matrix: the inverse matrix R^{-1} is obtained by reflecting the elements across the principal diagonal to form the transposed matrix R^T:

$$R(\theta)^{-1} = R(\theta)^T = \begin{pmatrix} \cos\theta & \sin\theta \\ -\sin\theta & \cos\theta \end{pmatrix}. \tag{21b}$$

Consider, for example, a $+45°$ rotation about $\mathbf{k}$. Expressing the amplitudes a_1, a_2 in terms of a_1', a_2' by the inverse of Eq. (20) leads to the product

$$a_1^* a_2 = \frac{1}{2}\left(|a_1'|^2 - |a_2'|^2\right) + \frac{1}{2}(a_1'^* a_2' - a_1' a_2'^*). \tag{22}$$

Since the second term is a pure imaginary number, $i\,\text{Im}\{a_1'^* a_2'\}$, the real part of $a_1^* a_2$ is recognized as the difference in intensities of orthogonal linear polarization states. From Eqs. (10a, b), the ratio ρ_1/ρ_0 is then clearly interpretable as the degree of linear polarization in a coordinate system rotated $+45°$:

$$\frac{\rho_1}{\rho_0} = \tau_L^{(45)} = \frac{|a_1'|^2 - |a_2'|^2}{|a_1'|^2 + |a_2'|^2}. \tag{23}$$

Likewise, the ratio

$$\frac{\rho_3}{\rho_0} = \tau_L^{(0)} = \frac{|a_1|^2 - |a_2|^2}{|a_1|^2 + |a_2|^2} \tag{24}$$

defines the degree of linear polarization in the original basis, denoted by a superscript for $0°$ rotation. Neither $\tau_L^{(0)}$ nor $\tau_L^{(45)}$ is an invariant, but combining the two according to Eq. (16) yields the invariant degree of linear polarization

$$\tau_L = \sqrt{(\tau_L^{(0)})^2 + (\tau_L^{(45)})^2}. \tag{25}$$

Besides rotation, it is often necessary to transform between states of linear polarization (LP) and circular polarization (CP). Since the convention in this regard is not uniform, let us first clarify what is meant by left and right circular polarizations. Optical physicists often (though not always) define the handedness or sense of a light wave with respect to the source; if the electric vector $\mathbf{E}$ rotates toward the left side of an observer facing the source, the light is LCP (and conversely for RCP). In other branches of physics, and especially within the engineering community, the sense of an electromagnetic wave is defined relative to the direction of propagation. The electric vector of LCP light in that case rotates toward the left side of an observer facing away from the source. To avoid confusion, I have found it useful to draw again from the quantum physicists' stock of terms and principles and classify the states of circular polarization not by handedness, but by helicity.

Circularly polarized light carries angular momentum. The helicity η of a circularly polarized photon refers to the projection of its spin angular momentum onto its linear momentum normalized by the product of the two magnitudes. This quantity assumes only two values: $+1$ if the component of photon spin is parallel to the propagation direction $\mathbf{k}$ (which corresponds to LCP in the convention of optical physics) and -1 if the spin component is antiparallel to $\mathbf{k}$ (RCP light in the same convention). Thus, the states of circular polarization can be labeled by $(\pm)$, instead of (L, R), to signify the appropriate helicity.

To express CP basis vectors and amplitudes in terms of corresponding LP quantities (and vice versa), one applies a unitary transformation U:

$$(\mathbf{e}_+ \quad \mathbf{e}_-) \equiv (\mathbf{L} \quad \mathbf{R}) = (\mathbf{x} \quad \mathbf{y})U \tag{26}$$

and

$$\begin{pmatrix} a_+ \\ a_- \end{pmatrix} = U^{-1} \begin{pmatrix} a_1 \\ a_2 \end{pmatrix} \tag{27}$$

in which

$$U = \frac{1}{\sqrt{2}} \begin{pmatrix} 1 & 1 \\ i & -i \end{pmatrix}. \tag{28a}$$

Eqs. (26) and (27) are of the same form as Eqs. (19) and (20) for a rotational transformation, and for precisely the same reason: changing between bases can have no physical effect on the field **E**. Thus, one could replace R by U in Eq. (18). There is a significant difference, however, between the matrices R and U. The inverse of a unitary matrix U is equal to the Hermitian adjoint matrix $U^\dagger$, where the latter is obtained by reflecting the complex conjugate of the elements across the diagonal:

$$U^{-1} = U^\dagger = \frac{1}{\sqrt{2}} \begin{pmatrix} 1 & -i \\ 1 & i \end{pmatrix}. \tag{28b}$$

The transformation expressed by matrix (28a) is indicative of another choice between two opposite conventions, in this case representation of the time dependence of a harmonic wave. Optical physicists usually express the temporal phase factor of a traveling wave by $e^{-i\omega t}$, and this is the convention I adopt. My engineering colleagues, however, prefer the form $e^{j\omega t}$, where the unit imaginary is $j = -i$. With the optical physics convention, the basis states expressed by Eq. (26) have the explicit form

$$\mathbf{e}_+ = \frac{1}{\sqrt{2}}(\mathbf{x} + i\mathbf{y}), \qquad \mathbf{e}_- = \frac{1}{\sqrt{2}}(\mathbf{x} - i\mathbf{y}). \tag{29}$$

To verify quickly that the designated helicity is correct, multiply $\mathbf{e}_+$, for example, by $e^{-i\omega t}$ and extract the real part of the expression to get the real physical waveform. (Maxwellian electrodynamics is an intrinsically real-valued field theory, and the use of complex numbers is only a convenience, not a necessity. This is not the case, by the way, in quantum mechanics where the equations of motion, like the Schrödinger equation, are intrinsically complex-valued relations.) The resulting wave, $\mathrm{Re}\{\mathbf{e}_+ e^{-i\omega t}\} = \frac{1}{\sqrt{2}}(\mathbf{x}\cos\omega t + \mathbf{y}\sin\omega t)$, rotates in a positive sense from **x** (at $t = 0$) to **y** (at $t = \pi/2\omega$) and is clearly LCP (for an observer facing the source) or helicity $+1$.

Now examine the product $a_1^* a_2$ expressed in terms of CP amplitudes. (We have already seen that the real part of this product contributes to $\tau_L^{(45)}$.) From the inverse of Eq. (27) it is not difficult to show that

$$a_1^* a_2 = i \left(\frac{|a_+|^2 - |a_-|^2}{2} \right) + \text{Im}\{a_+^* a_-\}. \tag{30}$$

Thus, the imaginary part of $a_1^* a_2$ is proportional to the difference in intensities of the CP components, and from Eqs. (10a, c) and the invariance of ρ_0 under a unitary transformation it follows that the ratio

$$\frac{\rho_2}{\rho_0} = \frac{|a_+|^2 - |a_-|^2}{|a_+|^2 + |a_-|^2} = \tau_C \tag{31}$$

indeed yields the degree of circular polarization (as expressed previously in Eq. [17] in terms of LCP and RCP labels.) Given the physical nature of circularly polarized states, it should be apparent that $\text{Im}\{a_1^* a_2\}$, and therefore τ_C, are invariant under a rotational transformation.

In the case of a coherent waveform for which no time-averaging is required, the density matrix formalism can be dispensed with, and the optical invariants expressed directly in terms of geometrical products of the field amplitude $\mathbf{E}$. The most familiar example is the intensity

$$I = \mathbf{E} \cdot \mathbf{E}^* = |\mathbf{E}|^2, \tag{32}$$

but comparable expressions can be found for the degree of linear polarization

$$\tau_L = \frac{|\mathbf{E} \cdot \mathbf{E}|}{\mathbf{E} \cdot \mathbf{E}^*} \tag{33}$$

and the degree of circular polarization

$$\tau_C = i\mathbf{k} \cdot \frac{\mathbf{E} \times \mathbf{E}^*}{\mathbf{E} \cdot \mathbf{E}^*}.^6 \tag{34}$$

That the vector products in Eqs. (32), (33), and (34) are independent of coordinate system and basis set is an immediate consequence of the properties of vectors; vector algebra was created, after all, to handle geometrical relations in a coordinate-free way. The equivalence of Eqs. (33) and (34) to previous expressions defined in terms of amplitude components is perhaps not transparent, but nonetheless demonstrable with a little algebra.

Expression (34) is especially interesting, for it is not, in my experience, a particularly well-known relation. Indeed, it is sufficiently obscure that in recent years an extensive scientific literature has developed examining in minute detail the far-reaching electrodynamic, quantum, and cosmological implications of a "new" nonlinear light interaction proportional to $\mathbf{E} \times \mathbf{E}^*$ (deduced by analogy to the Poynting vector $\mathbf{S} \propto \mathbf{E} \times \mathbf{H}^*$) and interpreted as a "longitudinal magnetic field" carried by the photon. Several books have even been written on the subject. Were any of this true, such a radical revision of Maxwellian electrodynamics would of course be highly exciting, but it is regrettably the chimerical product of self-delusion—just like the "discovery" of N-Rays in the early 1900s.[7] (During the period 1903–1906 some 120 trained scientists published almost 300 papers on the origins and characteristics of a totally spurious radiation first reported by the French scientist, René Blondlot.) The actual significance of the right side of Eq. (34) is that it is a generalization of the concept of helicity defined earlier for a single photon and now extended to a classical electromagnetic wave. If the wave is purely LCP (helicity +1), τ_C takes the value +1; similarly $\tau_C = -1$ for a pure RCP wave (helicity −1). For the general case of an elliptically polarized wave, the degree of circular polarization falls within the range $-1 \leq \tau_C \leq +1$. From relations (23) and (24), it is clear that the partial degrees of linear polarization $\tau_L^{(0)}$ and $\tau_L^{(45)}$ also fall within the same range, but τ_L (Eq. [25]) is a positive number between 0 and 1.

Since many optical problems are best approached by using a CP basis rather than a linear basis, it is useful to consider the effect of transformation (27) on the elements of the density matrix. These elements, labeled by indices ±, are obtained from the expression

$$\rho = U \overline{\binom{a_1}{a_2} (a_1^* \quad a_2^*)} U^\dagger. \tag{35}$$

The physically significant coefficients ρ_0 and ρ then become

$$\rho_0 = \frac{1}{2}(\rho_{++} + \rho_{--}), \tag{36a}$$

$$\rho_1 = \frac{i}{2}(\rho_{+-} - \rho_{-+}) = -\text{Im}\{\rho_{+-}\}, \tag{36b}$$

$$\rho_2 = \frac{1}{2}(\rho_{++} - \rho_{--}), \tag{36c}$$

$$\rho_3 = \frac{1}{2}(\rho_{+-} + \rho_{-+}) = \text{Re}\{\rho_{+-}\}. \tag{36d}$$

A glance at relation (36c) again shows immediately—in contrast to the corresponding relation (8c) in a LP basis—why ρ_2/ρ_0 is interpretable as the degree of circular polarization. From the defining equation (16) and relations (36b,d), the degree of linear polarization is seen to take the succinct form

$$\tau_L = \frac{|\rho_{+-}|}{\rho_0}. \tag{37}$$

Finally, in addition to the various degrees of polarization, two other useful experimental quantities sensitive to the structure of optical media are the linear and circular intensity differences

$$\delta_L \equiv \frac{I^{(1)} - I^{(2)}}{I^{(1)} + I^{(2)}} = \frac{\rho_0^{(1)} - \rho_0^{(2)}}{\rho_0^{(1)} + \rho_0^{(2)}} \tag{38}$$

and

$$\delta_C \equiv \frac{I^{(+)} - I^{(-)}}{I^{(+)} + I^{(-)}} = \frac{\rho_0^{(+)} - \rho_0^{(-)}}{\rho_0^{(+)} + \rho_0^{(-)}}. \tag{39}$$

The basic idea—although not necessarily the actual experimental procedure by which δ_L and δ_C are measured—is that one irradiates a sample first with light of one pure form of linear or circular polarization and then with light of the corresponding orthogonal polarization. Effectively, two different beams are employed. In each case the total resulting light intensity (irrespective of polarization) is detected, and the ratios recorded according to the defining expressions (38) and (39). The superscripts denote the pure polarization state of the first and second incident beam, and, from the invariance of the individual terms in numerator and denominator, it is clear that δ_L and δ_C are themselves invariant measures of the interaction between sample and light. The linear and circular intensity differences are to be distinguished from the degrees of linear and circular polarization. The latter refer to the differences in intensity of the LP or CP components of a single light beam.

The utility of the various measures of light polarization introduced in this chapter will be seen in the experimental studies to follow, especially those pertaining to reflection (chapter 11) and diffusive scattering (chapter 13).

9.5 THE ELLIPTICAL CONNECTION

An alternative way of describing the general (elliptical) polarization state of a coherent light wave is in terms of geometrical quantities that relate explicitly to the ellipse traced out by the vector **E** in a plane normal to the propagation direction. This geometry is illustrated in figure 9.2. The x- and y-axes define a convenient coordinate system in the laboratory relative to which the semimajor axis of the ellipse is oriented at an angle ϕ. The angle ψ, defined by

$$\tan \psi = \frac{\text{semiminor axis}}{\text{semimajor axis}} = \frac{b}{a}, \tag{40}$$

is the ellipticity, a measure of the degree of linear or circular polarization. For example, if $\psi = 0$, there is no amplitude component b, and the light is purely LP; if $\psi = \pi/4$, the amplitudes a and b are equal, and the light is purely CP. The sign of ψ, as will be seen shortly, conveys the sense (right or left) in which **E** is rotating. It should be evident from the figure that ψ is a purely geometric—and therefore invariant—property of the light, for the shape of the ellipse is in no way altered by the choice of the laboratory coordinate system. On the other hand, the orientation angle ϕ would appear to be a relative quantity whose value depends on the seemingly arbitrary choice of x-axis. This is not necessarily the case, however, for the reference axis is usually defined experimentally by the

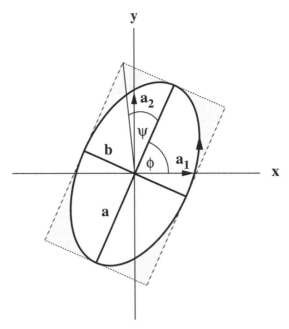

Figure 9.2. Geometrical invariants of elliptically polarized light: ellipticity (ψ), optical rotation (ϕ), amplitudes (a, b) along semi-major and semi-minor axes. a_1 and a_2 are (noninvariant) amplitudes along the laboratory axes.

linear polarization $\mathbf{e}_0$ of an incident light wave. The angle ϕ is then the "optical rotation" of the detected light $\mathbf{E}$ and, like ellipticity, expressible in the form of a geometric invariant

$$\cos \phi = \frac{\mathbf{E} \cdot \mathbf{e}_0}{|\mathbf{E}|}. \tag{41}$$

The science by which the geometric invariants of elliptically polarized light are measured is called ellipsometry. When, during my student days, I had resolved to teach myself optics—for I never had a formal course in the subject—I found the discussions of light polarization and ellipsometry to be unvaryingly dull, and my eyes would glaze over whenever they encountered pages black with trigonometric calculations and diagrams of ellipses. If I thought about the subject at all, it would be to lament the fate of those poor dreary souls whose livelihood depended on their learning such tedious material, and I could never have imagined the necessity of making such measurements myself. Indeed, even when, long afterward, I was totally immersed in the study of polarized light for purposes of testing aspects of electrodynamics and quantum physics, it did not occur to me that I was, in fact, doing ellipsometry—until I was invited one day to give a talk at an international ellipsometry conference. And there, effectively for the first time, I encountered a sizable fraction of the world's "ellipsometrists."

The conference was held in Paris, France—which is always a fine place for a conference—and the banquet took place in the fashionable restaurant below the "Pyramid" of the Louvre, tables bedecked with roses, and wine flowing freely. When the final strains of Boccherini's String Quartet in G minor faded away, and the routine expressions of gratitude to the conference organizers, sponsoring agencies, and others were over, I learned to my pleasant surprise how lively ellipsometrists are. Dreary souls indeed! A spontaneous urging suddenly overtook the participants to come up to the microphone, one at a time like the patrons in a karaoke bar, to pour forth the pleasures of ellipsometry. Some recounted their early experiences; some recited poetry; some told jokes. I particularly recall one specialist raising his wine glass and proclaiming: "To all of us here, ellipsometry is *not* a science, but a *religion!*" to which the seated audience voiced a loud cheer of approval. As an "outsider," my attachment to the discipline was (and still is) perhaps not quite so strong, but nonetheless I could appreciate, as I had never before, the devotion that ellipsometrists have for their work and the esprit de corps which binds them. From that time on, the sight of an ellipse (like figure 9.2) in an optics book awakens in me an altogether different association.

To relate the ellipsometric parameters of a coherent wave to the invariants developed in the preceding section, one starts by comparing the complex amplitudes a_1, a_2 in the laboratory coordinate system

$$a_1 = A_1 e^{i\alpha_1}, \qquad a_2 = A_2 e^{i\alpha_2} \tag{42}$$

to the corresponding amplitudes $a_{x'}$, $a_{y'}$ in the principal axis system of the ellipse

$$a_{x'} = A e^{i\alpha} \cos \psi, \qquad a_{y'} = A e^{i\alpha} \sin \psi. \tag{43}$$

There are four parameters in Eq. (42)—the real amplitudes A_1, A_2 and real phases α_1, α_2—and correspondingly four parameters in Eq. (43) if, in addition to the real amplitude A, global phase α, and ellipticity ψ, one recalls the angle ϕ which defines the principal axis x'. Only three parameters are ordinarily of physical significance; except for interference with an external reference wave, it is only the relative phase

$$\delta = \alpha_1 - \alpha_2 \tag{44}$$

that can be measured, and not the global phase

$$\alpha = \frac{1}{2}(\alpha_1 + \alpha_2) \tag{45}$$

(which, under the conditions of an interference experiment, could no longer be considered a global phase).

Upon transforming the amplitudes in Eq. (43) by a rotation $R(\phi)$ to express them in terms of the laboratory coordinate system, one obtains the equations

$$A_1 e^{i\alpha_1} = A(\cos\phi \cos\psi - i \sin\phi \sin\psi)e^{i\alpha}, \tag{46a}$$

$$A_2 e^{i\alpha_2} = A(\sin\phi \cos\psi + i \cos\phi \sin\psi)e^{i\alpha} \tag{46b}$$

from which all else follows. In particular, Eqs. (10a–d) lead to the coefficients of the density matrix

$$\rho_0 = \frac{1}{2}A^2, \tag{47a}$$

$$\rho_1 = \frac{1}{2}A^2 \sin 2\phi \cos 2\psi, \tag{47b}$$

$$\rho_2 = \frac{1}{2}A^2 \sin 2\psi, \tag{47c}$$

$$\rho_3 = \frac{1}{2}A^2 \cos 2\phi \cos 2\psi. \tag{47d}$$

The relative phase, given by

$$\tan\delta = \frac{-\tan 2\psi}{\sin 2\phi}, \tag{48}$$

ultimately follows from application of the trigonometric identity for reducing the tangent of the difference of two angles.

From relations (47a–d) the degrees of polarization, expressed in terms of ellipsometric parameters, take the especially simple forms

$$\tau_L = |\cos 2\psi|, \qquad \tau_C = \sin 2\psi. \tag{49}$$

The identity $\tau_L^2 + \tau_C^2 = 1$ signifies that the wave is 100% polarized.

9.6 OPTICAL SYSTEMS

The fundamental problem that the density matrix theory is designed to handle is of the following kind: What is the polarization state of an incident light wave $E^{(0)}$ that has passed sequentially through a chain of optical components? There is no restriction on the nature of the components (apart from consistency with physical laws). In passing through the system, the light can be polarized, depolarized, scattered, absorbed, amplified, and so on.

If the physical effect exerted on an incident light wave by the i-th component in a system of n components is represented mathematically by the two-dimensional matrix M_i, then the emerging light wave E (represented as a two-component column vector) takes the form

$$E = M_n \cdots M_2 M_1 E^{(0)} = M E^{(0)} \tag{50}$$

from which the system matrix M is defined. The order in which the component matrices M_i appear is important, for in general these matrices do not commute.

Using the definition of the density matrix in Eq. (1), one obtains the transformation

$$\rho = M \rho^{(0)} M^\dagger \qquad (51)$$

between the density matrix $\rho^{(0)}$ of the incident light and corresponding matrix ρ of the emerging light. The reader will perhaps recall that the Hermitian adjoint of a product of matrices is the product, in reverse order, of the adjoint of the individual matrices

$$M^\dagger = M_1^\dagger M_2^\dagger \cdots M_n^\dagger. \qquad (52)$$

Our basic objective is to determine the physically significant components ρ_0, ρ of the "output" density matrix in terms of the corresponding components $\rho_0^{(0)}, \rho^{(0)}$ of the "input" density matrix. To do so, we first expand M and $M^\dagger$ in terms of **1** and σ:

$$M = m_0 + \mathbf{m} \cdot \boldsymbol{\sigma}, \qquad M^\dagger = m_0^* + \mathbf{m}^* \cdot \boldsymbol{\sigma} \qquad (53)$$

where the second relation in Eqs. (53) has exploited the fact that the spin matrices are Hermitian (or self-adjoint): $\sigma_i^\dagger = \sigma_i$. The explicit connection between m_0, **m**, and the elements M_{ij} is precisely of the same form as that given for the expansion of ρ in Eqs. (8a–d). Next, multiply the expanded forms of the matrices in the right-hand side of Eq. (51) and reduce the resulting products of the spin matrices to expressions that involve either the unit matrix or the first power of σ. This reduction is greatly facilitated by the identity

$$(\mathbf{A} \cdot \boldsymbol{\sigma})(\mathbf{B} \cdot \boldsymbol{\sigma}) = \mathbf{A} \cdot \mathbf{B} + i \mathbf{A} \times \mathbf{B} \cdot \boldsymbol{\sigma} \qquad (54)$$

(derivable from Eq. [5]) where **A** and **B** are ordinary vectors. Finally, compare corresponding coefficients of **1** and σ on both sides of the equation.

The above program, which with the algebra of the spin matrices can actually be executed fairly easily, leads to a set of sixteen transformation coefficients that constitute one scalar quantity Σ_0^0, two 3-component vectors Σ^0 and Σ_0, and one 9-component tensor $\overline{\Sigma}$ as identified in the resulting expansion

$$\rho = \rho_0 + \boldsymbol{\rho} \cdot \boldsymbol{\sigma} = (\rho_0^{(0)} \Sigma_0^0 + \boldsymbol{\rho}^{(0)} \cdot \boldsymbol{\Sigma}^0) + (\rho_0^{(0)} \boldsymbol{\Sigma}_0 + \boldsymbol{\rho}^{(0)} \cdot \overline{\Sigma}) \cdot \boldsymbol{\sigma} \qquad (55)$$

where

$$\Sigma_0^0 = |m_0|^2 + |\mathbf{m}|^2, \qquad (56a)$$

$$\Sigma_i^0 = 2\left[\text{Re}\{m_0^* m_i\} - \varepsilon_{(ijk)} \text{Im}\{m_j^* m_k\}\right], \qquad (56b)$$

$$\Sigma_0^i = 2\left[\text{Re}\{m_0^* m_i\} + \varepsilon_{(ijk)} \text{Im}\{m_j^* m_k\}\right], \qquad (56c)$$

$$\Sigma_j^i = (|m_o|^2 - |\mathbf{m}|^2)\delta_{ij} + 2[\text{Re}\{m_i^* m_j\} + \varepsilon_{(ijk)} \text{Im}\{m_0^* m_k\}]. \qquad (56d)$$

In the labeling of the transformation coefficients, the upper index refers to the component ρ_μ ($\mu = 0, 1, 2, 3$) of the output matrix ρ; the lower index correlates with the component $\rho_\mu^{(0)}$ of the input matrix. Note, too, that I have placed the indices of the antisymmetric symbol ε_{ijk} in parentheses to indicate that, for a given index i, the remaining two indices are to be arranged in the *one* order for which the symbol takes the value $+1$. In other words, no summation over repeated indices in Eqs. (56b–d) is implied (as a reader familiar with the so-called Einstein convention might assume). In evaluating Σ_2^0, for example, the second term on the right-hand side of Eq. (56b) is $-\operatorname{Im}\{m_3^* m_1\}$, the components of **m** having been ordered so that $\varepsilon_{231} = +1$.

The procedure outlined above leads straightforwardly to Eq. (55) by vector algebra, and one can simply read the various Σ-coefficients from a single vectorial equation. Alternatively, by using the trace properties of the spin matrices, one can derive an operational expression

$$\Sigma_\nu^\mu = \frac{1}{2}\operatorname{Tr}\{\mathbf{M}^\dagger \sigma_\mu \mathbf{M}\sigma_\nu\} \tag{57}$$

for these coefficients directly in terms of the full system matrix M. The indices μ, ν in Eq. (57) assume the values 0, 1, 2, 3 with σ_0 identified as the unit matrix **1**. (It is a common convention among quantum theorists to use Greek indices to identify the components of a "4-vector" and Latin indices for a "3-vector.") To perform the matrix multiplications, expand M and $\mathbf{M}^\dagger$ according to Eq. (53) and make use of the following identities derivable from Eq. (5):

$$\frac{1}{2}\operatorname{Tr}\{\sigma_\mu\} = \delta_{\mu 0}, \tag{58a}$$

$$\frac{1}{2}\operatorname{Tr}\{\sigma_i \sigma_j\} = \delta_{ij}, \tag{58b}$$

$$\frac{1}{2}\operatorname{Tr}\{\sigma_i \sigma_j \sigma_k\} = i\varepsilon_{ijk}, \tag{58c}$$

$$\frac{1}{2}\operatorname{Tr}\{\sigma_i \sigma_j \sigma_k \sigma_m\} = \delta_{ij}\delta_{km} + \delta_{im}\delta_{jk} - \delta_{ik}\delta_{jm}. \tag{58d}$$

The two different, but equivalent, calculational methods must, of course, yield the same final expressions for the Σ elements, although each has its own advantages and drawbacks. I have found it useful to use both procedures in analyzing optical systems. Often, all that is measured in an optical experiment is the output light intensity I, in which case $\rho_0 = \frac{1}{2}\operatorname{Tr}\{\rho\}$ is the only density matrix component required. It then follows from Eq. (51) that the observable effects of the optical system on the light are fully rendered by the "output" matrix $\Sigma \equiv \mathbf{M}^\dagger\mathbf{M}$ or $\mathbf{M}\mathbf{M}^\dagger$ (order is unimportant in this case since $\operatorname{Tr}\{AB\} = \operatorname{Tr}\{BA\}$) whose elements Σ_{ij} are calculated directly by matrix multiplication. Since Σ is a two-dimensional matrix, its expansion in the basis

$(1, \sigma)$ is readily accomplished. I will illustrate such applications in the following sections.

Although the density matrix formalism employs strictly two-dimensional matrices (and takes advantage of what is technically called the algebra of the two-dimensional unitary group), one can, if needed, create a four-dimensional algebra by expressing the expansion coefficients of ρ as a 4-vector and arranging the sixteen Σ-coefficients (Eqs. [56a–d]) as the elements of a four-dimensional transformation matrix in the following way:

$$\rho = \begin{pmatrix} \rho_0 \\ \rho_1 \\ \rho_2 \\ \rho_3 \end{pmatrix} = \begin{pmatrix} \Sigma_0^0 & \Sigma_1^0 & \Sigma_2^0 & \Sigma_3^0 \\ \Sigma_0^1 & \Sigma_1^1 & \Sigma_2^1 & \Sigma_3^1 \\ \Sigma_0^2 & \Sigma_1^2 & \Sigma_2^2 & \Sigma_3^2 \\ \Sigma_0^3 & \Sigma_1^3 & \Sigma_2^3 & \Sigma_3^3 \end{pmatrix} \begin{pmatrix} \rho_0^{(0)} \\ \rho_1^{(0)} \\ \rho_2^{(0)} \\ \rho_3^{(0)} \end{pmatrix}. \tag{59}$$

Since the components of the density matrix 4-vector are proportional to the four Stokes's parameters of the output beam, Eq. (59) establishes the equivalence of the density matrix and Mueller calculus. But logical equivalence does not necessarily imply equivalent ease of application (Benzene!), and I have never, myself, had need of the four-dimensional matrices. To examine analytically the effect of matter on light, the two-dimensional density matrix, contrary to pervasive belief, provides no less complete a description of an optical system than the four-dimensional calculus and is much more easily utilized.

Having begun this chapter with reference to Wigner, I would like to digress briefly to point out that the preceding discussion is but a generalization of Wigner's group theoretical approach to rotation. Matrices that effect a coordinate rotation ($\mathbf{r}' = \mathbf{R}\mathbf{r}$) in three-dimensional space are elements of the mathematical group SO(3), the group of 3×3 orthogonal matrices of determinant $+1$. The same rotation, however, can be described by a two-dimensional unitary matrix u, also of determinant $+1$, according to the relation $R_{ij} = \frac{1}{2} \mathrm{Tr}\{u^{-1}\sigma_i u \sigma_j\}$—that is, precisely the form to which Eq. (57) reduces in the special case of a unitary system matrix $\mathbf{M}^\dagger = \mathbf{M}^{-1}$. Since the correspondence is quadratic in u, both u and $-u$ give rise to the same three-dimensional rotation matrix R. In group theoretical language one says that there is a twofold homomorphism between the group SU(2) (the group of 2×2 unitary matrices of determinant $+1$) and the group SO(3)—an association that is not especially noteworthy in classical mechanics, but that has significant implications in quantum theory.[8]

The general system matrix M is not endowed with special properties such as unitarity or unit determinant; it effects a general two-dimensional linear transformation and therefore has four independent complex-valued elements. In mathematical terminology it belongs to the general linear group over the complex numbers GL(2c). Correspondingly, the 4×4 matrix with elements

Σ_ν^μ in Eq. (59) does not represent a four-dimensional rotation because it is not necessarily an orthogonal matrix.

If the chain of components through which light passes is representable by a unitary system matrix M—a case applicable, for example, in the absence of absorptive or diffusive losses—then it is not difficult to demonstrate directly from Eq. (51) that the trace and determinant of ρ are equal to the trace and determinant of $\rho^{(0)}$. In this case the optical system, although it may well alter the state of polarization of the light, effects neither the output intensity nor the (full) degree of polarization. In the more general case the output intensity and the degree of polarization τ are

$$I = 2\rho_0 = 2(\rho_0^{(0)}\Sigma_0^0 + \rho^{(0)} \cdot \Sigma^0), \tag{60}$$

$$\tau = \frac{|\rho|}{\rho_0} = \frac{|\rho_0^{(0)}\Sigma^0 + \rho^{(0)} \cdot \overline{\Sigma}|}{\rho_0^{(0)}\Sigma_0^0 + \rho^{(0)} \cdot \Sigma^0}. \tag{61}$$

When the incident light is unpolarized, the vector $\rho^{(0)}$ vanishes (for its components signify the presence of linear and circular polarizations); the ratio of output to input intensity and the degree of polarization respectively take the simpler forms

$$\frac{I}{I^{(0)}} = \Sigma_0^0, \tag{62}$$

$$\tau = \frac{|\Sigma^0|}{\Sigma_0^0}. \tag{63}$$

To construct the individual matrices M_i for the most familiar optical components, one usually need apply, according to circumstances, only the transformations for rotation $R(\theta)$ and change of basis U.

Consider, for example, the matrix $M_{LP}(\alpha)$ representing the action of a linear polarizer with transmission axis (x') at angle α to the vertical. An incident light wave $\mathbf{E}^{(0)}$, expressed in a vertical and horizontal basis $(\mathbf{x}, \mathbf{y})$ with amplitudes a_1 and a_2 (Eq. [18]), takes the form

$$\mathbf{E}^{(0)} = (\mathbf{x}', \mathbf{y}')\begin{pmatrix} a_1 \cos\alpha + a_2 \sin\alpha \\ -a_1 \sin\alpha + a_2 \cos\alpha \end{pmatrix} \tag{64}$$

in a basis rotated by α about the propagation direction. The purpose of the polarizer, however, is to suppress the $\mathbf{y}'$ component in which case the wave $\mathbf{E}_p$ exiting the polarizer is

$$\mathbf{E}_p = (\mathbf{x}', \mathbf{y}')\begin{pmatrix} a_1 \cos\alpha + a_2 \sin\alpha \\ 0 \end{pmatrix}. \tag{65a}$$

Reexpressed in terms of the original laboratory basis, $\mathbf{E}_p$ becomes

$$\mathbf{E}_p = (\mathbf{x}, \mathbf{y})M_{LP}(\alpha)\begin{pmatrix} a_1 \\ a_2 \end{pmatrix} \tag{65b}$$

where the linear polarizer matrix is

$$M_{LP}(\alpha) = \begin{pmatrix} \cos^2\alpha & \cos\alpha\sin\alpha \\ \cos\alpha\sin\alpha & \sin^2\alpha \end{pmatrix}. \tag{66}$$

This matrix is symmetric about the principal diagonal with $\mathrm{Tr}\{M\} = 1$ and $\det\{M\} = 0$. Since the determinant is null, $M_{LP}(\alpha)$ has no inverse, which is to be expected on physical grounds since the transition from Eq. (64) to Eq. (65a) incurs loss of information.

A more concise way to arrive at $M_{LP}(\alpha)$ is to employ the quantum mechanical concept (and Dirac notation) of a projection operator $P(\alpha) \equiv |\alpha\rangle\langle\alpha|$ where the "ket" $|\alpha\rangle$ signifies the polarization state

$$|\alpha\rangle = |x\rangle\cos\alpha + |y\rangle\sin\alpha \tag{67}$$

and the "bra" $\langle\alpha|$ is the adjoint state. That is, if $|\alpha\rangle$ represents a column vector of amplitudes [e.g., $|x\rangle = \binom{1}{0}$], then $\langle\alpha|$ is the corresponding row vector of complex conjugate amplitudes [$\langle x| = (1 \quad 0)$]. $P(\alpha)$ projects an incident polarization state onto the transmission axis of the polarizer. Substituting the form (67) into the defining expression for $P(\alpha)$ leads to

$$\begin{aligned} P(\alpha) &= |x\rangle\langle x|\cos^2\alpha + |x\rangle\langle y|\cos\alpha\sin\alpha \\ &+ |y\rangle\langle x|\cos\alpha\sin\alpha + |y\rangle\langle y|\sin^2\alpha \end{aligned} \tag{68}$$

from which the matrix $M_{LP}(\alpha)$ (Eq. [66]) immediately follows.

As a second example, consider the matrix M_{LB} of a nonabsorbing linearly birefringent component

$$M_{LB} = \begin{pmatrix} e^{i\Delta_1} & 0 \\ 0 & e^{i\Delta_2} \end{pmatrix} = e^{i\bar\Delta}\begin{pmatrix} e^{2i\Delta} & 0 \\ 0 & e^{-2i\Delta} \end{pmatrix} \tag{69}$$

with principal axes (and associated retardations Δ_1 and Δ_2) along the vertical and horizontal directions of the laboratory frame. The second equality in Eq. (69) expresses the retardation in terms of an angle proportional to the birefringence

$$\Delta = \frac{(\Delta_1 - \Delta_2)}{2} \tag{70a}$$

and the mean retardation

$$\bar\Delta = \frac{(\Delta_1 + \Delta_2)}{2}. \tag{70b}$$

Unless the light beam is subsequently to be superposed coherently with another beam, the global phase factor $e^{i\bar\Delta}$ can ordinarily be ignored.

Finally, let us consider the matrix M_{CB} of a general circularly birefringent component. Such a matrix is diagonal in a CP basis with a form analogous to that of the first equality in Eq. (69), except with (complex) retardations denoted by Δ_+, Δ_-:

$$\Delta_\pm = \phi_\pm + i\chi_\pm. \tag{71}$$

By means of Eqs. (26) and (27) the wave $\mathbf{E}_{CB}$ exiting the device can be cast in the form

$$\mathbf{E}_{CB} = (\mathbf{e}_+ \quad \mathbf{e}_-) \begin{pmatrix} e^{i\Delta_+} & 0 \\ 0 & e^{i\Delta_-} \end{pmatrix} \begin{pmatrix} a_+ \\ a_- \end{pmatrix} = (\mathbf{x} \quad \mathbf{y})\mathbf{U} \begin{pmatrix} e^{i\Delta_+} & 0 \\ 0 & e^{i\Delta_-} \end{pmatrix} \mathbf{U}^{-1} \begin{pmatrix} a_1 \\ a_2 \end{pmatrix}$$

$$= (\mathbf{x} \quad \mathbf{y})\mathbf{M}_{CB} \begin{pmatrix} a_1 \\ a_2 \end{pmatrix} \tag{72}$$

which, upon substitution of Eq. (71), ultimately leads to

$$\mathbf{M}_{CB} = e^{i\bar{\Delta}} \begin{pmatrix} \cos(\phi + i\chi) & \sin(\phi + i\chi) \\ -\sin(\phi + i\chi) & \cos(\phi + i\chi) \end{pmatrix}. \tag{73}$$

The angle

$$\phi = \frac{(\phi_+ - \phi_-)}{2} \tag{74}$$

is the optical rotation, and

$$\chi = \frac{(\chi_+ - \chi_-)}{2} \tag{75}$$

is the circular dichroism. The global phase factor $e^{i\bar{\Delta}}$ with

$$\bar{\Delta} = \frac{(\Delta_+ + \Delta_-)}{2} = \frac{(\phi_+ + \phi_-) + i(\chi_+ + \chi_-)}{2} \tag{76}$$

is not necessarily ignorable if the component is absorbing, for the imaginary part of Eq. (76) leads to a diminution in output intensity. However, it can be neglected whenever one is concerned primarily with ratios of the components of the output density matrix such as, for example, the various degrees of polarization.

If the circularly birefringent component is transparent, χ vanishes, and the resulting form of $\mathbf{M}_{CB}$ (up to a global phase factor) is that of a rotation matrix (hence the term "optical rotation" for the chiral parameter ϕ). If the material absorbs LCP and RCP light inequivalently, but refracts the two polarizations equally, then ϕ vanishes and $\mathbf{M}_{CB}$ takes the form

$$\mathbf{M}_{CB} = e^{-\frac{1}{2}(\chi_+ + \chi_-)} \begin{pmatrix} \cosh\chi & i\sinh\chi \\ -i\sinh\chi & \cosh\chi \end{pmatrix}. \tag{77}$$

The elements of Eq. (73) can be expanded into real-valued sines and cosines of ϕ and χ, but there is usually little point in doing so until after one has evaluated the components ρ_μ of the density matrix and is ready to extract the experimentally measurable quantities.

9.7 A DEVICE FOR ALL SEASONS: THE PHOTOELASTIC MODULATOR

Measurement of light polarization is one of the principal means of investigating the interaction of light with matter. Although a full exposition of how light polarization is measured is well beyond the scope of this chapter, I would like to discuss a particular instrument of extraordinary versatility that I have used frequently in the course of my research, and whose understanding is facilitated by the theoretical formalism just developed. Known as a photoelastic modulator or PEM, the first models, to my knowledge, were constructed during the mid-1960s at the ESPCI in Paris by my colleague Jacques Badoz and his student Michel Billardon.[9] During the periods that I held the Joliot Chair of Physics at the ESPCI I had the pleasure of examining these original instruments, which—like the telescopes of Newton and the lenses and prisms of Fresnel— are historically important, for they have, in their own way, greatly extended the human capacity to see, interpret, and exploit optical phenomena.

The PEM is a simple, low-power device—small enough to fit in the palm of one's hand—that modulates the polarization state of light. Accompanied by synchronous (phase-sensitive) detection, this device has been used in applications that span broad fields of optics: polarimetry, ellipsometry, spectroscopy, interferometry, and optical imaging. Compared to other modulation methods (such as mechanically rotating polarizers and Pockels and Kerr cells) the PEM has a large aperture (as much as $\sim$5 cm) and wide angle of acceptance ($\sim$50° full cone angle) and is operational throughout the entire visible spectrum and substantial parts of the infrared and ultraviolet ($\sim$180 nm to $\sim$20 μm). With a PEM one can measure all the invariant attributes of light and numerous material properties—for instance, linear and circular birefringence and linear and circular dichroism—that provide keyholes through which to observe a world of diverse phenomena. Since its invention, the PEM has been employed in a greater variety of experiments than almost any other optical instrument I know. One can investigate strains in stressed materials, products of chemical reactions, magnetic fields of stars, light scattering in the interstellar medium, the flow of fluids, optical properties of semiconductors, phase changes in crystals, atomic collisions, and the detection of objects embedded in turbid media—to cite but a few.

In recent years the Swedish Academy has honored various "instrument makers," notably the inventors of electron microscopes and subatomic particle detectors, with the Nobel Prize in Physics. Less "flashy" perhaps, and undoubtedly much less well known throughout the full physics community, the PEM represents, in my opinion, an inventive achievement of comparable worth.

At the heart of the device, schematically illustrated in figure 9.3, is a bar of fused silica (quartz) to which is bonded a piezoelectric transducer (in essence, an oscillator). Although crystalline quartz is both birefringent and optically active, fused quartz is optically isotropic. Quartz is an example of a material whose

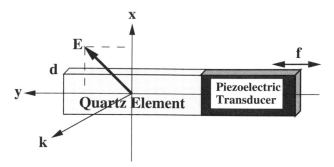

Figure 9.3. Schematic diagram of the operation of a PEM. Light propagating out of the page (direction **k**) is polarized at $45°$ to the fused quartz element of thickness d. The element is driven to oscillate at frequency f by a piezoelectric transducer, which results in an oscillating birefringence that modulates the phase of the transmitted wave.

optical activity derives, not from the structure of individual molecules, but from a helical molecular ordering which fusion destroys. Excited by the piezoelectric transducer, however, the bar vibrates longitudinally at an acoustic resonant frequency f (close to 50 kHz in my own PEMs) and thereby acquires a time-dependent, stress-induced linear birefringence. This dynamic birefringence results in a retardation

$$\varphi(t) = \varphi_m \sin(2\pi f t) \qquad (78)$$

between the components of an incident light field projected on the long (y) and transverse (x) axes of the quartz element. Thus, a light wave $\mathbf{E}_i \propto \mathbf{x} + \mathbf{y}$, linearly polarized at $+45°$ to the bar, emerges with polarization $\mathbf{E} \propto \mathbf{x} + \mathbf{y} e^{i\varphi(t)}$.

It is instructive to trace explicitly the time evolution of the light field $\mathbf{E}$ at different instants of the oscillation cycle for modulation amplitudes $\varphi_m = \pi/2$ and π:

Phase $(2\pi f t)$	0	$\pi/2$	π	$3\pi/2$	2π
$\varphi_m = \pi/2$	$\mathbf{x} + \mathbf{y}$	$\mathbf{x} + i\mathbf{y}$	$\mathbf{x} + \mathbf{y}$	$\mathbf{x} - i\mathbf{y}$	$\mathbf{x} + \mathbf{y}$
	LP	LCP	LP	RCP	LP
$\varphi_m = \pi$	$\mathbf{x} + \mathbf{y}$	$\mathbf{x} - \mathbf{y}$	$\mathbf{x} + \mathbf{y}$	$\mathbf{x} - \mathbf{y}$	$\mathbf{x} + \mathbf{y}$
	LP 1	LP 2	LP 1	LP 2	LP 1

In the first case ($\varphi_m = \pi/2$), $\mathbf{E}$ oscillates once per cycle between states of left and right circular polarization, whereas in the second case ($\varphi_m = \pi$) twice per cycle $\mathbf{E}$ flips between states of linear polarization at $+45°$ and $-45°$ to the PEM axis. At times other than those indicated, the light is elliptically polarized.

Many optical applications call for measuring a small difference with which a sample responds to incident orthogonal states of polarization. One could conceivably illuminate the sample with $\mathbf{x}$ (or $\mathbf{e}_+$) polarized light, record the signal, repeat with **y** (or e_-) polarized light, and take the difference of the two signals. Practically speaking, however, this is not a satisfactory way to proceed, for the small difference between two large signals would be lost amid the noise and system fluctuations during the time required to make separate measurements. The PEM provides light that evolves smoothly between desired polarization states and thereby probes the sample of interest over the brief time interval of approximately $1/f$ (~ 20 μsec for $f = 50$ kHz).

Since the silica element neither absorbs nor scatters light, the total intensity, $I = |\mathbf{E}|^2$, transmitted by the PEM equals the incident intensity and manifests no effect of the retardation $\varphi(t)$. Upon passage through a second polarizer, however, the orthogonal components of the transmitted wave are projected onto the same axis where they subsequently superpose and interfere, giving rise to a signal sensitive to $\varphi(t)$. Suppose, for example, the PEM is between crossed polarizers—that is, the second polarizer is at $-45°$ to the PEM axis. The transmitted light intensity (identically null if the quartz element is not excited) then becomes

$$I_d \propto \frac{1}{2}|1 - e^{i\varphi(t)}|^2 = 1 - \cos\varphi(t) \qquad (79)$$

up to a global numerical factor. Since light is ordinarily detected with a photo-multiplier tube, photodiode, or other "square-law" device whose photocurrent output is a measure of the incident light intensity, the symbol I_d represents equivalently both the light flux and the detector photocurrent. This photocurrent is sent to a lock-in amplifier, an instrument that decomposes I_d into harmonics of the reference frequency f provided by the PEM:

$$I_d(t) = I(0) + I(f)\cos(2\pi f t + \alpha_f) + I(2f)\cos(4\pi f t + \alpha_{2f}) + \cdots. \quad (80)$$

In many applications the phases ($\alpha_f, \alpha_{2f}, \ldots$) turn out to be either 0 or $\pi/2$ so that each harmonic varies in time as a pure cosine or sine function.

Explicit expressions for the coefficients $I(0), I(f), \ldots$ are deducible from the Fourier-Bessel series expansions

$$\cos(\varphi \sin \Omega t) = J_0(\varphi) + 2(J_2(\varphi)\cos 2\Omega t + J_4(\varphi)\cos 4\Omega t + \cdots), \quad (81a)$$

$$\sin(\varphi \sin \Omega t) = 2(J_1(\varphi)\cos \Omega t + J_3(\varphi)\cos 3\Omega t + \cdots). \quad (81b)$$

In practice, one does not usually need harmonics of I_d beyond $2f$. In the preceding example of a PEM between crossed polarizers the photocurrent of Eq. (79) gives rise to a mean dc component $I(0) \propto 1 - J_0(\varphi_m)$ and a component at the first harmonic $I(2f) \propto -2J_2(\varphi_m)$; there is no signal component at the fundamental frequency f.

An experimental configuration capable of yielding the degrees of linear and circular polarization of an arbitrarily polarized light beam is shown in figure 9.4. The light passes a PEM oriented at $45°$ to the optical table and then a vertical linear polarizer. For a general state of polarization the density matrix formalism is

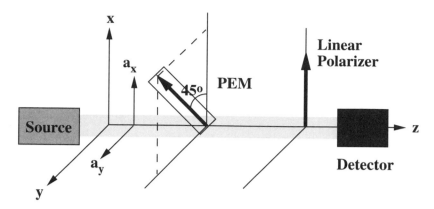

Figure 9.4. Optical configuration for determining degrees of linear and circular polarization with a PEM. Light (with component amplitudes a_x, a_y) passes the PEM (oriented at 45° to the optical table) and a vertical linear polarizer and is detected.

appropriate. Let the polarization of the light be defined with respect to the vertical (**x**) and horizontal (**y**) linear basis of the laboratory (i.e., optical table). The electric field **E** transmitted by the polarizer is related to the field $\mathbf{E}^{(0)}$ incident on the PEM by the following chain of operations (read from right to left)

$$M = M_{LP}(0)R\left(\frac{\pi}{4}\right)M_{LB}(\varphi)R\left(\frac{\pi}{4}\right)^{-1}. \tag{82}$$

$R(\frac{\pi}{4})^{-1} = R(-\frac{\pi}{4})$ rotates the laboratory basis vectors (**x y**) into the basis (**x′ y′**) defined with respect to the principal axes of the PEM quartz element; $M_{LB}(\varphi)$ represents passage of the wave through a linearly birefringent element with retardation $\varphi(t)$; $R(\pi/4)$ returns the basis to the laboratory frame; and $M_{LP}(0)$ lets pass only the vertical component. Multiplication of the component matrices, whose forms were derived in the preceding section, yields the system matrix

$$M = \begin{pmatrix} \dfrac{e^{i\varphi}+1}{2} & \dfrac{e^{i\varphi}-1}{2} \\ \dfrac{e^{i\varphi}-1}{2} & \dfrac{e^{i\varphi}+1}{2} \end{pmatrix} \tag{83}$$

connecting the output and input density matrices according to Eq. (51).

Since the single detected quantity here is the output light intensity $I = 2\rho_0$, only the output density matrix component

$$\rho_0 = \tfrac{1}{2}\operatorname{Tr}\{M\rho^{(0)}M^\dagger\} = \tfrac{1}{2}\operatorname{Tr}\{\rho^{(0)}M^\dagger M\} \tag{84}$$

is required, and this is readily calculable from the output matrix $\Sigma = M^\dagger M$:

$$\Sigma = \begin{pmatrix} \dfrac{1+\cos\varphi}{2} & \dfrac{i\sin\varphi}{2} \\ \dfrac{-i\sin\varphi}{2} & \dfrac{1-\cos\varphi}{2} \end{pmatrix} \tag{85}$$

whose expansion coefficients in the basis $(\mathbf{1}, \boldsymbol{\sigma})$ can be written down immediately by inspection

$$\Sigma_0 = \tfrac{1}{2}(\Sigma_{11} + \Sigma_{22}) = \tfrac{1}{2}, \tag{86a}$$

$$\Sigma_1 = \tfrac{1}{2}(\Sigma_{12} + \Sigma_{21}) = 0, \tag{86b}$$

$$\Sigma_2 = \tfrac{1}{2i}(\Sigma_{21} - \Sigma_{12}) = -\tfrac{1}{2}\sin\varphi, \tag{86c}$$

$$\Sigma_3 = \tfrac{1}{2}(\Sigma_{11} - \Sigma_{22}) = \tfrac{1}{2}\cos\varphi. \tag{86d}$$

To analyze simple experimental configurations, it is often easier to start from first principles and derive expansion coefficients like Eqs. (86a–d) directly, than to make formal substitutions into the general formulas developed in the preceding section. Since one employs only two-dimensional matrices, for which the manipulations can be performed in one's head, this strategy is particularly advantageous when only the total light intensity (and not the other output Stokes parameters) is sought.

The detected light intensity for the experimental configuration of Figure 9.4 therefore takes the form (up to a normalization constant)

$$I_d \propto \sum_{\mu=0}^{3} \rho_\mu^{(0)} \Sigma_\mu = \tfrac{1}{2}\rho_0^{(0)}(1 - \tau_C \sin\varphi + \tau_L^{(0)} \cos\varphi) \tag{87}$$

where the circular and (partial) linear degrees of polarization are defined by Eqs. (17) and (24). Note that a PEM oriented at $45°$ to the x- and y-axes yields the degree of linear polarization $\tau_L^{(0)}$ defined in terms of the corresponding $(\mathbf{x}\ \mathbf{y})$ basis. To determine the quantity $\tau_L^{(45)}$—which together with $\tau_L^{(0)}$ provides the invariant τ_L—one must measure the light intensity with the PEM oriented along the y-axis (i.e., horizontally) and the polarizer oriented at $45°$ to the horizontal. This $45°$ rotation of the entire system does not change the resulting value of τ_C because it is a rotational invariant.

Operationally, the incident light beam, whose polarization state is to be determined, is "chopped" (mechanically modulated) by a rotating disc with holes at a frequency (e.g., $F \sim 200$ Hz) much lower than the PEM frequency f. The chopping frequency F is low enough that the photocurrent component $I(F)$, measured by a lock-in amplifier, is an effective measure of the mean dc light level $I(0)$, but is also high enough for the amplifier to filter out noise, especially the all-pervasive 60 Hz noise of the electrical grid. The Fourier-Bessel series expansions (81a, b) lead directly to the harmonics of the photocurrent corresponding to Eq. (87):

$$I(0) \propto 1 + J_0(\varphi_m)\tau_L^{(0)}, \tag{88a}$$

$$I(f) \propto -2J_1(\varphi_m)\tau_C, \tag{88b}$$

$$I(2f) \propto 2J_2(\varphi_m)\tau_L^{(0)}. \tag{88c}$$

The degrees of polarization are then obtained straightforwardly from ratios of the fundamental and first harmonic components to the mean dc light level

$$\frac{I(f)}{I(0)} = \frac{2J_1(\varphi_m)\tau_C}{1 + J_0(\varphi_m)\tau_L^{(0)}}, \tag{89a}$$

$$\frac{I(2f)}{I(0)} = \frac{2J_2(\varphi_m)\tau_L^{(0)}}{1 + J_0(\varphi_m)\tau_L^{(0)}}. \tag{89b}$$

Setting the modulation amplitude to $\varphi_m = \pi/2$ radians (90°) or π radians (180°), the values used to illustrate the operation of the PEM, is by no means necessary even when it is desired to illuminate a specimen with light that oscillates respectively between orthogonal CP and LP states. The actual experimental value of φ_m is ordinarily chosen to simplify the resulting theoretical current ratios or to optimize a particular current component. Thus, substitution in expressions (89a, b) of the special value $\varphi_m = 2.405$ radians (138°)—the first zero of J_0—reduces the denominators to unity and thereby decouples the degrees of polarization from one another, τ_C being determined exclusively by $I(f)/I(0)$ and $\tau_L^{(0)}$ exclusively by $I(2f)/I(0)$. Furthermore, this is a reasonably good choice from the perspective of signal optimization; at the first zero of J_0 the Bessel functions J_1 and J_2 are at approximately 88% of their maximum values.

9.8 THE GREAT PEM PUZZLE

If the PEM functioned precisely as described in the preceding section—a description that I shall refer to as the "ideal model"—it would be a marvelously sensitive instrument indeed with which to probe subtle, if not exotic, optical properties of matter. That it did not, however, I first discovered in the midst of a project to measure the difference with which an optically active medium reflects LCP and RCP states of light (the differential circular reflection or DCR—equivalent to the generic function δ_C). The motivation for these experiments—in some ways a consummation of Fresnel's investigations of circular polarization, reflection, and optical activity—and an account of their ultimate success is the subject of a later chapter. For the present, however, it suffices to say only that the sought-for signal required an instrumental sensitivity of approximately one part in 10^5 or better.

The achievement of the requisite sensitivity should have posed no difficulty, for comparable sensitivities had been attained more or less routinely in PEM-based measurements of a related phenomenon, circular dichroism, observed in transmission rather than reflection. The reflection experiments, however, irrespective of whether they were performed in my laboratory in the United

States or at the ESPCI in Paris, led to perplexing results that could not be understood on the basis of the ideal model of PEM operation.

Recall that in the ideal model the modulator produces a sinusoidal (or cosinusoidal) retardation, Eq. (78),

$$\varphi_1(t) = \varphi_m \sin(2\pi f t) \tag{90}$$

between components of the transmitted wave parallel to, and perpendicular to, the oscillation axis. The reason for now designating this dynamic relative phase by a subscript "1" will become evident momentarily. Expansion in a Fourier-Bessel series of the theoretical expression for the resulting light flux in a given experiments leads to predictions of the dc, fundamental, and first harmonic components of the signal in terms of Bessel functions of the modulation amplitude (the maximum retardation) φ_m. The actual experimental components, or their ratios, are measurable with one or more lock-in amplifiers.

The ideal model, as has been recognized in the past, is not strictly accurate, for the modulator element, usually fused silica for visible light, may have a low-residual stress-induced *static* birefringence engendered perhaps during manufacture or afterward by pressure from its mount (since the element rests gently on two knife edges). This static birefringence—or rather what is generally believed to be the static birefringence—denoted here by the retardation ϕ_2', can be determined by orienting the PEM at 45° between crossed polarizers, as in the example of the preceding section, and measuring the current components $I(f)$ and $I(2f)$. For this configuration the photocurrent, Eq. (79), was shown to contain no fundamental component $I(f)$ in the absence of static birefringence. However, under the widely made assumption (to be challenged shortly) that the axes of dynamic and static birefringence are parallel and that the total retardation is simply the sum

$$\varphi(t) = \varphi_1(t) + \varphi_2', \tag{91}$$

Eq. (79) leads to the component $I(f) \propto 2J_1(\varphi_m) \sin \varphi_2'$ and consequently to the expression

$$\tan \varphi_2' \sim \varphi_2' = -\frac{J_2(\varphi_m)}{J_1(\varphi_m)} \frac{I(f)}{I(2f)}. \tag{92}$$

Examination in this way of a number of commercially available PEMs, as well as homemade ones, resulted in values of ϕ_2' of a few tenths of a degree ($\sim 10^{-3}$ radian). For such low values, the static birefringence has often been undetected or justifiably ignored in many experiments.

The PEM configurations employed in studies of reflection from chiral media were expressly designed (on the basis of the ideal model with static birefringence included) to lead to *no* signals (only noise) when the sample, either transparent or absorbing, was optically inactive.[10] This was not the case. There was observed, instead, a reflection curve—supposedly the DCR as a function of angle of incidence—in both optically active and inactive samples, qualitatively similar in form to my theoretical predictions[11] but with a peak value 10 to 100 times greater. In addition, measurements of the DCR as a function of modulation amplitude (at a fixed angle of incidence) did not exhibit a form governed by a simple Bessel function coefficient, $J_1(\varphi_m)$ or $J_2(\varphi_m)$, as one might naively expect, assuming that there should have been any signal at all (perhaps as a result of modulator or polarizer misalignment).

The measurement of the static birefringence itself in the crossed-polarizer configuration also led to anomalous results as a function of φ_m. For values of φ_m beyond $\sim$3.5 radians (200°), ϕ_2' was no longer small and independent of the modulation amplitude, but passed through what closely resembled singularities similar in form to an optical dispersion curve of a transparent medium.

This puzzling behavior was not the aberrant performance of a single instrument; all PEMs examined, irrespective of their origin, performed this way, and it was logical to conclude that the anomalous behavior was general, not isolated. That the operation of an instrument so widely used and long relied upon should actually be unpredictable and inexplicable could not be tolerated. If the PEM were to be exploited to its full potential in new kinds of experiments that probed the structure and interactions of matter, then it was imperative to know why the PEM functioned as it did and what steps must be taken to design the apparatus better. All work on chiral reflection stopped, and my colleagues and I devoted our full energies to understanding these discrepancies.

Our efforts were successful.[12] The theory we developed—which accounted for all observed anomalies—rested upon two basic assumptions: (1) the axes of static and dynamic birefringence in the modulator element are not necessarily collinear, but are at a small angle (to be designated γ) to one another; and (2) both the dynamic and static birefringence are distributed uniformly throughout the element.

The assumption, based on point (1) alone, that the static and dynamic birefringence may be treated as two phase plates of different retardations rotated with respect to each other leads to what I call the two-plate model. It might be anticipated on grounds of theoretical self-consistency that such a model is ambiguous and must be inadequate. The ambiguity is that the physical operations of the two plates on a light beam are not commutative; the state of the light after it first passes the static plate (constant retardation φ_2) and then the dynamic plate (time-varying retardation φ_1) is totally different from the corresponding state obtained by passage through the plates in reverse order. Physically, the symmetry of the PEM is such that there is no reason to prefer one ordering over the other. Indeed, calculations employing the two-plate model led to results in disagreement with experiment for a PEM between two polarizers of appropriate orientation.

An analysis of the PEM based on the uniform distribution of both static and dynamic birefringence—the distributed birefringence model—leads to phase

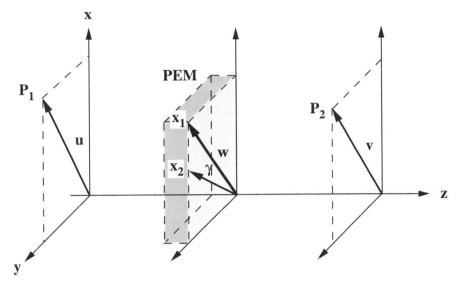

Figure 9.5. Distributed birefringence model of a PEM with modulation axis (x_1) at angle w to the vertical axis (x). Within each infinitesimal slab of the PEM element the axis of static birefringence (x_2) is oriented at an angle γ to the modulation axis. Angles u and v denote the transmission axes of polarizers P_1 and P_2. The configurations ($u = \pi/4$, $v = 0$, $w = \pi/2$) and ($u = 0$, $v = \pi/2$, $w = \pi/4$) pose stringent experimental tests of the model.

variations of the form $\cos \beta$ and $(\sin \beta)/2\beta$, where the relevant phase retardation is no longer φ_1 or $\varphi_1 + \varphi_2'$, but

$$2\beta - [(\varphi_1 + \varphi_2 \cos 2\gamma)^2 + (\varphi_2 \sin 2\gamma)^2]^{1/2}. \qquad (93)$$

The resulting expressions for detected light flux account without any further instrumental parameters for the observed anomalous signals and reproduce their geometric (configuration-dependent) and dynamic (power-dependent) characteristics. To deduce the effect of a distributed birefringence on the polarization of a transmitted light wave is in fact a nontrivial mathematical problem made tractable by the algebra of Pauli spin matrices. It is not a calculation I would care to perform with four-dimensional matrices.

Consider a PEM oriented between two polarizers P_1 and P_2, as shown in figure 9.5, with the modulation axis (x_1) and the axis of static birefringence (x_2) at angles w and $\Gamma \equiv w + \gamma$, respectively, to the vertical axis (x). The transmission axis of the first polarizer (P_1) is at an angle u to the vertical and that of the second polarizer (P_2) at an angle v. If the initial state of the light (normalized to unit amplitude) relative to the laboratory (x, y) coordinate system is represented by a column vector

$$\mathbf{E}^{(0)} = \begin{pmatrix} \cos u \\ \sin u \end{pmatrix}, \tag{94}$$

then the state of the light emerging from the modulator in the two-plate model (with dynamic birefringence preceding static birefringence) is determined from the matrix expression

$$\mathbf{E}_m = R(\Gamma)M_{LB}(\varphi_2)R(\gamma)^{-1}M_{LB}(\varphi_1)R(w)^{-1}\mathbf{E}^{(0)} \equiv \mathbf{M}\mathbf{E}^{(0)}. \tag{95}$$

In this chain of operations (read right to left) $R(w)^{-1}$ rotates the field amplitudes into the principal axis frame of the first plate; $M_{LB}(\varphi_1)$ generates the relative phase $\varphi_1(t)$ between the x_1 and y_1 components of the wave; $R(\gamma)^{-1}$ transforms to the principal axis frame of the second plate, $M_{LB}(\varphi_2)$ generates the relative phase φ_2 between the x_2 and y_2 components of the wave, and $R(\Gamma)$ transforms back to the laboratory coordinate system. Note that it is $R(\theta)^{-1} = R(-\theta)$, rather than $R(\theta)$, that appears in Eq. (95) since the field $\mathbf{E}$ is represented here as a column vector (see Eq. [20]).

As stated above, there is no physical justification in placing one birefringent plate before the other. The question of matrix ordering vanishes, however, for the case of infinitesimal transformations—transformations to first order in the differentials $d\varphi_1$ and $d\varphi_2$. Consider, then, an infinitesimally wide slice through the modulator element. To within an irrelevant global phase factor the transformation Eq. (95) engendered by the infinitesimal slab reduces to

$$\mathbf{E}^{(0)} + d\mathbf{E} = [1 + i R(\Gamma)dG R(-w)]\mathbf{E}^{(0)} \tag{96}$$

where the infinitesimal matrix

$$dG = \begin{pmatrix} \left(\dfrac{d\varphi_1 + d\varphi_2}{2}\right)\cos\gamma & -\left(\dfrac{d\varphi_1 - d\varphi_2}{2}\right)\sin\gamma \\ -\left(\dfrac{d\varphi_1 - d\varphi_2}{2}\right)\sin\gamma & -\left(\dfrac{d\varphi_1 + d\varphi_2}{2}\right)\cos\gamma \end{pmatrix} \tag{97}$$

represents the simultaneous effect of static and dynamic birefringence. To construct the finite transformation for a light wave that has propagated through the full width of the PEM element, first expand the matrices in Eqs. (96) and (97) in the basis $(\mathbf{1}, \boldsymbol{\sigma})$ to obtain the equivalent relation

$$\mathbf{E}(\beta + d\beta) = (\mathbf{1} + id\beta \cdot \boldsymbol{\sigma})\mathbf{E}(\beta) \tag{98}$$

in which the elements and magnitude of the finite parameter vector β are

$$\begin{aligned} \beta_1 &= \tfrac{1}{2}(\varphi_1 \sin 2w + \varphi_2 \sin 2\Gamma), \\ \beta_2 &= 0, \\ \beta_3 &= \tfrac{1}{2}(\varphi_1 \cos 2w + \varphi_2 \cos 2\Gamma), \end{aligned} \tag{99a}$$

$$\beta = (\beta_1^2 + \beta_2^2 + \beta_3^2)^{1/2} = \tfrac{1}{2}(\varphi_1^2 + \varphi_2^2 + 2\varphi_1\varphi_2 \cos 2\gamma)^{1/2}. \tag{99b}$$

The linear differential equation (98) is then readily integrated to yield the formally simple relation

$$\mathbf{E}_m(\beta) = e^{i\boldsymbol{\beta}\cdot\boldsymbol{\sigma}} \mathbf{E}^{(0)} = \left(\mathbf{1}\cos\beta + i\frac{\boldsymbol{\beta}\cdot\boldsymbol{\sigma}}{\beta}\sin\beta\right)\mathbf{E}^{(0)}. \tag{100}$$

The second equality in Eq. (100) follows from the Taylor series expansion of the exponential function, the algebraic properties of the spin matrices as embodied in Eq. (5), and the Taylor series expansions of the sine and cosine. (The same procedure can be used to evaluate the exponential of a general four-dimensional linear transformation, but the resulting expression is complicated and unsightly, not to mention tedious to apply. Such things are best left to general relativity and cosmology, where they may actually be needed.) Eq. (99b) is identical to Eq. (93), the origin of which is now clear. Evaluating Eq. (100) by means of the spin matrix representation of Eq. (6) leads to the following explicit expression for the components of the transmitted wave:

$$\mathbf{E}_m = \begin{pmatrix} \cos\beta\cos u + i\frac{\sin\beta}{2\beta}\{[\varphi_1\cos 2w + \varphi_2\cos 2\Gamma]\cos u + [\varphi_1\sin 2w + \varphi_2\sin 2\Gamma]\sin u\} \\ \cos\beta\sin u + i\frac{\sin\beta}{2\beta}\{-[\varphi_1\cos 2w + \varphi_2\cos 2\Gamma]\sin u + [\varphi_1\sin 2w + \varphi_2\sin 2\Gamma]\cos u\} \end{pmatrix}. \tag{101}$$

Since there is no absorptive or diffusive element in the PEM, the total flux through the device is still unity ($|\mathbf{E}_m|^2 = 1$). In addition, the field $\mathbf{E}_m$ reduces correctly (see note 10) in the special case of collinear static and dynamic birefringence ($\gamma = 0$).

Virtually everything one needs to know about the PEM follows from Eq. (101). Although full details can be found in the original literature (see note 12), it is instructive to consider briefly two experimental configurations that manifest a striking distinction between the distributed birefringence theory of the elastic modulator and the ideal model or two-plate model. In the first configuration, the PEM is situated between two polarizers with transmission axes at 45°; in the second the PEM is between crossed polarizers. These two configurations pose a stringent test of the new theory.

$\pi/4$-Polarizer Configuration

If the two polarizers and PEM are arranged so that $u = \pi/4$, $v = 0$, and $w = \pi/2$ (all angles measured relative to the vertical as shown in figure 9.5), then the exact relation for the transmitted light intensity can be reduced to the approximate expression

$$I_d = \tfrac{1}{2}\left[1 + (\varphi_2\sin 2\gamma)\left(\frac{1 - \cos 2\beta}{2\beta}\right)\right] \tag{102}$$

for which the Fourier decomposition (to first order in φ_2) yields

$$\frac{I(f)}{I(0)} = 2\varphi_2 \sin 2\gamma \left[\frac{1 - J_0(\varphi_m)}{\varphi_m} \right]. \tag{103}$$

According to the ideal model, the wave transmitted by the PEM should have the form $\mathbf{E}_m \propto \mathbf{x} + \mathbf{y}e^{i\varphi_1(t)}$, in which case subsequent transmission through a vertical polarizer ($\mathbf{x}$) would yield a constant photocurrent. The two-plate model—irrespective of which plate comes first—leads to a component $I(f)$ proportional to the Bessel function $J_1(\varphi_m)$ since a Fourier expansion is made of $\sin \varphi_1$ and $\cos \varphi_1$ (as in the ideal model), and not functions of $\beta(t)$. The distributed model predicts a component $I(f)$ that depends sensitively (i.e., as $\sin 2\gamma$) on the relative orientation of static and dynamic birefringence, and varies with modulation amplitude in an unusual way. In deriving Eqs. (102) and (103) it has been assumed that φ_2 and γ are small angles (compared to 1 radian), assumptions subsequently verified experimentally.

Figure 9.6 shows the experimental variation of $I(f)/I(0)$ with φ_m and corresponding theoretical predictions obtained from (1) numerical Fourier analysis of the exact expression from which Eq. (102) was derived, and (2) the approximate Eq. (103). The three curves are virtually identical and substantiate strongly the predicted $[1 - J_0(\varphi_m)]/\varphi_m$ power dependence. Also shown in the figure is the variation $J_1(\varphi_m)$, which does not resemble the observed results at all.

$\pi/2$-Polarizer Configuration

The configuration $u = 0$, $v = \pi/2$, and $w = \pi/4$ leads to a simple exact expression for the light intensity

$$I_d = \tfrac{1}{2}(1 - \cos 2\beta), \tag{104}$$

and approximate Fourier decomposition (to first order in φ_2) yields

$$\frac{I(f)}{I(2f)} \sim -\frac{J_1(\varphi_m)}{J_2(\varphi_m)} \tan(\varphi_2 \cos 2\gamma). \tag{105}$$

Eq. (105) is seen to be equivalent to Eq. (92) when φ_2' is identified with $\varphi_2 \cos 2\gamma$. Thus, measurement of the so-called static birefringence by the standard crossed-polarizer method does not yield the retardation angle φ_2 directly, but rather the product $\varphi_2 \cos 2\gamma$ (which is not sensitive to small γ).

Of particular interest in this case is the high-power behavior of the PEM, that is, the variation in $I(f)/I(2f)$ with φ_m for large modulation amplitude. The ideal model (as well as approximate relation [105]), predicts that the function $J_2(\varphi_m)/J_1(\varphi_m) \; I(f)/I(2f)$ be a constant, that is, independent of φ_m.

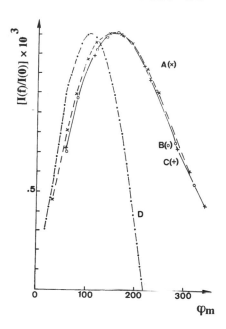

Figure 9.6. $\pi/4$ polarizer configuration. Variation of $I(f)/I(0)$ with modulation amplitude φ_m (degrees) as determined (A) experimentally, (B) by computer Fourier analysis of the unapproximated expression for light intensity, and (C) from the analytical approximation Eq. (103). Curve (D) shows the corresponding variation of Bessel function $J_1(\varphi_m)$.

Experimentally, this function is not constant, but reveals resonance-like singularities for values of φ_m at and beyond the first zero of $J_1(\varphi_m)$, as shown in figure 9.7. Computer simulation based on numerical Fourier analysis of the exact expression for the light flux, Eq. (104), quantitatively reproduces the observed features.

Physically, one can understand these observations as a consequence of the higher-harmonic contributions to the phase 2β. Carrying out a Fourier decomposition of Eq. (104) to second order in φ_2 leads to the result

$$\frac{J_2(\varphi_m)I(f)}{J_1(\varphi_m)I(2f)} \sim -\varphi_2 \cos 2\gamma + \frac{J_1(\varphi_m) - J_3(\varphi_m)}{J_1(\varphi_m)} \frac{(\varphi_2 \sin 2\gamma)^2}{\pi \varphi_m}, \qquad (106)$$

which explicitly shows a singularity at the first zero (~ 3.85 radians or $221°$) of $J_1(\varphi_m)$. More terms would need to be retained in the expansion of β and trigonometric functions of β in order to reproduce analytically the second observed singularity. Such an analysis is tedious and not necessary for the purpose of this discussion.

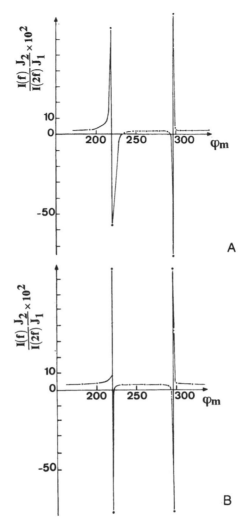

Figure 9.7. $\pi/2$ polarizer configuration. Variation of $J_2(\varphi_m)/J_1(\varphi_m)\ I(f)/I(2f)$ with modulation amplitude φ_m (in degrees) as determined by (A) experiment and (B) computer Fourier analysis of Eq. (104).

The point worth stressing, however, is that the distributed birefringence theory accounts quantitatively for the strong-modulation regime of the PEM in which higher harmonics in β contribute significantly to the signals measured at the fundamental and first harmonic frequencies. The Fourier coefficients of these contributions, although one may need to determine them numerically, are completely specified in an a priori way by the theory in terms of only two parameters φ_2 and γ.

One other point of interest is that the shape of the resonances, that is, whether the curve approaches the positive asymptote at 3.85 radians from above the abscissa (as shown in the figure) or crosses the abscissa and approaches the negative asymptote, depends sensitively on the relative locations of the points φ_m where J_2/J_1 and $I(f)/I(2f)$ cross the abscissa. The excellent agreement between predicted and measured resonances may therefore serve as a close check on PEM calibration, that is, establishing the correspondence between retardation φ_m and the voltage applied to the transducer.

From the measurement of $I(f)/I(0)$ in the $\pi/4$ configuration (Eq. [103]) and the measurement of $I(f)/I(2f)$ in the $\pi/2$ configuration (Eq. [105]), the angles φ_2 and γ can be determined separately. For the PEM used in obtaining the results of figures 9.6 and 9.7, these angles have been found to be $\varphi_2 \sim -1.43°$ and $\gamma \sim 2.2°$. In general, both angles are small and sensitive to the pressure exerted on the quartz element by its support.

Although small, the experimental implications of a nonvanishing axial inclination γ can be highly significant. As mentioned earlier, the investigation of light reflection from chiral media led to DCR values much greater than theoretically expected, even when the sample was not optically active and no signal was expected. The distributed model of the PEM accounts for this anomaly, for—as the reader may anticipate—the experimental configuration sensitive to the chiral structure of the sample corresponded precisely to one of the "critical" configurations for which the ideal PEM model failed to yield even approximately correct results. This was at first surprising, since these critical configurations require a polarizer and analyzer—and the reflection experiments employed only one initial polarizer.

The light flux, predicted by the distributed birefringence theory for the special "chiral reflection geometry" ($u = \pi/4$, $w = 0$), took the approximate form

$$I_d \sim 1 + \delta_L(\varphi_2 \sin 2\gamma)\frac{1 - \cos 2\beta}{2\beta} + \delta_C \sin 2\beta \qquad (107)$$

where the linear (δ_L) and circular (δ_C) intensity differences are defined in Eqs. (38) and (39). The latter (δ_C) is identical to the definition of DCR. One sees, therefore, that the detected light flux would have depended only upon the DCR in these reflection experiments, even if the PEM element had a residual static birefringence, so long as the axes of static and dynamic birefringence were parallel, that is, $\gamma = 0$. But $\gamma \neq 0$, and the component of the signal at the first harmonic

$$\frac{I(f)}{I(0)} = -2\delta_L(\varphi_2 \sin 2\gamma)\frac{1 - J_0(\varphi_m)}{\varphi_m} - 2\delta_C J_1(\varphi_m)\cos(\varphi_2 \cos 2\gamma) \qquad (108)$$

is dominated by the term proportional to δ_L, since $\delta_L \gg \delta_C$.

In the absence of optical activity, however, Eq. (107) is effectively equivalent to a two-polarizer configuration in which the extreme values of δ_L correspond to the setting of the second polarizer: $\delta_L = +1$ for vertical LP light ($v = 0$) and $\delta_L = -1$ for horizontal LP light ($v = \pi/2$). In effect, the experiment employed a two-polarizer $\pi/4$ configuration where the act of reflection constituted the "hidden" polarizer—a circumstance reminiscent of the interference colors of the cellophane bread wrapper lying on my kitchen table.

That such glaring anomalies apparently escaped the attention of scores of PEM users over the years is probably due to the fortuitous fact that few, if any, had occasion to use a PEM in the same experimental configuration. My colleagues and I know of no prior attempts to measure the weak chiral properties of matter by means of specular reflection.

Our theory of PEM operation now makes it possible to interpret a broader range of PEM-based experiments and to circumvent to a certain degree the problems engendered by the presence of stress-induced static birefringence. The ideal solution, of course, would be to fabricate silica elements free of stress, or for which the axes of static and dynamic birefringence are parallel—and, indeed, efforts along these lines are under way. But until such strain-free devices are produced, one can rely on theory to help design experiments more cleverly so that they need not be influenced by nonideal PEM behavior. The problem of chiral reflection, for example, was ultimately resolved in an elegant and satisfying way. Recognizing that the anomalous signal depended on the difference with which the material reflected orthogonal states of linearly polarized light, the experiment was carried out under conditions of total reflection for which δ_L vanished identically. It was then possible to measure directly the DCR of a chiral material whose refractive indices for right and left circular polarizations differed by only a few parts in ten million.[13]

I think Fresnel would have been pleased.

NOTES

1. Eugene Wigner, "Relativistic Invariance and Quantum Phenomena," in *Symmetries and Reflections* (Bloomington: Indiana University Press, 1967), pp. 51–81; quotation from pp. 53–54.

2. M. P. Silverman, "Interference Colors with 'Hidden' Polarizers," *Am. J. Phys.* 49 (1981): 881–82.

3. I discuss two-level quantum systems comprehensively in an earlier book: M. P. Silverman, *More Than One Mystery: Explorations in Quantum Interference* (New York: Springer-Verlag, 1995).

4. W. A. Shurcliff and S. S. Ballard, *Polarized Light* (Princeton, N.J.: van Nostrand, 1964), p. 92.

5. I give a nontechnical discussion of the phenomenon of quantum beats in M. P. Silverman, *And Yet It Moves: Strange Systems and Subtle Questions in Physics* (New

York: Cambridge University Press, 1993); the subject is treated in mathematical detail in Silverman, *More Than One Mystery*.

6. I. V. Lindell and M. P. Silverman, "Plane-Wave Scattering from a Nonchiral Object in a Chiral Environment," Helsinki University of Technology Report 211 (HUT, Espoo, Finland, 1995), and *J. Opt. Soc. Am. A* 14 (1997): 79–90.

7. R. T. Lagemann, "New Light on Old Rays," *Am. J. Phys.* **45** (1977): 281–84.

8. M. P. Silverman, "The Curious Problem of Spinor Rotation," *Eur. J. Phys.* **1** (1980): 116–23; see also Silverman, *And Yet It Moves*, section 2.3 ("Quantum Implications of Travelling in Circles").

9. M. Billardon and J. Badoz, "Modulateur de birfringence," *Compt. Rend. Acad. Sci. Paris B* **262** (1966): 1672–75; "Mesure du dichroïsme circulaire dans le visible et l'ultraviolet," *Compt. Rend. Acad. Sci. Paris B* **263** (1966): 139–42.

10. M. P. Silverman, N. Ritchie, G. M. Cushman, and B. Fisher, "Experimental Configurations Using Optical Phase Modulation to Measure Chiral Asymmetries in Light Specularly Reflected from a Naturally Gyrotropic Medium," *J. Opt. Soc. Am. A* **3** (1988): 1852–62.

11. M. P. Silverman, "Specular Light Scattering from a Chiral Medium," *Lettere al Nuovo Cimento* **43** (1985): 378–82; id., "Reflection and Refraction at the Surface of a Chiral Medium," *J. Opt. Soc. Am. A* **3** (1986): 830–37.

12. J. Badoz, M. P. Silverman, and J. C. Canit, "Wave Propagation through a Medium with Static and Dynamic Birefringence: Theory of the Photoelastic Modulator," *J. Opt. Soc. Am.* **7** (1990): 672–82.

13. M. P. Silverman, J. Badoz, and B. Briat, "Chiral Reflection from a Naturally Optically Active Medium," *Opt. Lett.* **17** (1992): 886–88.

Part Four

REFLECTION AND SCATTERING

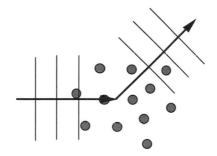

CHAPTER 10

The Grand Synthesis

Hence the velocity of light deduced from experiment
agrees sufficiently well with the value of v deduced
from the only set of experiments we as yet possess.
The value of v was determined by measuring the
electromotive force with which a condenser of
known capacity was charged, and then discharging
the condenser through a galvanometer, so as to
measure the quantity of electricity in it in
electromagnetic measure. The only use made of light
in the experiment was to see the instruments.
(James Clerk Maxwell, 1864)[1]

DESPITE THE enormous success of Fresnel's wave theory of light, there remained
a fundamental and thorny issue: If light were a kind of undulation, then what,
precisely, was "waving"?

The answer to this question was discovered by Scottish physicist James Clerk
Maxwell, not in optics but in the study of remotely connected phenomena of
electricity and magnetism. Deeply impressed by the experimental researches
of Michael Faraday, Maxwell set out to give mathematical structure to Faraday's geometrical conception of a continuum of lines of force permeating the
space between electrified and magnetized bodies. It is perhaps hard to imagine
today—when nearly all theoretical physics is a study in field theory—how bold
Faraday's idea was and how much resistance it incurred from contemporary
scientists and mathematicians. "I was aware," wrote Maxwell in the preface
to the first edition of his *Treatise on Electricity & Magnetism*, "that there was
supposed to be a difference between Faraday's way of conceiving phenomena
and that of the mathematicians, so that neither he nor they were satisfied with
each other's language. For instance, Faraday, in his mind's eye, saw lines
of force traversing all space, where the mathematicians saw centres of force
attracting at a distance: Faraday saw a medium where they saw nothing but
distance: Faraday sought the seat of the phenomena in real actions going on in
the medium; they were satisfied that they had found it in a power of action at a
distance impressed on the electric fluids."[2]

Like Fresnel, Maxwell lived a regrettably short time, dying of cancer at the
age of forty-eight in 1879. He began his investigations of electromagnetism in

the mid-1850s with the resolve to read Faraday's *Experimental Researches in Electricity*, and by the mid-1860s, through what is perhaps the most remarkable application of modeling by analogy of which I know, he arrived at his definitive dynamical theory of the electromagnetic field. In contrast to Ampère, whom he greatly admired and of whom he wrote "We can scarcely believe that Ampère really discovered the law of [magnetic] action by means of the experiments which he describes. We are led to suspect ... that he discovered the law by some process which he has not shewn us, and that when he had afterwards built up a perfect demonstration he removed all traces of the scaffolding by which he had raised it," Maxwell revealed all the interim steps in his own progress toward perfection.

In the first step, convinced of the conceptual fertility of Faraday's lines of force, Maxwell likened them to streamlines in a hydrodynamic model of in-compressible tubes of electric and magnetic fluids.[3] One insightful outcome of these considerations, which was to become a seminal part of the final theory, was Maxwell's distinction between a vectorial "quantity," associated with the flux of a field through an area, and a vectorial "intensity," associated with the circulation of a field about a closed path.

In the second step, he devised an extraordinary mechanical model whereby space-filling molecular vortices and idle wheels transmitted pressures and tensions representative of electrical and magnetic interactions, in particular the interaction between current-carrying wires (Ampère's law) and the induction of an electromotive force by a change in magnetic flux (Faraday's law).[4] From this hypothetical construction Maxwell deduced the existence of transverse waves propagating at a speed determined by the elasticity and density of the matter of the vortices (the ether)—which, when evaluated, very nearly equaled the speed of light as it was then known ($\sim 3.15 \times 10^8$ m/s).

In the final step, Maxwell displayed the architecture of his own edifice with the scaffolding removed (as Ampère had done from the outset without benefit of "blueprints") and presented in their awe-inspiring entirety the mathematical relations that account for all (nonquantum) electromagnetic phenomena. "The agreement of the results," wrote Maxwell, "seems to shew that light and magnetism are affections of the same substance, and that light is an electromagnetic disturbance propagated through the field according to electromagnetic laws" (see note 1). Thus, within the span of roughly ten years Maxwell effected a theoretical synthesis that not only brought electricity, magnetism, and optics under the same set of mathematical laws, but that was to provide the exemplar for virtually all other field theories of modern physics.

To the question "What is 'waving'?" Maxwell's theory provided the answer: an electromagnetic field. But of this, two things must be said.

First, though Maxwell's answer is correct and complete and nothing further is actually required, the question still remained troubling, for it was just as inconceivable to Maxwell and his contemporaries, as it had been to Huyghens,

Newton, Young, and Fresnel, that an undulation could propagate through space in the absence of a material medium. The search for a "luminiferous ether" consequently remained a pressing issue throughout Maxwell's lifetime and in fact until well after Einstein had effectively disposed of it in his 1905 paper on special relativity.[5] Imagine a substance that had to pass freely through the atoms of matter (to account for the aberration of starlight viewed through a telescope), yet must be a nearly incompressible elastic solid (if it were to transmit high-frequency transverse waves); a substance for which was claimed a negative modulus of compression so that it expanded under pressure and contracted when relaxed; a substance that now, to satisfy electrodynamics, was riddled with tubes of electric and magnetic flux. If you cannot, then you are in good company, for neither could anyone else.

Second, to the distress of many future generations of physics students, Maxwell's theory did not give rise to a *single* field—or even to two (electric and magnetic)—but rather to at least four, which in modern parlance are **E** (electric field), **D** (electric displacement), **B** (magnetic induction), and **H** (magnetic field). Actually, the symbolic designations are exactly those Maxwell had chosen (although he expressed them in Gothic letters), and the nomenclature is only slightly modified from the original; Maxwell termed **E** the "electromotive intensity" and **H** the "magnetic force."

Although **E** and **D**, and likewise **H** and **B**, are conflated in the absence of matter, their conceptual distinctions (as "intensities" and "quantities") are vital, and their properties in matter are sufficiently different that one must always exercise caution when referring to "transverse" waves of light. For example, in an anisotropic dielectric medium devoid of free charge and current, "quantities" **D** and **B** are transverse to the wave vector **k** (which is normal to the wavefront), whereas "intensities" **E** and **H** are transverse to the Poynting vector **S** (which gives the direction of power flux). **S** and **k**, however, need not be parallel to one another, with the consequence that the wavefronts are not transverse to the direction in which light energy is transported. Little things like this make crystal optics a fascinating subject for the devoted—or an ordeal to the geometrically impaired.

Having accomplished his immortal work, Maxwell retired from a professorship at King's College, London, in 1865 to write *A Treatise on Electricity and Magnetism* that should have "for its principal object to take up the whole subject in a methodical manner, and which should also indicate how each part of the subject is brought within the reach of methods of verification by actual measurement" (preface). It is this concern of Maxwell's, not only with theoretical foundations, but also with concrete experimental details (without which all science reduces to mere opinion), that has influenced so profoundly my own scientific education and research.

Although a significant fraction of my physics pursuits fall within the purview of electrodynamics, I have never had a course in the subject beyond an elementary introduction. Instead, as with most of the physics I learned, I studied the principles on my own—in this case with Maxwell's *Treatise* as both my inspiration and textbook. This is not an experience that I would necessarily recommend to others. For all his legendary gentleness, Maxwell is a demanding teacher, and his magnum opus is anything but coffee-table reading. He wrote at a time when vectors—introduced by William Rowan Hamilton as part of a long and largely opaque work on quaternions—were understood by only a handful of physicists. And, although Maxwell was among this select group and introduced vector terminology into his *Treatise*—indeed, it was he who created such familiar terms as "curl," "gradient," and "divergence"—he nonetheless preferred to express vectorial relations in their Cartesian components, each component distinguished by a different letter rather than a subscript. This did not make reading Maxwell any easier.

All the same, the experience was greatly rewarding in that I had come to understand, as I realized much later, aspects of electromagnetism that are rarely taught at any level today and that reflect the unique physical insight of their creator.

One of the most important of these, in regard to the relationship of electromagnetic waves to light, concerns the delicate subject of electromagnetic units. Few topics, I have found, seem more obscure or less interesting to students and professional physicists alike (except perhaps those at standards laboratories), and yet more likely to trigger heated discussion over preferences. At the introductory level, it would appear from perusing any number of general physics textbooks that the SI (*Système Internationale*) set of units has swept all others from the field, and so students begin their studies of electricity and magnetism by cluttering their minds with mysterious symbols—epsilon-naught (ε_0) and mu-naught (μ_0)—which are purely concocted numbers having, in fact, no basis in natural law. Some time later, perhaps, students will encounter in more advanced treatments the Gaussian form of Maxwell's equations overflowing with c's—but no explanation is ever given as to why the speed of light should occur in relationships between electric and magnetic fields, or between magnetic fields and currents. The absurdity of this situation must surely strike even the least perceptive student, for, as Maxwell wryly relates in the quotation that opens this chapter, the only role of light in the measurement of electric and magnetic parameters is to permit experimenters to see their instruments.

Why then should there be any surprise that Maxwell's equations yield electromagnetic waves propagating at the speed of light? In the first case (SI) one has simply contrived to make $1/\sqrt{\varepsilon_0\mu_0} = c$, and in the second case (Gaussian) an abundance of c's were inserted explicitly "by hand" at the outset. Students learning the subject from a modern textbook may well be excused if they are not impressed.

But I, who learned this marvelous result directly from Maxwell, was impressed indeed. To appreciate the fact that Maxwell's theory makes an extraordinary prediction, and not merely renders what is inserted beforehand, one must first understand how electric charge and current are measured—and I know of no account better than Maxwell's.

"The only systems of any scientific value," Maxwell states in his *Treatise* (2:266), "are the electrostatic and the electromagnetic systems."

According to the first (esu) system, a unit of charge is operationally defined by Coulomb's law: Two point charges q attracting or repelling one another with a force F of 1 dyne at a distance d of 1 cm each comprise 1 esu of charge. From $F = q^2/d^2$ and its equivalence to mass × acceleration, the dimension of charge, expressed in terms of the fundamental quantities of mass (M), length (L), and time (T), is readily seen to be

$$[q]_{esu} = [M^{1/2}L^{3/2}T^{-1}]. \tag{1}$$

It is not electric charge, however, but electric current that is primary in magnetism and consequently the basis for the second (emu) system of units. In this case the unit of current is operationally established by Ampère's law: Equal currents I in two straight segments of wire of length ℓ and separation d attract or repel one another with a force $F = I^2\ell/d$. It then follows that current has the dimension of the square root of force, or

$$[I]_{emu} = [M^{1/2}L^{1/2}T^{-1}]. \tag{2}$$

In the emu system, charge is a secondary quantity obtained by measuring the passage of current over a period of time; thus,

$$[q]_{emu} = [I]_{emu}[T] = [M^{1/2}L^{1/2}]. \tag{3}$$

Equations (1) and (3) illustrate that the two systems of units are incompatible, leading to designations of the same quantity (electric charge) that differ not only in their magnitude, but also in their *dimensions*. The number of esu units of charge that make up one emu unit of charge thus takes the form of a velocity

$$\frac{q_{esu}}{q_{emu}} = k[LT^{-1}] = k[v] \tag{4}$$

where the numerical factor k is a universal constant to be determined experimentally.

It is often quite surprising to those who encounter this basic feature of electromagnetism for the first time—and there are many, I suspect, who, not having read Maxwell, never encounter it. From grammar school onward one learns to express measurable quantities in different systems of units—for example, to convert between cgs and mks measures of mass, length, volume, speed, and so on, by a simple relocation of the decimal point. Those unfortunate enough to have been raised in the English system of units learn to convert between feet and miles, pints and quarts, bushels and pecks, ounces and pounds, and the like. But, so long as mechanical quantities only are involved, the conversion, however awkward, is simply a matter of a dimensionless numerical factor. A meter and a yard are both measures of length, and both have the dimension of length. Not so in electromagnetism; an esu of charge has an entirely different dimension from an emu of charge.

Unlike the SI parameters, the conversion factor $k[v]$ is not just a "pencil and paper affair" (to borrow the words of Nobelist Percy Bridgeman) but is amenable to direct measurement—and Maxwell, who, like Newton and Fresnel, was thoroughly conversant with experiment, described in his *Treatise* at least four ways to measure it. Of these, the method due to Weber and Kohlrausch is especially simple in principle.

A Leyden jar was charged with a certain quantity of electricity, determined in electrostatic measure as the product of the capacitance (C) of the jar and the potential difference (V) between its coatings. C was ascertained beforehand by comparison with the capacitance of a metal sphere suspended in an open space away from other bodies. In the esu system the capacitance of a sphere is given by its radius, and thus C for the Leyden jar could be expressed as a certain length. Correspondingly, V was measured by connecting the coatings to the terminals of a calibrated electrometer. The two measurements thus furnished $q_{esu} = CV$. The jar was subsequently discharged through the coil of a galvanometer. The transient current caused a small magnet to rotate, and from its extreme angular deviation the quantity of charge q_{emu} was deduced from the appropriate formula. The ratio of the two charges was found to be $k = 3.1074 \times 10^8$ m/s.

The system-dependent dimensions of charge have consequences for every electrical and magnetic quantity: for instance, current, potential permittivity, permeability, resistance, capacitance, inductance—and especially for the various electromagnetic fields. Within the esu system, the electric field **E** is defined by the force law **F** $= q$**E**; correspondingly, the force **F** $= I\ell \times$ **B** on a current-carrying wire of length ℓ can be used to define the magnetic induction **B** within the emu system.[6] Since the dimension of force $[MLT^{-1}]$ is that of mass $\times$ acceleration irrespective of the nature of the force, one has $[q]_{esu}[E]_{esu} = [I]_{emu}[B]_{emu}[L] = [q]_{emu}[B]_{emu}[v]$, or

$$[E]_{esu} = [B]_{emu} = [M^{1/2}L^{-1/2}T^{-1}]. \tag{5}$$

From the definitions of the fields and their interrelationships through Maxwell's equations, one can establish that each field (**E**, **D**, **B**, **H**) in the emu system is related to the corresponding field in the esu system by a velocity. For example, from Maxwell's expression of Faraday's law of induction it follows that $B_{emu} = k[v]B_{esu}$.

Now, in the formulation of the Maxwell equations, one is free to chose any system of units, so long as relations are expressed consistently. In the Gaussian system, for example, **E** is expressed in the esu system and **B** is expressed in the emu system, and the factor $k[v]$ enters each field equation as a conversion factor to maintain this consistency. The resulting wave equation (whose form is independent of the system of units) for a medium with permittivity ε and permeability μ—both constants being dimensionless numbers in the Gaussian system—leads to a phase velocity $u = k/\sqrt{\varepsilon\mu}$ that contains *not* the speed of light c, but rather the universal constant k giving the number of esu's of charge to one emu of charge. The fact that k turns out by measurement to have the same numerical value as c is wondrous indeed and strongly supports the belief that light is a form of electromagnetic wave. But nowhere is c "built into" Maxwell's equations at the outset, as one might infer from modern textbooks.

In the calculation that Maxwell himself made, the phase velocity of the resulting wave equation took the form $u = 1/\sqrt{\varepsilon\mu}$. For vacuum (or air, to good approximation), the esu values of the material parameters are $\varepsilon = 1$, $\mu = 1/k^2$, whereas the cmu values are just the reverse.

What is the best system of electromagnetic units to use? That depends on one's needs. I have always found the Gaussian system particularly suitable for theoretical analysis, for it manifests the intrinsic symmetry of Maxwell's equations, an especially attractive feature in light of relativity theory (unknown, of course, to Maxwell working in the 1860s). Thus, a phenomenon interpretable in terms of **B** in one inertial reference frame may be ascribed to the effect of **E** in another inertial frame. Since **E** and **B** are intimately related by a Lorentz transformation, it is physically significant that they have the same dimensions, as indicated in Eq. (5). On the other hand, the Gaussian system is not always convenient for practical work.

The reader might be interested to learn that Maxwell played a significant part in establishing the familiar set of electromagnetic units used in the laboratory— and he discussed this, too, in his *Treatise*. If one adopts the standard metric units of length (cm or m), time (sec), and mass (g or kg), then the units of resistance and electromotive force are too small to express laboratory measurements conveniently, and the units of charge and capacitance are correspondingly too large. The system of practical units (volt, ohm, farad, coulomb, ...) were initially based on selection of a unit of length of 10^6 m and a unit of mass of 10^{-14} kg. It is to achieve this convenience of scale, therefore, that ε_0 and μ_0 are inserted into the fundamental equations, dimensional consistency requiring that $1/\sqrt{\varepsilon_0\mu_0} = c$.

Students and physicist colleagues have occasionally asked me why the electromagnetic fields are designated by their particular letters. I have seen letters to the editors of various physics journals also pose this question from time to time. In a few cases the choice is self-evident, as in **E** for electric field and **D** for displacement field. But what is one to make of **B** and **H** for magnetic fields? Is there some scientifically significant language (Latin, Greek, German, French, ...) for which the names of these fields correlate with the choice of symbol? I suspect not, but speculate instead that Maxwell simply designated all his electromagnetic variables in an alphabetical order: **A** (vector potential), **B** (magnetic induction), **C** (electric current), **D** (displacement), **E** (electric field—Maxwell's electromotive intensity), **F** (mechanical force), **H** (magnetic field—Maxwell's magnetic force), and so on. If so, herein lies another significant insight into Maxwell's thinking—an insight that until recent times has largely been lost in modern presentations of the subject.

That the first letter of the alphabet is assigned to the vector potential is not, I believe, an arbitrary choice, but signifies instead the signal importance which Maxwell attached to this function. **A** is the mathematical embodiment of what Faraday, in his qualitative but perceptive reasoning, had termed the "electrotonic state," a "peculiar electrical condition of matter" whereby an isolated circuit remains unaffected by a constant electromagnetic field, but produces a current if the same state of the field were brought into existence suddenly. "The whole history of this idea in the mind of Faraday," Maxwell wrote, "is well worthy of study. By a course of experiments, guided by intense application of thought, but without the aid of mathematical calculations, he was led to recognise the existence of something which we now know to be a mathematical quantity, *and which may even be called the fundamental quantity in the theory of electromagnetism*" (*Treatise*, 2:187; my italics).

To Maxwell this fundamentality lay in the distinct dual purposes **A** served in the dynamical theory of the electromagnetic field. First, he introduced **A** as the potential function (hence the name vector potential) from which the magnetic induction was obtained by **B** = curl **A**, in analogy to a scalar magnetic potential of which **H** was the gradient. Later in the *Treatise*, upon investigating induction in a secondary circuit by current changes in a primary circuit, Maxwell showed that **A**—termed at that point the "electrokinetic momentum"—was related to the time integral of the electric field. In this capacity **A** not only determined an induced electric field by the relation $\mathbf{E} = -\partial \mathbf{A}/c\partial t$ (in the Gaussian system) but was also interpretable as a linear momentum of the electromagnetic field which could be communicated to a charged particle in the secondary circuit if the primary current were suddenly stopped. This dual significance of **A**, as both a potential and a form of momentum, was to have profound implications in the quantum theory of matter and electromagnetic fields.

In the form of Maxwell's theory that one presently encounters, condensed and simplified by Heinrich Hertz and Oliver Heaviside with full employment of the vector analysis of Willard Gibbs, it is the four electromagnetic fields, not the electromagnetic potentials, that are conceptually fundamental. The fields are regarded as physically real, for they are directly related to observable forces. The vector and scalar potentials, by contrast, are merely auxiliary functions from which these fields are determined by derivative operations. The potentials, unlike the forces, are not unique, but can be modified by a so-called gauge transformation that leaves Maxwell's equations and the force laws invariant.[7]

The twentieth-century discovery that electrons have wavelike properties, however, has brought to light (no pun intended) an amazing physical possibility, the implications of which lend strong support to Maxwell's own views of the interpretation and fundamentality of the vector potential. Is there any way the state of motion of a charged particle passing through a region of space where **A** is nonvanishing, but **E** and **B** are strictly null, can be influenced? To this question classical physics yields an unequivocally negative answer, for under such circumstances there can be no force on the charged particle. Neverthe-less, as recognized first by British electron microscopists W. Ehrenberg and R. E. Siday in 1949,[8] and independently by quantum theorists D. Bohm and Y. Aharonov ten years later,[9] such a phenomenon is indeed conceivable.

The Aharonov-Bohm (or AB) effect, as it is known today, entails the modi-fication of an electron self-interference pattern by passage of a coherently split electron "wave" around (but not through) an excluded region of space within which is confined a magnetic field **B** (as illustrated in figure 10.1). In the figure, the magnetic field is produced by an ideally infinitely long solenoid perpendic-ular to the page; the only electromagnetic influence at the location of a moving electron is the external vector potential **A**. Quantum theory predicts—and a number of independent experiments have confirmed[10]—that the fringes of the electron interference pattern received on a distant screen should be shifted by the angle $\alpha = \Phi/\Phi_0$ relative to the pattern for $\mathbf{B} = 0$, where

$$\Phi = \oint \mathbf{A} \cdot \mathbf{ds} = \iint \mathbf{B} \cdot \mathbf{dS} \qquad (6)$$

is the magnetic flux through the excluded domain and $\Phi_0 = hc/e = 3.9 \times 10^{-9}$ gauss-cm^2 is the quantum unit of magnetic flux, or fluxon. The contour of the line integral of **A** in Eq. (6) can be any closed path around the excluded domain connecting the points of electron emission and detection.

The startling nature of the effect is illustrated in the displaced fringe pattern of the figure, which indicates a value of **B** such that no electrons at all are received in the forward direction (i.e., along the optic axis). From a classical perspective, the trajectories of the electrons seem to have been shifted despite the absence, under the circumstances, of any Lorentz force. Quantum mechanics, however, does not permit us to think of electron trajectories in the context of an interference experiment. Correspondence with classical physics is established by the fact that the overall beam, as defined by the single-slit diffraction pattern (and not the two-slit interference pattern), is unaffected by the isolated magnetic field.

As a quantum physicist I have given much thought to the implications and novel experimental extensions of the AB effect, which I have discussed in earlier books.[11] Concerning the fundamental quantities of electrodynamics, however, one can say the following. Although the AB fringe shift is dependent on the confined magnetic flux Φ—which, according to Eq. (6), can be expressed in terms of either **A** or **B**—the derivation of the AB effect, indeed the fundamental starting point for treating all electromagnetic interactions of a quantum particle,

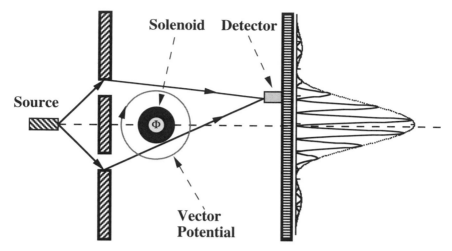

Figure 10.1. Geometry of the two-slit Aharonov-Bohm effect. Electrons diffract around a long solenoid (perpendicular to the page) and are counted at the detector, giving rise to interference fringes whose locations depend on the magnetic flux Φ confined to the solenoid interior. In the exterior region there is only a cylindrically symmetric vector potential field. The single-slit diffraction envelope is not displaced since, under the circumstances, there is no Lorentz force acting on the electrons.

begins with potentials, not fields or forces. It is the vector potential **A** and scalar potential V that enter the Lagrangian and Hamiltonian formulations of quantum mechanics that prescribe the time evolution of a quantum system. **A**, in particular, occurs in association with the canonical linear momentum **p** in expressions that reveal its dual role as both a potential and momentum-like quantity—as, for example, in the Hamiltonian operator

$$H = \frac{1}{2m}\left(\mathbf{p} - \frac{q}{c}\mathbf{A}\right)^2 + qV \tag{7}$$

of a single nonrelativistic charged particle. It is worth noting in this regard that it was Maxwell who introduced Hamiltonian ideas (based on energy and potential) into electrodynamics, thereby sidestepping methods based on force which would have required detailed knowledge of the electromagnetic medium (ether) and nature of electric charge (electron). Maxwell, it should be remembered, knew nothing of the electron, whose discovery took place well after his death.

A question which frequently arises, especially in view of the AB effect, is whether or not the vector potential is a measurable quantity. Maxwell, I believe, certainly thought so, as he associated it with a transferable linear momentum. In quantum mechanics, quantities designated as "dynamical observables"—that is, accessible to measurement—must satisfy two requirements. First, they must be representable by Hermitian (i.e., self-adjoint) operators, since these have real-valued eigenvalues, and the outcome of measurement can only be expressed by real numbers.[12] Second, they must be invariant to a gauge transformation, since the latter is analogous to selecting a local coordinate system, and the outcome of a physical measurement cannot depend on the arbitrariness of such a choice. Since $\mathbf{A}$ and V are not gauge invariant, they are not, strictly speaking, measurable. (In electrostatics, it is always a potential difference, ΔV, that is measured.)

Having said this, however, it is important to note that a gauge transformation is not entirely arbitrary, for the result of any such transformation must not change the physical presence of the magnetic field $\mathbf{B}$ = curl $\mathbf{A}$. Now any vector field can be decomposed into "transverse" and "longitudinal" components $\mathbf{A} = \mathbf{A}_{\perp} + \mathbf{A}_{\parallel}$ such that div $\mathbf{A}_{\perp} = 0$ and curl $\mathbf{A}_{\parallel} = 0$. The actual decomposition takes the form

$$\mathbf{A}_{\perp} = \frac{1}{4\pi} \nabla \times \left(\nabla \times \int \frac{\mathbf{A}(\mathbf{r}') \, d^3 r'}{|\mathbf{r} - \mathbf{r}'|} \right), \tag{8a}$$

$$\mathbf{A}_{\parallel} = -\frac{1}{4\pi} \nabla \int \frac{\nabla' \cdot \mathbf{A}(\mathbf{r}') \, d^3 r'}{|\mathbf{r} - \mathbf{r}'|}, \tag{8b}$$

and one can show with a little effort that the sum of Eqs. (8a) and (8b) does indeed reduce to $\mathbf{A}(\mathbf{r})$. A gauge transformation modifies only the component $\mathbf{A}_{\parallel}$, leaving the component $\mathbf{A}_{\perp}$ and the physical field $\mathbf{B}$ = curl $\mathbf{A}_{\perp}$ unchanged. The transverse part of the vector potential is a measurable quantity.

Maxwell did not live to see the confirmation of his greatest prediction—the existence of electromagnetic waves—which were demonstrated experimentally by Heinrich Hertz in 1889, some ten years after Maxwell's death. Hertz's investigations are ingeniously simple and of profound significance—an inspiration to anyone who, like me, takes joy in small-scale "tabletop" experiments. And yet I cannot recall ever seeing a discussion of these seminal experiments in any of the optics or electrodynamics textbooks from which I have studied or taught. Fortunately, early in my career I picked up for a pittance a collection of Hertz's papers[13] in a second-hand bookshop and have had the pleasure of following Hertz's thoughts and actions in his own words.

Hertz, interestingly enough, did not set out to find electromagnetic waves. On the contrary, his investigations were initially motivated by an entirely different objective brought to his attention by his mentor, the leading German physicist Hermann von Helmholtz. Hertz writes, "The general inducement was this. In the year 1879 the Berlin Academy of Science had offered a prize for research on the following problem:—To establish experimentally any relation between electromagnetic forces and the dielectric polarisation of insulators—that is to say, either an electromagnetic force exerted by polarisations in non-conductors, or the polarisation of a non-conductor as an effect of electromagnetic induction" (1).

The term "polarisation" here has no bearing on light, but refers instead to the separation of electrical charge (although, again, it must be remembered that in 1879 discrete units of electrical charge were still unknown). In brief, the focus of attention of the Berlin Academy, and ultimately of Hertz, was on Maxwell's predicted displacement current, the existence of which was by no means generally accepted at the time even in Britain, let alone in Germany. Representable (in the Gaussian system) by a density

$$\mathbf{J}_d = \frac{\varepsilon}{4\pi} \frac{\partial \mathbf{E}}{\partial t} \tag{9}$$

dependent exclusively on the temporal variation of a *neutral* "substance" (the electric field), the displacement current raised troublesome questions concerning the nature of electricity and the closure of electrical circuits.

According to Eq. (9), one might be able to detect the effects of a displacement current in an insulator if the means were at hand to generate rapidly oscillating electric fields. In this endeavor Hertz had good fortune, for in the collection of physical instruments at the Technical High School at Karlsruhe, where he carried out his investigations, he had earlier found—and used for lecture purposes—a pair of so-called Riess or Knochenhauer spirals. The discharge of a small Leyden jar through one of the spirals, Hertz discovered, amply sufficed to produce sparks in the other, "provided it had to spring across a spark gap" (2). Upon optimizing conditions, he eventually succeeded "in obtaining a method of exciting more rapid electric disturbances than were hitherto at the disposal of physicists" (3). Thus did Hertz auspiciously embark upon his researches.

But the work did not proceed well—that is, as Hertz hoped it would. Actually, it proceeded only too well, although he did not at first recognize it. Having a means of generating rapidly oscillating electric sparks between the terminals of a spark gap in a primary circuit, Hertz attached to each terminal a conducting plate, inserted between the two plates an insulator, and endeavored to determine the effect of the insulator on electrical oscillations induced between the gap of a separate loop antenna as schematically shown in figure 10.2.

The expectation was that "when the block was in place very strong sparks would appear in the secondary, and that when the block was removed there would only be feeble sparks" (5). The basis for this expectation was that electrostatic forces—forces derivable from a potential and which ordinarily diminish rapidly with distance from the source—could not induce sparking in a nearly closed secondary circuit (since their integral over a nearly closed contour ought to be vanishingly small). Any sparking, therefore, would have to be induced by the displacement current in the dielectric block. But this was not what occurred. "The experiment," Hertz wrote, "was frustrated by the invariable occurrence of strong sparking in the secondary conductor" (5) whether the insulator was present or not.

Gradually it became clear to Hertz that he was not dealing with static or quasi-static fields, as had been commonly the case among electrical investigators up

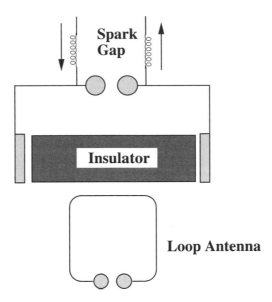

Figure 10.2. Diagram of Hertz's apparatus designed to test the effect of an insulator on the sparks induced in a secondary circuit (loop antenna) as a result of electrical oscillations in a primary circuit (spark gap and conducting plates).

to that time, but with a field of such high frequency that only the laws of a true electrodynamics would be applicable. "I perceived that I had in a sense attacked the problem too directly" (5), Hertz concluded somewhat understatedly. More importantly, he also perceived that the particular problem of the Academy, which until then had served as his guide, could be approached in an entirely different and more productive way. Since air—and indeed empty space—according to Maxwell's theory ought to behave like all other dielectrics, there was no need really to look for the effects of the displacement current generated in a solid dielectric. A more worthy and attainable goal, Hertz decided, would be to look for the direct transmission of an electrical signal through the air and to measure its rate of propagation.

The rest, as one says, is history. Connecting a powerful induction coil to a spark gap between two large square brass conducting plates (providing capacitance), Hertz probed the presence of a transmitted field at various distances along a horizontal "baseline" perpendicular to the gap and the plane of the plates (figure 10.3). Employing as his detector a circular or square loop antenna with spark gap resonant with the primary circuit, Hertz records, "I was able to observe the sparks [in the antenna] along the whole distance (12 metres) at my disposal, and have no doubt that in larger rooms this distance could be still farther extended" (109). At any given distance, however, the induced sparking

Induction Coil

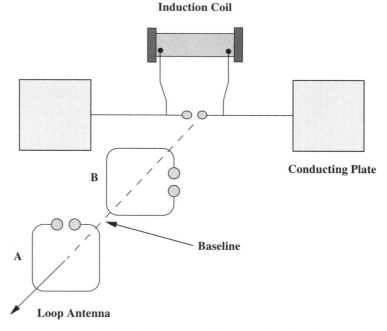

Conducting Plate

B

Baseline

A

Loop Antenna

Figure 10.3. Schematic of Hertz's apparatus demonstrating the existence of electromagnetic waves produced by a spark-gap oscillator comprising an induction coil and conducting plates. Waves were detected by the loop antenna in orientation A (with spark gap in a horizontal plane above the baseline), but not in orientation B (with spark gap in a horizontal plane through the baseline).

could be terminated by rotating the antenna; indeed, it is part of the ingenious simplicity of the experiment that not only did Hertz detect the transmission of electromagnetic waves, but he simultaneously confirmed their transverse polarization by appropriately orienting the plane of the secondary circuit. "The reason is obvious," Hertz wrote of the cessation of sparking: "the electric force is at all points perpendicular to the direction of the secondary wire" (109).

There remained the intriguing question of the speed of propagation of these newly discovered waves. To this end, Hertz modified his apparatus, adding an additional brass plate parallel to, and a short distance behind, one of the original plates, and extending from that plate a long straight copper wire parallel to the "baseline." The wire passed through the window of his laboratory for a distance of some 60 meters and ended freely in the air. By passing one of his tuned loop antennas nearby along the length of the wire, Hertz observed the periodic increase and decrease of sparking characteristic of a standing wave pattern. From the electrical properties of the primary circuit he estimated the period of oscillation to be 0.14 ns, later shown by Poincaré to be an overestimate by

$\sqrt{2}$. Placing paper riders at the nodal (no-sparking) positions on the wire, Hertz determined a wavelength of 2.8 m. Dividing the wavelength by the corrected period of 0.10 ns led to a wave velocity of 2.8×10^8 m/s.

With evident satisfaction and pleasure, Hertz concluded:

> it is clear that the experiments amount to so many reasons in favour of that theory of electromagnetic phenomena which was first developed by Maxwell from Faraday's views. It also appears to me that the hypothesis as to the nature of light which is connected with that theory now forces itself upon the mind with still stronger reason than heretofore. Certainly it is a fascinating idea that the processes in air which we have been investigating represent to us on a million-fold larger scale the same processes which go on in the neighbourhood of a Fresnel mirror or between the glass plates used for exhibiting Newton's rings.[14]

What more can one say?

Actually, there *is* one more thread to Hertz's story that must be mentioned, an ironic and adventitious twist of fate such as occurs rarely, but nevertheless does occur. Shortly after devising his system of producing rapid electrical oscillations, but before he could apply it to the examination of displacement current in insulators, Hertz had first to free himself from an earlier and somewhat frustrating investigation: "Soon after starting the experiments I had been struck by a noteworthy reciprocal action between simultaneous electrical sparks. I had no intention of allowing this phenomenon to distract my attention from the main object which I had in view," he lamented, "but it occurred in such a definite and perplexing way that I could not altogther neglect it. For some time, indeed, I was in doubt whether I had not before me an altogether new form of electrical action-at-a-distance" (4).

The puzzling effect which riveted Hertz's attention was the apparent diminution in intensity of the spark in his loop antenna when nonconducting materials (glass, paraffin, ebonite, etc.) were interposed between it and the primary oscillator. This shielding effect persisted irrespective of the distance between the two circuits. By contrast, coarse metal gratings, which in principle are excellent electrostatic screens, showed no shielding effect at all.

After groping in the dark (somewhat literally), Hertz established that it was absorption of the ultraviolet component of the light from the primary spark that degraded the intensity of the induced spark; reciprocally, the ultraviolet light from the secondary spark sustained the spark of the primary circuit when the latter was adjusted sufficiently close to misfiring.

Hertz, who died in 1884, never understood the implications of these observations, experiments undertaken as a side study to the more important task set by the Berlin Academy. He had discovered, in fact, the photoelectric effect—and therefore, even *before* his definitive confirmations of the existence of electromagnetic waves, had provided the first (albeit unrecognized) experimental evidence of the granular nature of light.

NOTES

1. J. C. Maxwell, "A Dynamical Theory of the Electromagnetic Field," reprinted from *Trans. Roy. Soc.* 155 (1865), in *The Scientific Papers of James Clerk Maxwell*, ed. W. D. Niven (New York: Dover, 1952): 526–97; quotation from p. 580.

2. J. C. Maxwell, *A Treatise on Electricity & Magnetism*, 2 vols. (New York: Dover, 1954), republication of unabridged 3d ed., published by the Clarenden Press in 1891. Quotations from Vol. 1, pages viii–ix and Vol. 2, pages 175–76.

3. J. C. Maxwell, "On Faraday's Lines of Force," reprinted from the *Trans. Cambridge Phil. Soc.* 10, part 1 (1855–56), in Maxwell, *Scientific Papers*, 155–229.

4. J. C. Maxwell, "On Physical Lines of Force," reprinted from *Phil. Mag.* 21 (1861–62), in Maxwell, *Scientific Papers*, 451–513.

5. A. Einstein, "On the Electrodynamics of Moving Bodies," *Annalen der Physik* 17 (1905): 891–921.

6. The magnetic induction can also be defined in terms of the Lorentz force on a moving charged particle. Maxwell did not refer to the Lorentz force, which was introduced after his death. Rather, he defined **B** and **H** in terms of the force on a hypothetical unit magnetic pole, a construct that as far as we know, still has no realization in nature.

7. A gauge transformation of electromagnetic potentials consists of the following. From a given pair of vector and scalar potentials $(\mathbf{A}, V)$, one constructs a new pair $(\mathbf{A}', V')$ by means of an arbitrary gauge function $\Lambda(x, t)$: $\mathbf{A}' = \mathbf{A} + \nabla\Lambda$; $V' = V - \partial\Lambda/c\partial t$. For the case of fields coupled to charged particles, as treated within the framework of quantum mechanics, a gauge transformation also entails a unitary transformation of the particle wave function. See, for example, J. J. Sakurai, *Advanced Quantum Mechanics* (Reading, Mass.: Addison-Wesley, 1967), pp. 14–16.

8. W. Ehrenberg and R. E. Siday, "The Refractive Index in Electron Optics and the Principles of Dynamics," *Proc. Phys. Soc. London B* 62 (1949): 8–21.

9. Y. Aharonov and D. Bohm, "Significance of Electromagnetic Potentials in the Quantum Theory," *Phys. Rev.* 115 (1959): 485–91.

10. See, for example, N. Osakabe et al., "Experimental Confirmation of the Aharonov-Bohm Effect Using a Toroidal Magnetic Field Confined by a Superconductor," *Phys. Rev. A* 34 (1986): 815. In this experiment the desired field configuration is achieved with a toroidal ferromagnet covered with a superconducting outer layer. The Meissner effect expels the magnetic flux from the layer, thereby confining it to the toroidal interior.

11. M. P. Silverman, *And Yet It Moves: Strange Systems and Subtle Questions in Physics* (New York: Cambridge University Press, 1993); *More Than One Mystery: Explorations of Quantum Interference* (New York: Springer-Verlag, 1995).

12. One might inquire why a complex number—in effect, a coupled pair of real numbers—cannot serve as a measurement outcome. The answer is that two measurements would be required to furnish the two numbers, and the order of these observations would matter, since quantum measurements are not, in general, commutative.

13. Henrich Hertz, *Electric Waves* (New York: Dover, 1962), an unabridged republication of the work first published in 1893 by Macmillan. Quotations in the text are taken from the introduction, pages 1–20, and from his papers "On the Finite Velocity of Propagation of Electromagnetic Actions," p. 109, and "On Electromagnetic Waves in Air and Their Reflection," p. 136.

14. H. Hertz, *Electric Waves*, p. 136.

CHAPTER 11

New Twists on Reflection

> I carefully separated the crystals which were
> hemihedral to the right from those hemihedral to the
> left, and examined their solutions separately in the
> polarising apparatus. I then saw with no less surprise
> than pleasure that the crystals hemihedral to the right
> deviated the plane of polarisation to the right, and
> that those hemihedral to the left deviated it to the left;
> and when I took an equal weight of each of the two
> kinds of crystals, the mixed solution was indifferent
> towards the light in consequence of the neutralisation
> of the two equal and opposite individual
> deviations.... The two kinds of crystals are
> isomorphous.... But the isomorphism presents itself
> with a hitherto unobserved peculiarity; it is the
> isomorphism of an asymmetric crystal with its
> mirror image.
> (*Louis Pasteur, 1850*)[1]

11.1 THE PERVASIVE HAND OF CHIRAL ASYMMETRY

ALTHOUGH MOST people probably accord no special notice to the fact that they are right- or left-handed, chiral[2] asymmetry in nature has profound and far-reaching implications.

At the uppermost echelon of complexity is life itself. The capacity to create and consume molecules of preferential handedness—that is, only one of two possible mirror-image or enantiomeric[3] forms—is virtually the hallmark of the living state. Amino acids, for example, that comprise the proteins of all living things from viruses to humans are predominantly L-amino acids.[4] To a living organism the wrong enantiomer may not be just a useless molecule, but quite possibly a toxin or mutagen with deadly consequences. The sweetener limonene comes in two enantiomeric forms: The left-handed form imparts the flavor of lemons, the right-handed form that of oranges; consuming a mixture of them is innocuous. There are likewise two enantiomers of thalidomide; one is a mild drug that helps settle the stomach, but the other is a teratogen leading to gross limb malformation in fetal development.

How this "homochirality" of life evolved—whether it was an accident of evolution or the ineluctable outcome of physical law, whether it was a precursor to life or the result of it—is not known. There has been no shortage of speculation, however, ever since Louis Pasteur first discovered (~1897) the existence of naturally occurring mirror-image molecules and wondered whether the chiral ("dissymmetric" in Pasteur's terminology) agent leading to their unequal prevalence was light, electricity, magnetism, heat, the rotation of the Earth, or some combination. Over the years researchers, including Pasteur, have tried to impress a chiral asymmetry upon otherwise achiral chemical synthesis or decomposition by, among other things, stirring, performing the reactions in magnetic fields, illuminating the reactants with circularly polarized light, or bombarding the reactants with polarized electrons. Most of the interventions that have been tried cannot work, for they do not possess the correct attributes of true chirality.

The nature of the symmetry inherent in a truly chiral system is exemplified by the phenomenon of optical activity, of which one of the most familiar manifestations is the optical rotation of linearly polarized light.[5] Consider a light beam directed from left to right through an optically active medium (such as corn syrup) resulting in a rotation of the electric vector $\mathbf{E}$ from its initially vertical orientation to an orientation directed out of the page—that is, a counterclockwise (ccw) rotation to an observer facing the light source. As shown in figure 11.1 (part A), the mirror image of the preceding process is another process whereby the light beam propagates from right to left through the sample with a clockwise (cw) optical rotation (again with respect to an observer facing the light source). Spatial inversion has reversed the "sign" of the optical rotation; in the language of quantum physics such behavior is referred to as "parity-odd." Actually, the total parity must be inferred from the results of mirror reflection across three mutually perpendicular planes. Since reflection is a discrete operation—that is, implementable only in a single, abrupt step, rather than in gradual increments as in the case of a rotation or translation—parity is a multiplicative index of symmetry; the total parity is then the product of the parities of these three reflections. The outcome is still the same: Optical rotation is parity-odd.

Someone familiar with the phenomenon of optical activity may be inclined to think at first that if the initial process (ccw rotation) occurs, then the mirror-image process does not, for experimentally it should also have led to a ccw (and not cw) optical rotation. The sense of rotation in true optical activity—as opposed to the fundamentally different phenomenon of the Faraday effect—is always defined with respect to the propagation vector of the light. There is, in fact, no other relevant direction within a homogeneous, isotropic, optically active medium free of external static fields.

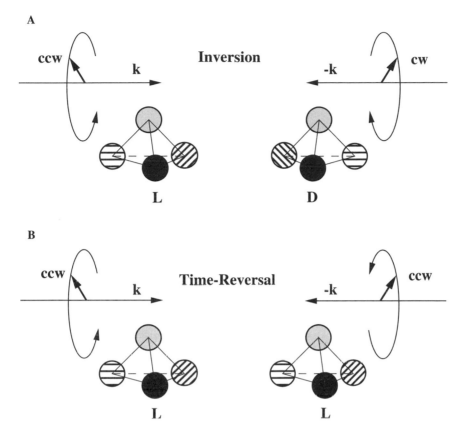

Figure 11.1. Symmetry of optical rotation: (A) odd under space inversion (ccw → cw) and (B) even under time reversal (ccw → ccw). In each case the transformation of an allowed process is also an allowed process.

Nevertheless, the mirror image of the first optical rotation *is* in fact an allowable process. Symmetry operations apply not just to the light beam, but to the whole system of light + sample, in which case the mirror reverses not only the direction of light propagation but also the handedness of the chiral molecules. Thus, in total conformity with physical law, if the L-form of a chiral molecule rotates light counterclockwise, then the corresponding D-form generates a clockwise rotation. Because both processes are allowed, one says that natural optical activity is invariant to spatial inversion.

There is another symmetry of fundamental importance: time reversal. Figure 11.1 (part B) shows the effect on the original process of reversing time in all dynamical quantities that depend on it, in this case the propagation vector (or linear momentum) of the light and the sense of rotation of the electric field.

The static molecular framework of the chiral sample is unaffected. The optical rotation (ccw) of the time-reversed process, as shown in the figure, is the same as in the original process; that is, optical rotation is "time-even." Thus, if the original process occurs, so must the time-reversed process, and the phenomenon of natural optical activity is said to be invariant to time reversal.

Processes designed to create molecules of preferential handedness must, like natural optical activity, be parity-odd and time-even. Carrying out a chemical synthesis in the presence of a powerful magnet, for example, will not—if our understanding of electrodynamics is correct—produce anything other than an equal mixture of enantiomeric forms (termed a racemic[6] mixture), since a magnetic field is parity-even and time-odd. (Mathematically, the magnetic field is represented by an axial vector with the space- and time-inversion properties of angular momentum, in contrast to a polar vector, which represents an electric field.) Likewise, stirring—even vigorous centrifugation—of a chemical brew will not engender chiral asymmetry, since stirring is parity-even and time-odd. The rotation of the Earth, therefore, could not in itself lead to biological homochirality. What might work then?

One possibility is simply a photochemical reaction with circularly polarized light; the light should interact differently with left- and right-handed forms of any chiral reactants. However, the overall equality of LCP and RCP light fluxes would seem to eliminate this prospect. Nevertheless, there is a credible chance that this simplest of mechanisms (with neglect of electric, magnetic, and centrifugal forces) may yet suffice. According to published claims,[7] the morning skylight contains a greater degree of left circularly polarized light than right, a condition which reverses in the afternoon. Although the composition of circularly polarized (CP) light averaged over the full period of daylight is effectively the same, the chiral asymmetry engendered by the morning illumination is not necessarily undone by day's end because of one additional feature: Excluding the polar regions of the globe, the ambient temperature at sunset is approximately 10°C higher than at sunrise. Since most chemical reactions are known to proceed more quickly at higher temperatures, the chirally asymmetric irradiation is conceivably linked to a kinetic asymmetry in the rate of molecular synthesis.

The matter is by no means settled—if, indeed, it ever will be; terrestrial evolution is not an experiment that can be replicated in the laboratory. In contrast to mechanisms dependent upon the accidents of circumstance, there are other mechanisms of biological homochirality based on the intrinsic chiral asymmetry of physical laws. Thus, at the opposite end of the scale of complexity from living organisms are the elementary particles whose activities, ironically, may be no less complex.

Prior to about 1956 most physicists would have scoffed at the proposition that the laws of physics were not invariant under mirror reflections. That year, however, T. D. Lee and C. N. Yang published their critical appraisal of the question of parity conservation,[8] coming to the conclusion that all previous experiments purportedly ruling out the nonconservation of parity were basically irrelevant. They proposed, instead, a number of possible tests including the β-decay of polarized Co^{60} nuclei. The latter experiment was performed shortly afterward by C. S. Wu and her collaborators and showed unmistakably that at the level of the weak nuclear interactions mirror reflection was not a universal symmetry of nature. Wolfgang Pauli, upon first learning of Lee and Yang's paper, expressed disbelief that "God was a weak left-hander." The final arbiter in the matter, however, was experiment, not Pauli.

Although β-decay is chirally asymmetric, it is a charge-changing interaction that destroys nucleons and transmutes atomic nuclei—hardly the place to look for the origin of an enantiomeric excess of stable molecules. In the 1960s, however, the development of gauge-field theories and the "electro-weak unification" wrought a theoretical synthesis not seen since Maxwell's unification a century earlier. Now, under one theoretical framework were to be found all of electromagnetism (including optics) and the weak nuclear interactions.[9] Although the electro-weak theory is more the playground of the high-energy or elementary particle physicist than the researcher concerned with tabletop optics, two significant points pertinent to the present discussion and the general theme of this book can be made regarding it.

The first involves symmetry and the conceptual foundations of optics. With respect to its theoretical structure, Maxwell's electrodynamics proceeded from the phenomenological researches of Coulomb, Ampère, and Faraday—and the resulting gauge invariance (which to my knowledge was unknown to Maxwell) was long afterward seen largely as an incidental curiosity, useful to the extent that it permitted the introduction of vector and scalar potentials that afforded a certain calculational simplicity. In its bare essentials, then, one postulated the form (inverse square) of the basic electric and magnetic interactions, and at the end of a chain of reasoning leading to the wave equation there emerged electromagnetic waves of light.

The logical progression in the gauge field theory of electromagnetism and the weak interactions (and indeed in other gauge theories as well) is quite different. Here, the essential is to preserve local gauge invariance—that is, invariance of the fundamental quantum equations of motion (ordinarily in Lagrangian form) under a particular kind of phase change of the particle wave function. The designation "local" signifies that the parameters of the transformation can vary with the coordinates of the observer (i.e., the field point). To impose this tightly restrictive symmetry, one must couple to the wave function of the particles additional fields that, in the case of electromagnetism alone, turn out to be precisely the vector and scalar potentials. In a manner of speaking, one preserves gauge invariance by introducing light (i.e., photons)—and the theory then *predicts* the basic electromagnetic interactions.

The second point bears directly on chiral asymmetry. To put both electromagnetism and the weak interactions under one roof, so to speak, and still maintain local gauge invariance required coupling not only the electromagnetic field to the electron wave function, but additional fields, including the "weak" counterpart to the photon, a neutral particle labeled the Z^0 boson. In marked contrast to the massless photon, however, the Z^0 is approximately 100 times as massive as a proton—and its range of action, defined by the Compton wavelength,[10] barely extends to one ten-millionth the size of a typical atom. As a neutral particle, however, the Z^0 can participate in charge-preserving, parity-violating interactions that do not transmute atoms, but manifest themselves, instead, in a weak (both senses of the word) optical activity. This optical activity has been observed.[11]

Thus, an individual unbound atom—which electrodynamically is a spherical system superposable on its mirror image—is revealed by the electroweak theory to have enantiomeric forms. In what way could the enantiomeric forms of an atom possibly differ? Although well-defined particle orbits are generally not pertinent to the quantum description of atoms, it is not altogether incorrect to conceive of certain electron states—or coherent superpositions of states—as characterized by a minuscule precessional movement in either a cw or ccw sense. However, unlike a sugar molecule or a quartz crystal—both forms of which in principle can be synthesized even if one form did not occur naturally—only one of the two atomic enantiomers theoretically exists. An atom of the opposite handedness could not be made of normal matter, but only of the corresponding antiparticles.

The prospect that every atom in existence has a built-in chiral asymmetry has naturally raised the question of whether this feature may have played a role in the subsequent evolution of chirally asymmetric life forms. Numerous calculations have apparently established that, based on the weak nuclear interactions alone, the enantiomeric forms of molecules may differ in energy, and hence stability, although by amounts so small that their consequences for biochemical evolution are hard to envision. Nevertheless, evolution has had, for all practical purposes, unlimited time (billions of years) in which to proceed. Does an "infinite" number of "infinitesimal" steps lead to a finite result in this curious mix of physics, chemistry, and biology, as it does in a calculus exercise? The fact is, one simply does not know.

From elementary particles exchanging photons and Z^0 bosons to human beings who marvel at it all—one feels the hand of nature's handedness. As both a chemist and a physicist, I have long wondered about the persistent, intricate, and far-reaching connections of nature, so rhapsodically captured in Feynman's poetic musings:[12]

> Ages on ages . . . before any eyes could see . . . year after year . . . thunderously pounding the shore as now. For whom, for what? . . . on a dead planet, with no life to entertain.

Never at rest ... tortured by energy ... wasted prodigiously by the sun ... poured into space. A mite makes the sea roar.

Deep in the sea, all molecules repeat the patterns of one another till complex new ones are formed. They make others like themselves ... and a new dance starts.

Growing in size and complexity ... living things, masses of atoms, DNA, protein ... dancing a pattern ever more intricate.

Out of the cradle onto the dry land ... here it is standing ... atoms with consciousness ... matter with curiosity.

Stands at the sea ... wonders at wondering ... I ... a universe of atoms ... an atom in the universe.

(There are no elisions; the lines are as Feynman wrote them.)

Among the many things that have piqued my curiosity throughout the years, the origin, nature, and practical consequences of chiral interactions and objects figure prominently. An erstwhile organic chemist, I have pondered the distinctive behaviors of the complex molecules of life, "masses of atoms ... dancing a pattern ever more intricate." As an atomic physicist, I have examined the detailed anatomy of atoms, probing the most basic of elements ever more precisely with light beams and microwaves for some minute aberration in the dance of an electron, the exciting sign that "a new dance starts."

It was not at the limits of the ultra-elemental or transcendentally complex that I first encountered for myself the capricious subtleties of chirality. Rather, in setting out to investigate the reflection of light from naturally optically active crystals and liquids, I came to a surprising realization. Here, firmly within the domain of classical optics well trod by the likes of Arago, Biot, and Fresnel, this phenomenon of chiral light reflection had never been observed—quite possibly never even looked for. Furthermore, its previous theoretical treatment was incomplete and incorrect.

The problem of how to measure and account for the minute differences with which a chiral medium reflects LCP and RCP light marks a confluence of two major directions in the work of Fresnel. It was, first of all, Fresnel who explained the striking effects of optical activity observed by Biot and Arago in light transmission by hypothesizing the existence of CP light and subsequently separating the distinct LCP and RCP forms by differential refraction of an incident linearly polarized beam through a concatenation of quartz prisms. And, of course, it was Fresnel again who first arrived at the "Fresnel amplitudes" for light reflection from a plane achiral dielectric surface.

Apart from interwoven strands of history, which were not uppermost in my mind when I began my own investigations, the chiral reflection of light is, itself, a conceptually rich and practically useful phenomenon. For well over a century, the optical rotation or circular dichroism of light transmitted through bulk crystals or liquids has provided information on the identity, concentration, and stereochemistry of materials. Indeed, the very existence of optical activity informed chemists—even before a consensus on the reality of atoms was firmly established in the twentieth century by the experiments of Jean Perrin—that at least some molecules had to have a three-dimensional structure, since any planar geometry is superposable on its mirror image. Chiral reflection extends this valuable probe to new and broader classes of materials than can be conveniently studied by transmission. In addition, chiral reflection amplitudes carry more information than the optical rotation of a transmitted beam.

Experimentally, the problem presented a number of technical difficulties, for, depending on wavelength and material, the difference in LCP and RCP indices of refraction (i.e., the circular birefringence) of many naturally chiral molecules of interest is very small, at most a few parts in a million. In contrast to optical rotation or circular dichroism, which increases linearly with the pathlength through a sample, specular reflection from a surface is effectively a "single-shot" affair; and chiral asymmetries in the reflected light are ordinarily of the same order of magnitude as the circular birefringence itself. To detect the manifestations of "ordinary" molecular optical activity in reflection can be comparable in difficulty to observing atomic optical activity induced by weak nuclear interactions. Experimental success, therefore, required devising various ways to enhance very small chiral asymmetries.

It is not only the intricate web of physical connections, such as may link the electron dance in atoms to the spiral twists of snail shells, that is intriguing, but the fortuitous crossings of human world lines as well. Many years ago, after completing at Harvard University a lengthy study of the hydrogen atom, I was invited to the Ecole Normale Supérieure (ENS) in Paris. By happenstance, I arrived at a time when the first experiment to search for atomic optical activity was in preparation under the direction of M. A. Bouchiat, and I made the aquaintance of Lionel Pottier, who was to pursue this research for his doctoral thesis. My own work at the ENS focused on the structure of Rydberg atoms and did not concern chirality at all, but I learned interesting things about it in talking with Lionel. Perhaps a "germ" was planted in my mind; I cannot say. Eventually, I left the ENS, and we had no further communication.

Over ten years later, during a lengthy stay in Japan as a guest of the Hitachi Advanced Research Laboratory, I received a letter (in English) from a French physicist that began, "Sir, I was deeply interested in your paper on reflection and refraction at the surface of a chiral medium" and continued "may be would you like to know why I am interested in your work. Since a long time we use measurements of rotatory power or better of circular dichroism to study optical activity either natural or induced by a magnetic field. We measure this activity through transmission.... However, it would be sometimes interesting to do this through the light reflected by the medium under study. Your work seems to me to be a milestone on this way." The letter concluded with one of those ritualized French endings that are charmingly incomprehensible when rendered literally into English: "With my thanks will you agree my best regards, J. Badoz." Professor Jacques Badoz was then director of the Laboratoire d'Optique Physique at the Ecole Supérieure de Physique et Chimie Industrielles in Paris (the ESPCI or "PC," as it is affectionately called)—coincidentally, just across the street from the ENS. We did not know one another then, and I could never have foreseen it at the time, but that letter was to be the starting point of a long and fruitful collaboration between us with numerous visits on both sides of the Atlantic.

All good things eventually end, however, and my French colleague had to retire. On the occasion of my final visit to the PC to wrap up our research together and to say farewell, whom should I encounter occupying the office a few doors from my own, but Lionel Pottier; we had not seen one another in over twenty years. Working now at the PC, he had long since left optical activity for other interests. Meeting both these men at the same time and same place—a stone's throw from where I began research in France—symbolized for me the beginning and a kind of conclusion, a coming full circle, in my personal and scientific affairs with chiral asymmetry.

11.2 LEAPS AND BOUNDARIES

Although the generation and detection of electromagnetic waves by Hertz had to wait a quarter century after Maxwell first predicted their existence, dynamical evidence for the electromagnetic identity of light could have been provided almost immediately by a phenomenon familiar to physicists for decades: light reflection.

The derivation of the so called Fresnel amplitudes for specular reflection at the interface between two homogeneous dielectrics (e.g., air and glass) is one of the archetypical applications of classical electromagnetism found in virtually every optics and E&M text I know. In contrast to kinematical features of reflection and refraction such as Snel's law or the "law of the mirror," which are derivable from any wave theory, the specific form of the Fresnel reflection and transmission amplitudes are a unique consequence of Maxwell's theory of the electromagnetic field.

Assuredly Maxwell must have been aware of the importance of reflection as a test of his theory. And yet it was with surprise bordering on consternation that I searched in vain years ago, first through Maxwell's *Treatise*[13] and subsequently through his collected scientific papers,[14] without finding even a trace of the Fresnel relations, let alone the analysis I looked for. Since such an analysis proceeds straightforwardly from the boundary conditions of the fields at the interface, I can only assume that Maxwell must have been unsure of the appropriate boundary conditions for high-frequency fields as he had, himself, established these conditions in the case of static fields. Thus, the first derivation of the Fresnel amplitudes from Maxwell's theory was made by H. A. Lorentz in the 1870s.

As is well known today, the boundary conditions for static and temporally varying fields are identical—the standard derivations with rectangular loops and little pillboxes making no distinction between them—and generally take the following form:

$$(E_2 - E_1) \times n = 0, \tag{1}$$

$$(H_2 - H_1) \times n = 0, \tag{2}$$

$$(D_2 - D_1) \cdot n = 0, \tag{3}$$

$$(B_2 - B_1) \cdot n = 0, \tag{4}$$

in the familiar case of a stationary boundary. (If one medium moves relative to the other, then the boundary conditions are more complicated, and such problems are perhaps best handled by means of Einstein's theory of special relativity, which, in fact, was expressly created to treat the electrodynamics of moving bodies, as indicated by the title of Einstein's 1905 paper.)[15] In the preceding relations, n is the unit normal vector at the interface directed from medium 1 into medium 2 as illustrated in figure 11.2. According to Eqs. (1) and (2), the tangential component (E_t, H_t) of the electric and magnetic fields is continuous across the interface, whereas, from Eqs. (3) and (4), the normal component (D_n, B_n) of the electric displacement and magnetic induction is continuous.

Students encountering Maxwell's theory for the first time are frequently mistrustful of the derivation of the boundary conditions, feeling somehow that they have arisen more out of mathematical chicanery than from some deeper-lying physical necessity. That same sentiment, I suspect, may have underlain the unwarranted freedom with which electrodynamic boundary conditions were ignored—and hence wrong conclusions drawn—in the fundamental problem of reflection from a chiral medium to be discussed later in this chapter. However, the conditions imposed upon the fields at discontinuities do indeed have a physical significance: They are tantamount to ensuring the conservation of energy (or, more precisely, energy flow).

The formal expression for the balance of radiant energy, which follows naturally from Maxwell's equations without any additional ad hoc assumptions, is the differential equation[16]

$$\nabla \cdot \mathbf{S} + \frac{\partial u}{\partial t} = -\mathbf{J} \cdot \mathbf{E} \tag{5}$$

where the Poynting vector (in the Gaussian system of units)

$$\mathbf{S} = \frac{c}{4\pi} \mathbf{E} \times \mathbf{H} \tag{6}$$

is interpretable as the power flux (i.e., the energy transported per unit of time across a unit area normal to the light beam), and u is the energy stored in the

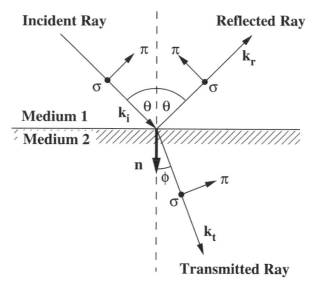

Figure 11.2. Geometric configuration of the electric fields of the incident, reflected, and transmitted beams. σ-polarized fields (black circles) are directed out of the page; π-polarized fields lie in the plane of incidence. Magnetic fields (not shown) are oriented so that each cross product $\mathbf{E} \times \mathbf{H}$ is parallel to the corresponding wave vector $\mathbf{k}$.

electric and magnetic fields per unit volume of space. The right-hand side of Eq. (5) expresses the rate of work done by the electric field $\mathbf{E}$ on charged particles comprising the current density $\mathbf{J}$. Thus, in the absence of free charges ($\mathbf{J} = 0$) and when the incident light beam is steady (u independent of time), the conservation of electromagnetic energy in two contiguous media is simply expressible by the equality of the flux of radiant power across each point of the boundary

$$(\mathbf{S}_1 - \mathbf{S}_2) \cdot \mathbf{n} = 0. \tag{7}$$

When the Poynting vector in each medium is expressed in terms of the fields of that medium according to Eq. (6), and the fields then related by the boundary conditions (1) and (2), the equality (7) is readily established. Thus, the Maxwellian boundary conditions are a mathematically integral and physically plausible part of Maxwell's theory. We shall see again shortly, in an entirely different context, the intricate way in which they are connected to the conservation of energy.

In deriving the Fresnel relations from Maxwell's equations for homogeneous dielectric media, one ordinarily applies only the boundary conditions (1) and (2); relations (3) and (4) are not explicitly needed because they do not furnish additional information. However, to obtain the appropriate wave equation for whose free-space solutions the boundary conditions are required, one must first know the so-called constitutive or material relations that connect, in Maxwell's terminology, "quantities" and "intensities." For the case at hand, in which each medium is completely characterized electromagnetically by a scalar dielectric constant ε and magnetic permeability μ, these relations are simply

$$\mathbf{D} = \varepsilon\mathbf{E}, \tag{8}$$

$$\mathbf{B} = \mu\mathbf{H}. \tag{9}$$

The associated phase velocity of light—that is, the speed with which the crest of a monochromatic wave advances through the medium (and which is not necessarily the speed at which an information-bearing signal can propagate)— is $v = c/\sqrt{\varepsilon\mu}$; correspondingly, the index of refraction of the medium predicted by Maxwell is

$$n = \frac{c}{v} = \sqrt{\varepsilon\mu}. \tag{10}$$

For most nonmagnetic media of optical interest, the permeability μ differs insignificantly from the value 1 and can be so designated. Moreover, to the extent that plane waves (with wave vector $\mathbf{k}$ and angular frequency ω) provide an adequate model of the incident light—and this is frequently the case (especially in textbooks)—Maxwell's equation for Faraday's law of induction furnishes the magnetic field in terms of the electric field in the following familiar way:

$$\mathbf{H} = \frac{\mathbf{k}}{k_0} \times \mathbf{E} = n\hat{\mathbf{k}} \times \mathbf{E} \tag{11}$$

where $k_0 \equiv \omega/c$ and the unit vector $\hat{\mathbf{k}}$ designates the direction of propagation. Equation (11) shows that (in the Gaussian system of units) the ratio of field amplitudes in a homogeneous dielectric medium is $|\mathbf{H}|/|\mathbf{E}| = n$, the refractive index.

In the standard problem of reflection at the interface between two achiral dielectric media, it is customary to consider separately the cases in which the electric vector of the incident light is either parallel (π-polarization) or perpendicular (σ-polarization, from the German *senkrecht*) to the plane of incidence defined by the incident wave vector and the normal to the reflecting surface. These two orthogonal polarizations are fundamental since, under the stated conditions, the processes of reflection and refraction do not mix them. An incident light wave of arbitrary polarization can be decomposed uniquely into σ- and π-components whose reflected (and transmitted) waves can then be superposed.

Let us designate by r and t, respectively, the amplitudes of the reflected and transmitted waves engendered by an incident wave of unit amplitude. As in chapter 2, θ is chosen to be the angle of incidence (in medium 1), and ϕ the angle of transmission (in medium 2); the relative refractive index of the two media is $n_{21} = n_2/n_1$. Then, as deduced from the geometry of figure 11.2, the boundary conditions for the two basic polarization states can be summarized as follows:

	σ-polarization	π-polarization
continuity of E_t	$1 + r_\sigma = t_o$	$(1 - r_\pi) \cos \theta = t_\pi \cos \phi$
continuity of H_t	$(1 - r_\sigma) \cos \theta = n_{21} t_\sigma \cos \phi$	$1 + r_\pi = n_{21} t_\pi.$

$$(12)$$

(It is readily demonstrable that the continuity of the normal components $H_n^{(\sigma)}$ and $D_n^{(\pi)}$ leads to the same conditions as the continuity of the tangential components $E_t^{(\sigma)}$ and $H_t^{(\pi)}$, respectively.) Now, although there is no motivation for doing so at present, let us multiply together the two continuity conditions for each polarization to obtain the relation

$$(1 - r^2) \cos \theta = n_{21} t^2 \cos \phi \tag{13}$$

in which polarization labels have been suppressed since Eq. (13) is the same for both σ and π waves. I will return to the significance of this equation shortly.

The remaining steps of the solution are trivial; one has merely to solve the two separate pairs of coupled linear algebraic equations to arrive at Fresnel's renowned relations

$$r_\sigma = \frac{\cos \theta - n_{21} \cos \phi}{\cos \theta + n_{21} \cos \phi} = -\frac{\sin(\theta - \phi)}{\sin(\theta + \phi)}, \tag{14a}$$

$$r_\pi = \frac{n_{21} \cos \theta - \cos \phi}{n_{21} \cos \theta + \cos \phi} = \frac{\tan(\theta - \phi)}{\tan(\theta + \phi)}. \tag{14b}$$

To evaluate the amplitudes as explicit functions of the incident angle one invokes Snel's law

$$\sin \theta = n_{21} \sin \phi. \tag{15}$$

Even for simple homogeneous materials, Eqs. (14a, b) embody a rich content of reflective phenomena depending upon the dielectric properties of the two media and the angle of light incidence. For full details, the reader can consult any appropriate optics text. Suffice it to say that for the conditions envisioned here of a nonabsorbing incident medium (like air) of lower refractive index than the transparent reflecting medium (like glass), the amplitudes are real-valued over the full angular range $0 \leq \theta \leq \pi/2$; the reflectances $R \equiv |r|^2 = r^2$ are generally low at normal incidence $[R(0) = (\frac{n_{21}-1}{n_{21}+1})^2]$ approaching 100% at grazing incidence $[R(\pi/2) = 1]$. For σ-polarized light, the increase is monotonic; by contrast, for π-polarized light, R_π first decreases to 0 at the Brewster (or polarizing) angle θ_B given by $\tan \theta_B = n_{21}$.

I often wondered, when I began my studies of the electromagnetic theory of light, how it was that Fresnel arrived at relations (14a, b) without benefit of

Maxwell's theory. He derived these expressions sometime before 1823—a half century, more or less, in advance of Lorentz—on the basis of a model of the ether as an elastic solid.[17] But how? And how could it be that a starting point founded upon principles so very different from those of electrodynamics could lead to the same result? And if the results are indeed the same, does this not contradict the uniqueness of Maxwellian electrodynamics to predict correctly and self-consistently phenomena that no other competing theory could satisfactorily account for?

Some years later, in the course of a long and productive collaboration in France, I held at various times an invited chair at the PC. This is a school with a rich scientific history. Founded in 1882 with the mandate to train students simultaneously in physics and chemistry, the PC could count among its distinguished faculty Pierre Curie (who discovered there the piezoelectric effect) and Paul Langevin (who investigated magnetism, Brownian motion, and special relativity), and among its students Frédéric Joliot (eventual Nobelist together with his wife, Irène Curie, for discovery of artifical radioactivity) and Georges Claude (noted for his cryogenic work in liquefying gases). In recent years, Pierre de Gennes, recipient of the Nobel Prize for theoretical studies of liquid crystals, has been the director of the school.

Such names had special significance for me, for, during my lengthy stays in Paris, my family and I lived on the school premises in apartments designated the "Apartement Joliot" or "Apartement Georges Claude." (The incomparable worth of an apartment associated with an academic chair—a practice virtually unheard of in the United States—could be fully appreciated only by someone who has had to search for accommodations in Paris during the peak tourist season.) My lodgings were merely a short walk through several narrow corridors to the Laboratoire d'Optique Physique, where I shared the office and laboratory of my colleague J. Badoz.

From the window of my office I could look out at the site (today a parking lot) where Pierre and Marie Curie once processed tons of uranium ores for precious bits of radium. This office, however, held greater treasure than vanished glory—for in the large, glass-doored cabinet behind my worktable, dustily arrayed on the topmost shelf, seemingly untouched for decades, were the complete works of Fresnel. Fresnel was a "polytechnicien" (graduate of the Ecole Polytechnique) who died long before the portals of the ESPCI were opened. All the same, a large portrait of him, taped to the glass doors of the cabinet, looked down wistfully over my shoulder as I worked—and it was during these years that I came to realize to what extent the research I had initiated for entirely different reasons was, in some ways, a synthesis and completion of his own. I will elaborate on this point later.

With respect to Fresnel's thinking, however, I was able at last to satisfy my curiosity, for there on page 767 began the "Memoire on the Law of Modifications That Reflection Impresses upon Polarized Light."[18] Despite the musty smell of antiquity in the close-printed pages that had confined the thoughts of the young Fresnel for well over a century, the article made most edifying reading.

The ether, according to Fresnel, consisted of molecules which filled the entire universe and whose vibrations constituted the passage of light. The elasticity of the ether was everywhere the same, but its density ρ varied with the material it permeated. In the elastic theory of solids the speed of a wave in a medium is proportional to the square root of the ratio of elasticity to density. Thus, in the passage of light from one substance to another, the ratio of the wave speeds in each medium is

$$\frac{c_1}{c_2} = \sqrt{\frac{\rho_2}{\rho_1}} = \frac{n_2}{n_1} = \frac{\sin\theta}{\sin\phi}. \qquad (16)$$

The second equality in Eq. (16) reflects in modern notation the definition of the index of refraction, and the third equality is simply Snel's law.

In Fresnel's model the ether did not sustain slippage, and therefore the velocity (v) of "particle" vibrations parallel to the interface between two media had to be continuous. Fresnel took v as a measure of the amplitude of light—what we today recognize as the counterpart to Maxwell's electric field E. Thus, in Fresnel's theory, the continuity of the tangential component v_t is equivalent to the Maxwellian boundary condition (1) on E_t.

Next, the energy (termed "force vive" or "live force") of the vibrating ether had to be conserved as the wave passed from one medium to another. If the incident wave shook a mass m_1 of ether with speed v_i, and this energy subsequently went into a reflected wave shaking the same mass m_1 with speed v_r and a transmitted wave shaking mass m_2 with speed v_t, the conservation of mechanical energy required

$$m_1(v_i^2 - v_r^2) = m_2 v_t^2 \qquad (17a)$$

or

$$1 - r^2 = \frac{m_2}{m_1} t^2 \qquad (17b)$$

in which the reflection and transmission amplitudes are defined by the ratios $r = v_r/v_i$ and $t = v_t/v_i$. The masses appearing in Eq. (17b) are each the product of the appropriate density and volume of shaken ether, the volumes V_1, V_2 in each medium being proportional to the corresponding shaded areas in figure 11.3: $V_1 \propto \overline{AB}^2 \sin\theta\cos\theta$ and $V_2 \propto \overline{AB}^2 \sin\phi\cos\phi$. It therefore follows from Eqs. (15) and (16) that

$$\frac{m_2}{m_1} = \frac{\rho_2}{\rho_1} \cdot \frac{V_2}{V_1} = \left(\frac{\sin\theta}{\sin\phi}\right)^2 \cdot \frac{\sin\phi\cos\phi}{\sin\theta\cos\theta} = n_{21}\frac{\cos\phi}{\cos\theta}. \qquad (18)$$

Upon substitution of Eq. (18) into Eq. (17b) one obtains a very interesting result: precisely Eq. (13), the product of the Maxwellian boundary conditions.

It is now clear why Fresnel was able to deduce correct expressions for the amplitudes of reflection. Although his mechanical model is, strictly speaking,

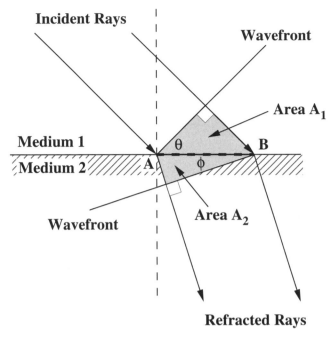

Figure 11.3. Geometry of Fresnel's elastic solid model of reflection. A part of the incident wavefront in medium 1 propagates into medium 2 to become the refracted wavefront. The shaded areas A_1 and A_2 are proportional to the corresponding volumes of ether set into vibration by the passing wave.

incompatible with an electromagnetic theory of light, his starting assumptions—the continuity of the tangential component of **v** and the conservation of kinetic energy—are equivalent to the Maxwellian boundary conditions (1) and (2). Fresnel's theory, however, is not a self-consistent one, as Maxwell, himself, emphasized in a review of the ether in the *Enclopaedia Britannica*.[19]

Of the three difficulties Maxwell listed, perhaps the most devastating is that physically real elastic solids must admit of vibrations *normal* to the interface. By ignoring this degree of freedom at the outset and assuming light to be a transverse vibration, Fresnel had inconsistently applied the requisite boundary conditions. Like many a student on an examination, Fresnel got the right answer by the wrong calculation. "The only way," wrote Maxwell, "of accounting for the fact that the optical phenomena which would arise from these [normally polarized] waves do not take place is to assume that the aether is incompressible." Yet, if the elasticity of the ether is null, then so must be the velocity of wave propagation. Only Maxwell's theory leads directly and consistently to transverse waves of light.

Apart from resolving a point of history that perplexed me, I find Fresnel's analysis noteworthy for the fact that the Maxwellian boundary conditions represented in Eq. (13) again emerge from consideration of energy conservation (Eq. [17b]), albeit in this case mechanical kinetic energy. Clearly Eq. (13) must also be deducible directly from electromagnetic energy conservation—and, indeed, with a little effort one can show that it is identical to relation (7) applied to the configuration of fields in figure 11.2. Here, however, one must be a bit careful, for the Poynting vector in the incident medium

$$\mathbf{S} = \frac{c}{4\pi}(\mathbf{E}_i + \mathbf{E}_r) \times (\mathbf{H}_i + \mathbf{H}_r) = \mathbf{S}_i + \mathbf{S}_r + \mathbf{S}_{ir} + \mathbf{S}_{ri} \qquad (19)$$

is not simply the sum of the Poynting vectors of the incident ($\mathbf{S}_i$) and reflected ($\mathbf{S}_r$) waves as defined by Eq. (6), but contains terms, $\mathbf{S}_{ir} = \frac{c}{4\pi}\mathbf{E}_i \times \mathbf{H}_r$ and $\mathbf{S}_{ri} = \frac{c}{4\pi}\mathbf{E}_r \times \mathbf{H}_i$, that mix incident and reflected fields. The geometry of reflection is such, however, that the sum of these terms, projected onto the normal to the boundary, vanishes—and one recovers from Eq. (7) the seemingly (but not actually) transparent result

$$(\mathbf{S}_i + \mathbf{S}_r) \cdot \mathbf{n} = \mathbf{S}_t \cdot \mathbf{n}, \qquad (20)$$

which leads precisely to the product of Maxwellian boundary conditions, Eq. (13).

As a final point, it is instructive to note that, just as Snel's law of refraction was deducible from a quantum particle model (demonstrated in chapter 2), so too is the energy-conservation and boundary-condition relation (13). The amplitudes of reflection r and transmission t no longer represent physical vibrations of a mechanical medium or oscillations of an electromagnetic field, rather, they are probability amplitudes whose absolute magnitude squared gives the probability of an incident photon appearing in the reflected or transmitted beam. In keeping with the foregoing interpretation, let us set $r^2 = N_r/N_i$ and $t^2 = N_t/N_i$ where N_i, N_r, and N_t are the numbers of photons per second in the incident, reflected, and transmitted beams.

Now, the energy $\mathcal{E}$ of a single photon can be expressed in terms of the refractive index n of the medium by the chain of steps

$$\mathcal{E} = pc = \frac{hc}{\lambda} = \frac{nhc}{\lambda_0} \qquad (21)$$

where the first equality comes from application of special relativity to massless particles of momentum p, the second equality from substitution of the de Broglie relation, and the third equality from the relation between wavelength in vacuum (λ_0) and in matter (λ). For a stream of N photons per second impinging upon a surface at angle α to the vertical, the flux of radiant energy transported normal to the surface is therefore $(Nnhc/\lambda_0)\cos\alpha$. Neglecting factors common to all terms, the balance of energy of incident, reflected, and transmitted fluxes across the interface takes the form

$$n_1 N_i \cos\theta - n_1 N_r \cos\theta = n_2 N_t \cos\phi \qquad (22)$$

which, with the quantum definitions of r^2 and t^2, is again exactly Eq. (13).

It should be stressed that the preceding argument does not lead to a quantum derivation of the Fresnel amplitudes, for, examined closely, there are interpretative difficulties. Unlike the electron, which obeys the Schrödinger or Dirac equation and for which a physically meaningful conserved probability density is derivable, the photon obeys a massless Klein-Gordon equation whose corresponding probability density can be negative. Moreover, with r and t defined as scalar probability amplitudes, it is difficult to justify the separate boundary conditions, Eqs. (1) and (2). For a consistent quantum treatment of reflection and refraction, one must employ, QED, the quantized generalization of Maxwell's theory.

11.3 CONSTITUTIVE RELATIONS

The determination of the specular reflection and transmission amplitudes (the generalized Fresnel relations) for a chiral medium requires the solution of Maxwell's equations augmented by the appropriate constitutive or material relations. An ordinary dielectric medium is characterized by a scalar dielectric constant and magnetic permeability. By contrast, a homogeneous chiral medium requires a third material constant, the chiral or gyrotropic[20] parameter.

Optical activity arises from the spatial dispersion of an electromagnetic wave over a noncentrosymmetric molecule (such as a carbon atom with four different substituents attached) or molecular grouping (as in the helical arrangement of achiral SiO_2 molecules in the unit cell of a quartz crystal). To avoid cumbersome locutions in the following discussion, let us regard the elementary chiral constituent of a medium as a molecule, rather than a molecular aggregate. There is, of course, an important physical distinction: the optical activity of chiral molecules persist in all phases (gas, liquid, and solid), whereas it vanishes upon melting of a solid whose handedness derives from the chiral ordering of essentially achiral molecules.

From the perspective of quantum physics, linearly polarized light incident on a nonabsorbing chiral molecule induces virtual electric dipole (E1) and magnetic dipole (M1) transitions between the ground and excited states. These states can undergo both E1 and M1 transitions because parity is not a sharp quantum number in a chiral system. Expressed in terms of the electric and magnetic fields of the light wave, the induced electric and magnetic dipole moments of the molecule take the form[21]

$$\mu_E = \alpha E - \frac{\beta}{c}\frac{\partial H}{\partial t},$$

$$\mu_M = \gamma H + \frac{\beta}{c}\frac{\partial E}{\partial t} \tag{23}$$

in which α is the familiar polarizability

$$\alpha = \frac{2}{3h}\sum_b \frac{\nu_{ba}|\langle a|\mathbf{d}|b\rangle|^2}{\nu_{ba}^2 - \nu^2}, \tag{24}$$

and γ is the structurally similar (but ordinarily less important) magnetizability

$$\gamma = \frac{2}{3h}\sum_b \frac{\nu_{ba}|\langle a|\mathbf{m}|b\rangle|^2}{\nu_{ba}^2 - \nu^2}. \tag{25}$$

In the preceding two equations, $\mathbf{d}$ and $\mathbf{m}$ are, respectively, the electric and magnetic dipole operators, h is Planck's constant, ν is the frequency of the light wave, and $\nu_{ba} = (E_b - E_a)/h$ is the Bohr frequency corresponding to the difference in energies of states b and a. The additional microscopic parameter for a chiral medium—which, for want of a better term, I shall call the "dissymmetry" in remembrance of Pasteur—expresses directly the parity nonconserving nature of the molecular transitions

$$\beta = \frac{2}{3\pi h}\sum_b \frac{\text{Im}\{\langle a|\mathbf{d}|b\rangle \cdot \langle b|\mathbf{m}|a\rangle\}}{\nu_{ba}^2 - \nu^2}. \tag{26}$$

From a classical perspective, the relative phase of the light-induced electric and magnetic moments of one enantiomer is opposite to that of its mirror image. As a consequence, incident and forward-scattered light waves superpose to produce a transmitted wave with linear polarization that has been rotated by the same amount in either a cw or ccw sense depending on the handedness of the molecules. Analogous reasoning underlies other manifestations of optical activity, such as circular dichroism (the differential absorption of CP light).

The macroscale constitutive relations, to which the quantum Eq. (23) gives rise, therefore contain light-induced polarization $\mathbf{P}$ and magnetization $\mathbf{M}$ terms— or, equivalently, terms that link the displacement $\mathbf{D}$ (as well as the magnetic induction $\mathbf{B}$) to both the electric field $\mathbf{E}$ and magnetic field $\mathbf{H}$ as follows:

$$\mathbf{D} = \varepsilon\mathbf{E} - g\frac{\partial\mathbf{H}}{\partial t},$$

$$\mathbf{B} = \mu\mathbf{H} + g\frac{\partial\mathbf{E}}{\partial t}, \tag{27}$$

in which $g = 4\pi N\beta/c$ is the "gyrotropic" parameter and N is the number of molecules per volume of sample. In the case of monochromatic plane waves,

Eq. (27) can be reformulated by use of Maxwell's equations into linear relations between **D** and **E** and between **B** and **H**:

$$\mathbf{D} = \varepsilon \left[(1 - f^2)\mathbf{E} + i\frac{f}{nk_0}(\mathbf{k} \times \mathbf{E}) \right],$$

$$\mathbf{B} = \mu \left[(1 - f^2)\mathbf{H} + i\frac{f}{nk_0}(\mathbf{k} \times \mathbf{H}) \right]. \tag{28}$$

In the above expressions the parameter $f = \omega g/n$, with $n = \sqrt{\varepsilon\mu}$, is a convenient measure of the intrinsic chirality of the medium. Since the tensorial relationship between **D** and **E** is identical to that between **B** and **H**, I will refer to the set of material relations in Eq. (28) as the "symmetric set."

It may seem surprising—although it ought not to—that **B** is an anisotropic function of **H** even though the medium is intrinsically nonmagnetic. The neglect of this fact has been the fatal flaw in a number of theoretical studies of optical activity in which the anisotropic term proportional to the chiral parameter f was dropped from the magnetic material relation. I will refer to the resulting truncated material relations as the "asymmetric set." Several justifications have been offered for the adoption in optics of the asymmetric set.

One argument was based on symmetry. Maxwell's equations are rich in symmetries. Besides gauge invariance, which I have already discussed, the electrodynamic equations of an intrinsically nonmagnetic material are also invariant under a linear transformation that intermixes **P** and **M**, thereby permitting one to dispense with magnetization entirely and incorporate in the new polarization all the effects of bound charges and currents. Consider, for example, the following transformation of **P** and **M**:

$$\mathbf{P}' = \mathbf{P} - \xi \operatorname{curl} \mathbf{M},$$

$$\mathbf{M}' = \mathbf{M} + \frac{\xi}{c}\frac{\partial \mathbf{M}}{\partial t} \tag{29}$$

where ξ is the transformation parameter. These fields lead to new displacement and magnetic fields

$$\mathbf{D}' = \mathbf{E} + 4\pi\mathbf{P}' = \mathbf{D} - 4\pi\xi \operatorname{curl} \mathbf{M},$$

$$\mathbf{H}' = \mathbf{B} - 4\pi\mathbf{M}' = \mathbf{H} - \frac{4\pi\xi}{c}\frac{\partial \mathbf{M}}{\partial t} \tag{30}$$

that satisfy Maxwell's equations if the original fields **D** and **H** do. For plane waves of frequency ω and the choice $\xi = c/i\omega$, **M**' vanishes, thereby resulting in a scalar constitutive relation $\mathbf{H}' = \mathbf{B}$, and a tensorial relation between **D**' and **E**. The optical properties of the medium are then effectively contained in the dielectric tensor alone, the permeability simply being unity. Thus, some would say that reducing the magnetic material relation from a tensor to a scalar relation is not even an approximation, but the result of an exact symmetry transformation.

A second argument was simply that this theoretical procedure seemed to account adequately for the observed properties of wave propagation through isotropic chiral media; that is, both the symmetric and asymmetric sets of relations led to chiral refractive indices of the form $n_\pm = n(1 \pm f)$. In his classic treatment of optical activity, Max Born noted that neglect of the magnetic anisotropy merely led to a rescaling of the chiral parameter f.[22] This equivalent description of light transmission encouraged a belief that the two sets of constitutive relations were in all ways equivalent.

Born's note is technically correct, but the widely assumed implication that a physically unobservable rescaling in the refractive index is the only consequence of simplifying the constitutive relations is not correct. My analyses of reflection from isotropic (as well as birefringent) chiral media[23] have shown that the symmetric and asymmetric sets of constitutive relations lead to significantly different physical outcomes and can be readily distinguished, for example, by the unequal reflection of left and right circularly polarized light as a function of incident angle or wavelength. Moreover, under conditions of total reflection for one or both circular polarizations the asymmetric set, but not the symmetric set, results in violations of energy conservation (Poynting's theorem for chiral media).

Why does the presence or absence of a small light-induced magnetization—the key feature distinguishing the symmetric and asymmetic constitutive relations—have no significant effect on optical rotation or circular dichroism, but a major impact on the Fresnel reflection amplitudes? The critical difference is that light reflection, but not transmission, requires implementation of electrodynamic boundary conditions. The fallacy of the foregoing symmetry argument embodied in Eqs. (29) and (30), which purportedly demonstrate the equivalence between the original fields $\mathbf{D}$, $\mathbf{H}$ and the transformed fields $\mathbf{D'}$, $\mathbf{H'}$, is that the latter do not generally satisfy the Maxwellian boundary conditions. For the particular transformation given, the normal component of $\mathbf{D'}$ and the tangential component of $\mathbf{H'}$ are not continuous across the boundary unless the original magnetization $\mathbf{M}$ is normal to the boundary, a generally invalid restriction.

Given the importance of chiral media, the question of their appropriate electrodynamic boundary conditions has far-reaching implications. In the past, implicit or explicit adoption of the asymmetric set of constitutive relations has at times been accompanied by proposals for supplementing or transforming the standard boundary conditions and changing the basic expressions for power flux and energy conservation. None of this is necessary or desirable. The standard electrodynamic boundary conditions derive in a general and model-independent way from the macroscopic Maxwell equations; these conditions, together with the equations of motion (Maxwell's equations) and material relations, should suffice for obtaining internally consistent, complete solutions to well-posed electrodynamic problems. The need for supplementary boundary conditions may arise only when a problem encompasses a broader theoretical framework than that of classical electrodynamics alone.

Maxwell's equations do not, of course, prescribe a general form for the constitutive relations; these relations depend on the particular material in question and are not necessarily unique. But here, again, there are general symmetry considerations that may prove useful, if not decisive. Maxwell's equations are invariant not only under the linear transformation of **P** and **M**, but more generally under what is termed a "duality transformation" of fields expressed by

$$\begin{pmatrix} \mathbf{F}_1 \\ \mathbf{F}_2 \end{pmatrix} = \begin{pmatrix} \cos \tau & \sin \tau \\ -\sin \tau & \cos \tau \end{pmatrix} \begin{pmatrix} \mathbf{F}'_1 \\ \mathbf{F}'_2 \end{pmatrix} \tag{31}$$

where τ is the transformation parameter and the paired fields $(\mathbf{F}_1, \mathbf{F}_2)$ are $(\mathbf{E}, \mathbf{H})$, $(\mathbf{B}, \mathbf{D})$, and $(\mathbf{P}, \mathbf{M})$. For $\tau = \pi/2$, for example, one recovers the familiar transformation $\mathbf{E} \rightarrow \mathbf{H}$, $\mathbf{H} \rightarrow -\mathbf{E}$ by which a second solution to the wave equation is generated from a known solution. This duality symmetry has a deep geometric significance in the relativistic formulation of electrodynamics and implications for the relationship between electric and magnetic charge.[24] The symmetric set of constitutive relations represented by Eqs. (27) or (28) is invariant under a duality transformation, but the asymmetric set is not.

Conceivably, the imposition of invariance under a duality transformation may be a needed criterion for limiting a priori the fundamental forms of physically acceptable constitutive relations. Ultimately, of course, the adequacy of any phenomenological description of optical media must be determined by experiment. It was a surprising realization to me, however, at the time when the foregoing considerations first entered my mind, that the chiral Fresnel amplitudes which I derived had never been tested experimentally—even though natural optical activity was first observed more than a century and a half earlier.

11.4 HELICITY WAVES AND CHIRAL REFLECTION

The constitutive relations, together with Maxwell's equations, determine the polarization states and refractive indices of the characteristic waves of a medium; imposition of appropriate boundary conditions on these waves then leads to the reflection and transmission amplitudes at the interface of two contiguous media.

An essential point of the previous section is that a part of the polarization **P** of an isotropic chiral medium is generated by $\partial \mathbf{H}/\partial t$, and, reciprocally, a part of the magnetization **M** is generated by $\partial \mathbf{E}/\partial t$. Substituting the (plane-wave) material relations (28) into Maxwell's equations yields relations between **H** and **E** that differ from Eq. (11) for ordinary dielectric media by terms proportional to the chiral parameter

$$\mathbf{H} = \frac{\mathbf{k} \times \mathbf{E}}{\mu k_0} + i \frac{nf}{\mu} \mathbf{E},$$

$$\mathbf{E} = \frac{-\mathbf{k} \times \mathbf{H}}{\varepsilon k_0} - i \frac{nf}{\varepsilon} \mathbf{H}. \tag{32}$$

To maintain full symmetry throughout the present analysis, an arbitrary permeability μ has again been assumed. The wave equation which follows from Eq. (32),

$$\left[\left(\frac{k}{nk_0} \right)^2 - (1 - f^2) \right] \mathbf{E} - 2if \left(\frac{k}{nk_0} \right) \times \mathbf{E} = 0, \tag{33}$$

has helicity eigenwave solutions, designated $(+, -)$, of the form

$$\mathbf{E}_\pm(\mathbf{r}, t) = E \left[\frac{\mp i\beta_\pm}{n_\pm}, 1, \frac{\pm i\alpha_\pm}{n_\pm} \right] \exp[ik_0(\alpha_\pm x + \beta_\pm z)] \exp(-i\omega t),$$

$$\mathbf{H}_\pm(\mathbf{r}, t) = \mp i \left(\frac{n_\pm}{\mu_\pm} \right) \mathbf{E}_\pm(\mathbf{r}, t) \tag{34}$$

in which the wave vector

$$\mathbf{k}_\pm = k_0(\alpha_\pm, 0, \beta_\pm) \tag{35}$$

is here taken without any loss of generality to lie in the x-z plane of a Cartesian coordinate system (as shown in figure 11.4).

The helicity waves in Eq. (34) are characterized by chiral refractive indices

$$n_\pm = n(1 \pm f) \tag{36a}$$

related in the expected way

$$n_\pm = \sqrt{\varepsilon_\pm \mu_\pm} \tag{36b}$$

to the chiral material constants

$$\varepsilon_\pm = \varepsilon(1 \pm f),$$
$$\mu_\pm = \mu(1 \pm f) \tag{37}$$

associated with the resulting displacement and induction fields

$$\mathbf{D}_\pm = \varepsilon_\pm \mathbf{E}_\pm,$$
$$\mathbf{B}_\pm = \mu_\pm \mathbf{H}_\pm. \tag{38}$$

It should be noted that Eq. (34) is valid for both freely propagating traveling waves as well as surface waves damped in a direction normal to the interface,

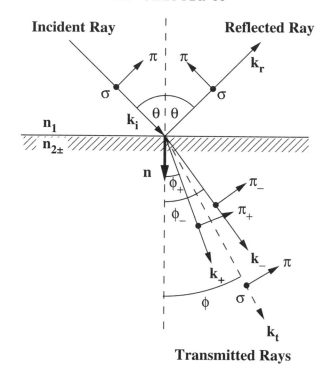

Figure 11.4. Geometry of chiral reflection. An incident linearly polarized wave in a nonchiral medium gives rise to two waves of opposite helicity and different transmission angles ($\phi_{\pm}$) in the optically active medium. The dashed wave vector $\mathbf{k_t}$ at angle ϕ indicates the single transmitted wave in the absence of chirality.

such as those arising in reflection beyond a critical angle. In the first case the wave vector is real and in the chosen coordinate system takes the form

$$\mathbf{k}_{\pm} = n_{\pm}k_0(\sin\phi_{\pm}, 0, \cos\phi_{\pm}) \tag{39}$$

where the $(+, -)$ solutions are LCP and RCP waves, respectively. In the second case, the component of $\mathbf{k}$ normal to the interface (the $z = 0$ plane) is imaginary:

$$\beta_{\pm} = i\sqrt{\alpha_{\pm}^2 - n_{\pm}^2}. \tag{40}$$

The $(+, -)$ solutions are then elliptically polarized and inhomogeneous—that is, planes of constant phase do not coincide with planes of constant amplitude. Equation (40) is equivalent to the trigonometric identity for the cosine when the sine is greater than unity, a substitution routinely made in the treatment of total reflection from optically inactive media.

Consider a plane wave of unit amplitude originating in a nonchiral medium (of refractive index n_1 and permeability μ_1), incident at an angle θ upon the surface of an optically active medium (mean index n_2, chiral parameter f, and permeability μ_2), producing reflected and transmitted waves as shown in figure 11.4. For the present, both media are transparent, homogeneous, and isotropic. The optically active medium is circularly birefringent with chiral indices $n_{2\pm}$ given by Eq. (36a); upon entering the medium, the light decomposes into helicity waves refracted at angles $\phi_\pm$ in accordance with Snel's law

$$n_1 \sin\theta = \alpha_\pm = n_{2\pm} \sin\phi_\pm. \tag{41}$$

The second equality in Eq. (41) applies, strictly speaking, to freely propagating waves but can apply as well to damped waves, although it is then understood that $\sin\phi_\pm$ are simply complex numbers with $\phi_\pm$ no longer interpretable as angles.

In contrast to reflection from an ordinary dielectric medium, which preserves incident states of σ- or π-polarization, the wave reflected from a chiral medium contains in general (except at normal and grazing incidence) both these components. Thus, there are four possibly complex-valued reflection amplitudes r_{ij}:

$$\begin{aligned} \mathbf{r}_1 &= r_{11}\hat{\mathbf{e}}_1 + r_{21}\hat{\mathbf{e}}_2, \\ \mathbf{r}_2 &= r_{12}\hat{\mathbf{e}}_1 + r_{22}\hat{\mathbf{e}}_2 \end{aligned} \tag{42}$$

where the subscript on the reflected electric vector $\mathbf{r}_i$ signifies incident σ- ($i = 1$) or π- ($i = 2$) polarizations. An incident wave of polarization $\hat{\mathbf{e}}_i$ gives rise to a reflected wave of polarization $\hat{\mathbf{e}}_j$ with amplitude r_{ji}. Since the intrinsic chiral parameter (f) is small for most naturally optically active media, one would expect the amplitudes r_{ii} to attain much larger values than the amplitudes r_{ij} ($i \neq j$) which vanish for nonchiral media.

Within the chiral medium only helicity waves with polarizations $\hat{\mathbf{e}}_\pm$ can propagate; the total transmitted field is then expressible as the superposition

$$\mathbf{t}_i = t_{i+}\hat{\mathbf{e}}_+ + t_{i-}\hat{\mathbf{e}}_- \tag{43}$$

where the transmission amplitudes $t_{i\alpha}$ ($i = 1, 2, \alpha = +, -$) represent a transformation between linear and circular polarization states.

The exact generalized Fresnel amplitudes that one obtains upon implementing the standard boundary conditions—continuity of the tangential components of $\mathbf{E}$ and $\mathbf{H}$ across the interface—are fairly complicated expressions which are collected in the appendix to this chapter. Of more interest to us here is to examine their general form and physical content, particularly in the special (but widely applicable) case of an intrinsically small chiral parameter in the optical domain. In that case r_{11} and r_{22} are virtually identical to the familiar Fresnel amplitudes r_σ and r_π of Eqs. (14a, b). The chiral components r_{12} and r_{21} are related generally (for the homogeneous medium under consideration) by $r_{21} = -r_{12} = ib$, where the amplitude b (real-valued for reflection without critical angles) is proportional to $\beta_+ - \beta_-$, which varies linearly with f for $f \ll 1$.

Having the reflection amplitudes for incident σ- and π-polarized waves, one can construct the reflected wave for any incident polarization and determine from this the fractional reflected intensity, or reflectance. In this regard, it is useful to examine the difference with which a chiral medium reflects incident LCP and RCP waves. Expressed in terms of the above amplitudes, the reflectances of incident CP light are

$$R_\pm = \tfrac{1}{2}[|r_{11} \mp ir_{12}|^2 + |r_{22} \pm ir_{21}|^2]. \tag{44}$$

The difference in LCP and RCP reflectances, normalized by the sum, constitutes the differential circular reflection or DCR

$$\delta_C \equiv \frac{R_+ - R_-}{R_+ + R_-} \sim \frac{2(r_\sigma + r_\pi)b}{r_\sigma^2 + r_\pi^2} \tag{45}$$

which is equivalent to the function δ_C defined earlier in Eq. (6.39) for light of any provenance, not just reflection. Although the weak chiral component b (and therefore the chiral parameter f) enters R_+ and R_- quadratically, the DCR, as shown by Eq. (45), is linearly proportional to b and consequently much more sensitive to the chiral structure of the medium.

At normal and grazing incidences the angle of transmission is the same for both forms of circularly polarized light, in which case the amplitude b vanishes. Thus, the DCR vanishes at normal incidence, increases with incident angle to a maximum value roughly in the vicinity of Brewster's angle (as demonstrated in the papers listed in note 23), and then decreases to zero at grazing incidence. Figure 11.5 shows the exact variation as a function of incident angle, calculated without approximation from the amplitudes in the appendix. For purposes of comparison, the figure also shows the corresponding prediction derived from the asymmetric constitutive relations. The result is maximum at normal incidence and remains nearly constant over a wide angular range before dropping off monotonically to zero at grazing incidence. It is clear that the two phenomenological descriptions of a chiral medium can, at least in principle, be distinguished by experiment.

Since the asymmetric relations are now known to provide an incorrect description of a truly chiral medium, there is no point in recording here the formulas for the associated reflection and transmission amplitudes; these can also be found in note 23. In one respect worth commenting upon, however, it would seem that the invalid amplitudes lead to a more "sensible" result. Why should a medium which has an intrinsic chiral structure reflect incident left and right circularly polarized light to the same extent at normal incidence?

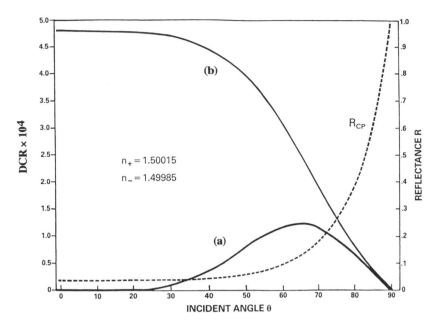

Figure 11.5. Differential reflection of LCP and RCP waves as predicted from the (a) symmetric and (b) asymmetric constitutive relations of a chiral medium. The dashed line shows the reflection curve for CP light. Individual LCP and RCP curves are indistinguishable at this scale.

A heuristic explanation of this puzzling feature may be sought in the model of reflection justified by the so-called Ewald-Oseen extinction theorem.[25] According to this theorem, the incident light beam does not interact with the reflecting medium at the surface only, but propagates into the medium and is extinguished by the molecular dipoles, which are induced to radiate secondary wavelets that superpose coherently to form the reflected wave. At normal incidence this interaction is equivalent to a penetration of the wave by some characteristic depth followed by reflection as from a mirror. Reversal of wave helicity upon reflection results in an opposite chiral effect on the outgoing wave—as in the case of optical rotation of a beam that has made an even number of passes through an isotropic chiral medium. The result is that the DCR vanishes. At larger angles of incidence the incident and reflected beams no longer overlap, and such perfect cancellation of chiral asymmetry does not occur.

The foregoing idea suggests the possibility of enhancing the DCR by multiple reflections from a chiral medium at a sufficiently large angle of incidence— as illustrated in figure 11.6. Detailed study of this question has shown that, although this does not occur for ordinary reflection from a transparent medium,

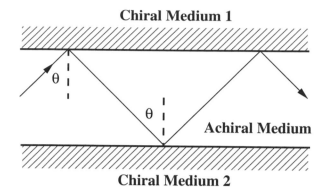

Figure 11.6. Multiple reflection of light between two chiral media separated by an achiral dielectric can lead to significant enhancement of chiral asymmetry.

a significant enhancement can occur in the case of total reflection in the vicinity of critical angle, particularly at frequencies that fall within the absorption band of the sample.[26] In that case a DCR several orders of magnitude greater than in the case of single-pass ordinary reflection should be possible. To a good approximation, the theoretical signal should increase linearly with the number of reflections (up to a certain maximum number, after which growth occurs more slowly), and reciprocally with the "detuning" of the relative refractive index of the two media from unity.

Although discussion to this point has focused on transparent media for which the material parameters ε, μ, g, n, f are all real, no such restriction need be made. The constitutive relations take the same form for nontransparent media but with complex-valued parameters

$$\tilde{\varepsilon} = \varepsilon_r + i\varepsilon_i, \qquad \tilde{n} = n + i\kappa = \sqrt{\tilde{\varepsilon}\mu},$$

$$\tilde{g} = g_r + ig_i, \qquad \tilde{f} = f_r + if_i = \frac{\omega\tilde{g}}{\tilde{n}} \qquad (46)$$

(with μ still a real parameter for nonmagnetic optical media). That the chiral parameters $(\tilde{g}, \tilde{f})$ as well as the mean dielectric constant and refractive index are complex in this case is made plausible by examining the dissymmetry parameter β, Eq. (26), to which they are proportional. Here, β contains sums of quantum mechanical matrix elements divided by factors of the form $\nu_{ab}^2 - \nu^2$. Excited molecular states are ordinarily unstable states; the energy eigenvalues are then complex-valued, the imaginary part representing the level width (reciprocal of the lifetime). Thus, the chiral parameters become complex-valued because the terms ν_{ab} are complex. The reflection and transmission amplitudes summarized in the appendix remain valid for nontransparent chiral media upon substitution of the above material parameters.

In a series of experiments begun with my undergraduate students[27] and concluded with colleagues at the PC, the DCR from a naturally chiral medium has been observed for the first time.[28] The experiments, which in a historical sense complement Fresnel's separation of circularly polarized light by chiral refraction, test the generalized Fresnel amplitudes and confirm the predicted chiral enhancements under conditions involving multiple reflection, critical angle, and weak absorption.

In the optical configuration which ultimately proved successful, a probe beam, derived from a xenon-arc lamp, propagated back and forth in an achiral medium (fused silica) between two parallel compartments of chiral solution (camphorquinone in methanol) as illustrated in figure 11.7. Since the signal to be observed was quite small even with enhancement, it was extracted from the output current of the photodetector by means of optical phase modulation and synchronous detection (as described in chapter 9). The light, initially polarized at 45° to the quartz element of the PEM (which lay in the plane of incidence), traversed the PEM, multiply reflected from the chiral medium, and was synchronously detected by a photomultiplier and lock-in amplifier. The DCR was then deducible from the photocurrent components $I(0)$ and $I(f)$ at the modulation frequency $f = 50$ kHz, according to the relation

$$\delta_C = \frac{I(f)/I(0)}{2J_1(\varphi_m)} \tag{47}$$

derivable from Eq. (9.107) where φ_m is the modulation amplitude. (Experimentally, for the reasons explained in chapter 9, φ_m was set to ~ 2.4 radians, the first zero of the Bessel function J_0.)

The outcome is shown in figure 11.8. $I(f)/I(0)$ was recorded across the absorption band of camphorquinone as a function of wave number for the following three cases: (1) compartments C1 and C2 filled with pure methanol, (2) C1 filled with methanol and C2 filled with chiral solution, and (3) C1 and C2 both filled with chiral solution. The DCR (designated D) for single chiral reflection was determined from cases 2 and 1; the signal (designated 2D) for double chiral reflection was determined from cases 3 and 1. Maximum signals were in the expected ratio of 1:2.

The theoretical DCR was calculated from the chiral Fresnel amplitudes of this chapter and optical constants were deduced from measurements of absorption, rotary power, and dichroism as a function of wave number. As discussed above, the complex chiral refractive indices take the form

$$\tilde{n}_\pm = n_\pm + i\kappa_\pm = (n + i\kappa)[1 \pm (f_r + if_i)]. \tag{48}$$

At 21,000 cm^{-1} (476 nm), close to the maximum of the absorption band of the sample, the optical constants were found to have the following values: mean

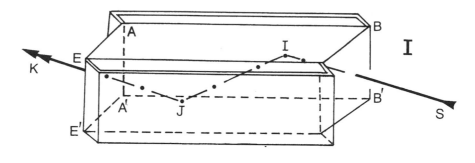

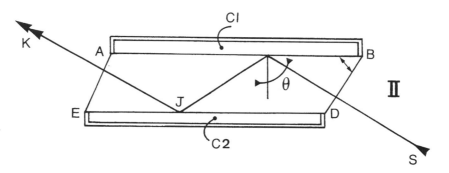

Figure 11.7. Side view (I) and top view (II) of the two-compartment (C1, C2) fused-silica reflection cell by means of which the differential circular reflection of light from a chiral sample was first observed. The angle ABD of the rhombus is 68°.

refractive index $n = 1.3327$, mean absorption parameter $\kappa = 1.46 \times 10^{-4}$, and real and imaginary chiral parameters $f_r = -0.94 \times 10^{-7}$ and $f_i = -6.02 \times 10^{-7}$. The refractive index of the fused silica was $n_1 = 1.4637$.

The solid curve in figure 11.8 traces the theoretical prediction of the DCR (for two reflections) as a function of wave number for light incident upon the sample at $\theta = 66.50°$. Agreement of experiment and theory is seen to be excellent. In accordance with theory, the differential reflection increases rapidly as θ approaches the critical angle θ_c. At 21,000 cm^{-1} a value of $D = 17 \times 10^{-5}$ with rms noise of 10^{-6} was measured when θ was set as close as experimentally possible to $\theta_c = 65.58°$. (This is to be compared with $D = 6 \times 10^{-5}$ for $\theta = 66.50°$.)

It is worth noting explicitly here that the differences in chiral refractive indices (circular birefringence) and absorption constants (dichroism) are numbers of the order of 10^{-7}. Under conditions of ordinary single-pass reflection this would also have been the order of magnitude of the DCR (and other manifestations of

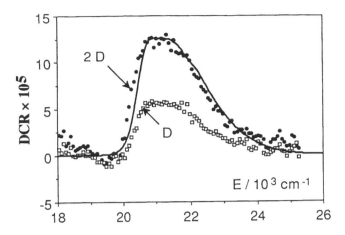

Figure 11.8. Differential circular reflection (DCR $\times 10^5$) as a function of wave number E for incident angle $\theta = 67°$. D denotes single chiral reflection; $2D$ denotes two chiral reflections. The solid curve is the theoretical $2D$ determined from the chiral Fresnel amplitudes and optical constants of the solution.

chiral scattering). Thus, the observed maximum DCR of approximately 10^{-4} represents a thousandfold enhancement of chiral asymmetry.

11.5 THE CHIRAL FABRY-PEROT INTERFEROMETER

Apart from a pioneering test of principle, the capacity to measure weak chiral asymmetries in matter by light reflection opens up new avenues in the exploration of materials and interactions. Among the advantages of specular reflection are the following: (1) As a form of coherent light scattering, the reflected "carrier" wave can be sufficiently intense to permit detection of optical activity by direct voltage and current measurements (in contrast to Rayleigh and Raman scattering, for example, for which photon counting would be required); (2) it can be employed under circumstances for which the chiral imprint on a transmitted light beam, or the transmitted beam itself, is very weak as in the case of chiral thin films or strongly opaque substances; (3) it can be readily implemented over a broad spectral range, in particular from the IR through UV; and (4) it furnishes through the Fresnel coefficients more information than obtainable by transmission measurements of rotatory power or circular dichroism.

Until recently the quantitative detection of optical activity in naturally occurring chiral substances by light reflection had remained an elusive goal because of the weakness of the phenomenon and the occurrence of instrumental artifacts. Through the use of multiple reflection under conditions of total reflection that goal has now been attained. However, other more versatile configurations are needed if the full potential of a reflection-based probe of chiral matter is to be achieved. The advantage of measuring optical activity by light transmission, it will be recalled, is that the signal increases with the optical pathlength through the sample—until various processes of extinction, which also increase with pathlength, render the signal too weak to be detected. Is there a way to take advantage of the pathlength enhancement afforded by light propagation, yet observe the chiral signal in reflected light? An affirmative answer is provided by the chiral analog of a Fabry-Perot interferometer (FPI).[29]

The FPI configuration (figure 11.9) is totally different from the one of the preceding section, employing the interference of arbitrarily many beams multiply reflected within (rather than outside of) a chiral medium of arbitrary opacity. The system—a chiral layer between two achiral media—can be expected to display optical activity in reflection as a result of two basically different processes. First, there is the chiral asymmetry intrinsic to the reflection and transmission amplitudes at each interface. Second, there is the asymmetry in retardation incurred by passage of LCP and RCP waves across the chiral layer. Although chiral effects intrinsic to reflection and transmission are ordinarily very weak, those attributable to propagation can be substantial even for thin films.

Heuristically, one might imagine the optically active layer to "rotate" the polarization of a linearly polarized incident wave penetrating the surface—the larger the number of internal reflections, the greater the optical rotation—which, upon subsequent retransmission across the first interface, superposes with the externally reflected wave (and other successive internally reflected waves) to give rise to a large optical rotation. This expectation is indeed borne out although, when analyzed rigorously, a reflected linearly polarized wave displays ellipticity as well as optical rotation, both effects depending sensitively on the angle of incidence and layer thickness.

To be of practical use, it is necessary that the resulting reflectance be adequate—that is, it would do little good to have a large chiral effect impressed upon a wave of vanishingly small intensity. By judicious choice of the material properties of the upper and lower media, reasonably large reflectances—sometimes reaching 100%—can be obtained. For example, large chiral asymmetries at high reflectance occur for thin films under conditions of (1) total reflection or (2) metallic reflection at the rear chiral-achiral interface. Moreover, under appropriate circumstances, ordinary reflection from macroscopically thick chiral samples can produce a "cross-polarized" component—for example, a component reflected with polarization e_1 when the incident wave is of orthogonal polarization e_2—large enough to be seen with one's eyes. Recall that this component vanishes in the absence of chirality and would yield a reflectance proportional to $f^2 \ll 1$ for single-pass external reflection.

With a chiral FPI one could in principle study a wide variety of structures ranging from chiral thin films to macroscopic bulk samples. Among many

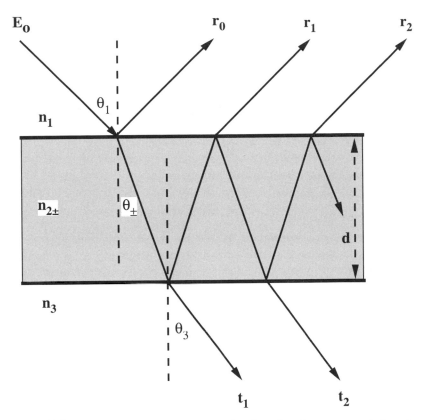

Figure 11.9. Schematic diagram of a chiral Fabry-Perot interferometer consisting of an isotropic optically active medium (indices $n_{2\pm}$) of thickness d between achiral media with indices n_1 and n_3. The incident wave $\mathbf{E}_0$ leads to net reflected and transmitted waves formed by the linear superposition of a suite of partial waves. Each internally reflected ray gives rise to two rays (not shown) at angles $\theta_\pm$.

potential applications is the exciting possibility of monitoring in real time the selective creation and degradation of enantiomeric molecular layers of biological significance in an effort to examine different mechanisms for the origin of biological homochirality. Recent theories suggest that the molecular precursors of life formed in thin layers on solid catalytic surfaces like clay, rather than in an organic "soup."

As indicated in figure 11.9, the net reflected wave $\mathbf{r}$ in the incident medium is the vector sum of the primary (externally reflected) wave $\mathbf{r}_0$ and the series of secondary waves $\mathbf{r}_1$, $\mathbf{r}_2$, and so on, where $\mathbf{r}_j$ has been internally reflected $2j - 1$ times. Were the middle layer an isotropic achiral dielectric, the calculation of the reflected wave could be effected immediately by use of the well-known formula for summing a geometric series. There are, however, two important aspects of chiral reflection that complicate the analysis of the optically active FPI. First, at each encounter of a circularly polarized wave with an interface, in general reflected and transmitted waves of both helicities are produced; and second, for the same angle of incidence LCP and RCP components reflect at *different* angles within the optically active medium. Here is an example—in case the reader had never considered the prospect—of a system for which the law of the mirror does not hold:[30] For a single angle of incidence, there are two angles of reflection corresponding to two orthogonally polarized CP waves. These polarization transformations necessitate the treatment of vector, rather than scalar, waves in a chiral FPI.

The intrinsically vectorial nature of the problem makes it an interesting one from the standpoint of mathematical method, as well as physical content. Following familiar procedure for a boundary-value problem, one could in principle solve for the reflected, transmitted, and internal waves by imposing appropriate boundary conditions at each interface and directly integrating Maxwell's equations. Although seemingly straightfoward, this is not the best way to proceed; the approach leads to large matrices (8 × 8 in the simplest case of a three-layer system) for which the inversion must certainly be executed numerically and which can, in fact, be ill-defined when the transmittance is weak (as in the case of total reflection).

There is an alternative and far simpler procedure (see note 29) which generalizes the standard geometrical series solution of a scalar Fabry-Perot. This approach, which involves only 2 × 2 matrices, provides a relatively simple explicit formula for the reflected and transmitted waves separately, under all conditions, that is, for both ordinary and total reflection. Moreover, the method is applicable to any kind of optical medium once one knows the single-pass reflection and transmission coefficients at each interface.

Let us designate respectively by $r_{\alpha\alpha'}$, $r'_{\alpha\alpha'}$, $r''_{\alpha\alpha'}$ (where α, α' can each be $+$ or $-$) the amplitudes for conversion of an incident circularly polarized wave of polarization α into a wave of polarization α' by (1) external reflection (in achiral medium 1) at interface 1-2, (2) internal reflection (in chiral medium 2) at interface 1-2, and (3) internal reflection at interface 2-3. Similarly, $t_{\alpha\alpha'}$, $t'_{\alpha\alpha'}$, $t''_{\alpha\alpha'}$ designate the corresponding amplitudes for polarization conversion effected by transmission (1) from medium 1 to 2, (2) from medium 2 to 1, and (3) from medium 2 to 3. Thus, a LCP wave originating in medium 1 generates an externally reflected wave $\binom{r_{++}}{r_{-+}}$ in medium 1 and a transmitted wave $\binom{t_{++}}{t_{-+}}$ in medium 2, where $\binom{1}{0}$, $\binom{0}{1}$ are the helicity $(+)$ (LCP) and $(-)$ (RCP) basis states. For an incident wave $\mathbf{E}_0$ of arbitrary polarization, the externally reflected and internally transmitted waves take the form $\mathbf{RE}_0$ and $\mathbf{TE}_0$ where

$$R = \begin{pmatrix} r_{++} & r_{+-} \\ r_{-+} & r_{--} \end{pmatrix}, \qquad T = \begin{pmatrix} t_{++} & t_{+-} \\ t_{-+} & t_{--} \end{pmatrix} \tag{49}$$

are the relevant chiral reflection and transmission matrices. The corresponding matrices (R', T'), (R'', T'') for internal reflection and external transmission at the 1-2 and 2-3 interfaces, respectively, have the same form as relation (49), but employ the appropriate elements. Exact expressions for all the matrix elements (in CP and LP bases) are not particularly enlightening here and are therefore relegated to the appendix.

It is useful to note that the matrices defined above satisfy the following identities (where **1** is the unit 2×2 matrix)

$$R^2 + T'T = R'^2 + TT' = 1 \tag{50}$$

corresponding to the scalar relationship $tt' = 1 - r'^2$ for the Fresnel amplitudes of an achiral layer. The chiral matrices, however, are noncommutative. Also, the relation $R' \sim -R$ is not an identity (as is the case for the corresponding achiral scalar amplitudes) but holds rigorously true only at normal and grazing incidence and approximately at any other incident angle to the extent that the chiral parameter f is small.

Within the chiral medium LCP and RCP waves of vacuum wavelength λ propagate without change in polarization, but incur for each one-way passage across the chiral layer of thickness d respective phase shifts

$$\delta_\pm = \frac{2\pi n_{2\pm} d}{\lambda} \cos\theta_\pm. \tag{51}$$

The propagation of an abitrarily polarized wave through the layer can then be represented by the phase matrix

$$\Lambda = \begin{pmatrix} \exp(i\delta_+) & 0 \\ 0 & \exp(i\delta_-) \end{pmatrix}. \tag{52}$$

Although relations (51) and (52) are the apparently natural extensions to chiral reflection of the corresponding phase shift within an achiral layer, the simplicity of the result actually masks an underlying subtlety arising from an important, nontrivial distinction between the scalar (achiral) and vector (chiral) Fabry-Perot interferometers.

In the standard analysis of an achiral FPI, a ray that has made one round-trip passage (at an angle ϕ to the normal) through a layer of index n_2 incurs a retardation of $4\pi n_2 d/\lambda \cos\phi$—that is, with the cosine factor in the denominator. However, a second ray, which is externally reflected from the 1-2 interface (at an angle θ to the normal) at the point of emergence of the first ray, has traveled relative to the latter an extra distance in medium 1 (e.g., air) and therefore incurs a retardation of $(4\pi n_1 d/\lambda) \tan\phi \sin\theta$. It is the difference between these two phase shifts—one inferred from propagation within the layer and the other from propagation outside the layer—that leads to the net relative phase $(4\pi n_2 d/\lambda) \cos\phi$ for a round trip through the layer or, equivalently, half that value for a one-way trip.

Were the preceding argument to be applied without caution to the case of a chiral Fabry-Perot, the apparent generalization might seem to necessitate two propagation matrices of diagonal form (52), one characterizing phase shifts incurred within the chiral layer and the other characterizing phase shifts in achiral medium 1. Since these matrices would be separated by noncommuting transmission and reflection matrices, it is by no means evident that, in the resulting multiple products of matrices, the phase shifts would combine to produce the simple result of Eqs. (51) and (52). In fact, they do not.

My derivation of relations (51) and (52) is actually based on an alternative point of view—described long ago by Max Born for scalar waves and useful in the theory of X-ray interference[31]—whereby propagation only *within* the chiral medium need be considered. Examination of all possible round-trip pathways resulting in (1) no polarization conversion (e.g., a component $t'_{++}r''_{++}t_{++}$), (2) polarization conversion on transmission (e.g., $t'_{+-}r''_{--}t_{-+}$), and (3) polarization conversion on reflection (e.g., $t'_{+-}r''_{-+}t_{++}$) leads readily to the above relations as the correct description of the (one-way) propagation phase shifts. The details of Born's argument generalized to chiral media can be found in note 29.

The net amplitude in medium 1 of N contributing reflected waves is then expressible by the series

$$\mathbf{r}_{[N]} = \mathbf{r}_0 + \mathbf{r}_1 + \mathbf{r}_2 + \cdots + \mathbf{r}_{N-1} = M_{[N]}\mathbf{E}_0 \tag{53}$$

in which the system reflection matrix $M_{[N]}$ is defined by

$$M_{[N]} = R + T(1 + S + S^2 + \cdots + S^{N-1})(\Delta R''\Delta)T \tag{54a}$$

with

$$S \equiv \Delta R''\Delta R'. \tag{54b}$$

To infer the sequence of events (transmission, propagation, reflection) for each contributing wave, one reads the matrix series in Eq. (54a) from right to left.

To perform the sum, first diagonalize the matrix S and then close the resulting geometric series that constitute the diagonal elements. The outcome takes the form

$$M_{[N]} = R + T'Z(\Delta R''\Delta)T \tag{55}$$

where

$$Z \equiv \frac{1 - S^N}{1 - S} = UZ_DU^{-1}. \tag{56}$$

The first relation in Eq. (56) is a symbolic representation of the matrix summation, the practical evaluation of which is indicated by the second relation containing diagonal matrix Z_D,

$$Z_D = \begin{pmatrix} \dfrac{1 - s_+^N}{1 - s_+} & 0 \\ 0 & \dfrac{1 - s_-^N}{1 - s_-} \end{pmatrix}, \tag{57}$$

a function of the eigenvalues $s_\pm$ of S:

$$s_\pm = \tfrac{1}{2}\left[\mathrm{Tr}(S) \pm \sqrt{[\mathrm{Tr}(S)]^2 - 4\det(S)}\right].\tag{58}$$

The transformation matrix U that diagonalizes S and Z can be expressed in terms of the elements s_{ij} and eigenvalues $s_\pm$ of S in the following way:

$$U = \begin{pmatrix} s_+ - s_{22} & s_{12} \\ s_{21} & s_- - s_{11} \end{pmatrix}.\tag{59}$$

If S is initially diagonal, then U is taken to be the unit matrix **1**.

In a similar way one can deduce the net transmitted wave in medium 3. The details are left to the reference, but the final result is

$$\mathbf{t}_{[N]} = M'_{[N]}\mathbf{E}_0\tag{60}$$

in which

$$M'_{[N]} = T''\frac{1 - S'^N}{1 - S'}\Delta T\tag{61a}$$

with

$$S' = \Delta R'\Delta R''.\tag{61b}$$

The concrete evaluation of the symbolic expression in Eq. (61a) is performed in the same way as that of matrix Z in Eq. (56). Note that S and S' are distinguished by the order of matrix multiplication and are not in general the same for a nonvanishing chiral parameter.

In the case of a chiral layer of infinite extent the number of contributing beams $N \rightarrow \infty$, and, assuming that the matrix series converge, the reflection and transmission matrices reduce, respectively, to

$$M_{[\infty]} = R + T'\left(\frac{1}{1 - S}\right)(\Delta R''\Delta)'T,\tag{62}$$

$$M'_{[\infty]} = T''\frac{1}{1 - S'}\Delta T.\tag{63}$$

(It is under these circumstances that one can also treat the system as a boundary-value problem at two interfaces, as previously discussed.)

Equations (62) and (63) are the principal analytical results of this section. Given an incident wave $\mathbf{E}_0$ of known polarization, one can deduce from Eq. (53) the LP or CP components of the reflected wave and consequently any experimentally relevant quantity such as the DCR, optical rotation, or ellipticity. Suppose, for example, the σ- and π-components of the reflected wave are represented by (real) amplitudes a and b and relative phase δ_0. These quantities then determine the ellipticity and optical rotation of the wave through the following expressions:

$$e = \left[\frac{(a^2 + b^2) - [a^4 + b^4 + 2a^2b^2\cos 2\delta_0]^{1/2}}{(a^2 + b^2) + [a^4 + b^4 + 2a^2b^2\cos 2\delta_0]^{1/2}}\right]^{1/2} \times \mathrm{sgn}(\delta_0)\tag{64}$$

and

$$\tan 2\phi = \frac{2ab\cos\delta_0}{b^2 - a^2}. \tag{65}$$

The ellipse orientation has been chosen so that $\phi = 0$ represents a pure π-polarization; $\phi = \pi/2$ then connotes a pure σ-polarization.

For ordinary reflection—that is, reflection at incident angles below the critical angles of the 1-2 and 2-3 interfaces—the reflected wave in medium 1 is determined largely by the first two partial waves in the series (53). The chiral asymmetry is then generally weak—except in the vicinity of an incident angle at which the reflectance vanishes. Near these locations, the small angular displacement of RCP and LCP reflectance nodes leads to chiral effects numerically large compared with the size of the chiral parameter(s). However, since the detector is scarcely illuminated, these large chiral asymmetries are not practically observable.

An interferometer configuration with indices $n_1 > (n_\pm \sim n_2) > n_3$ leads to total reflection at the 2-3 interface for incident angles (within the chiral medium) beyond the critical angle $\theta_{2c} = \sin^{-1}(n_3/n_2)$. The condition $n_1 > n_2$ avoids total internal reflection at the 1-2 interface where high transmittance is desired. The condition $n_2^2 > n_1 n_3$ assures that the critical angle $\theta_{1c} = \sin^{-1}(n_2/n_1)$ for total external reflection in medium 1 at the 1-2 interface is greater than θ_{2c}. In this way one can expect over the range of incident angles between θ_{2c} and θ_{1c} a strongly reflected wave in medium 1 that has undergone many internal reflections within the chiral medium 2.

Experimentally, such a configuration can be realized by coupling light through a quartz prism into a chiral layer in air. Theoretical modeling suggests that a thin, weakly absorbing chiral film can lead to peak values of the DCR some 5000 times the circular birefringence, with comparable results for the ellipticity and optical rotation. Moreover, the overall reflectance is quite high, varying between approximately 60% and 80%.

Particularly striking are the results of chiral reflection from an optically thick layer of several thousands of wavelengths or more. The chiral sample can be surrounded by air on both sides; no special arrangements for total or metallic reflection are necessary. Figure 11.10 shows the variation with incident angle of the σ-component of the reflected light flux when the incident wave in medium 1 (air) is π-polarized. This is the cross-polarization referred to above which, in the absence of chirality, would be strictly null. In an actual experiment the rapid variation in reflectance between extremals would be averaged over by the finite divergence of the beam, finite size of the detector, wavelength spread of the light source, and slight deviation of the surfaces of the sample from exact parallelism.

For the sample of the figure with a thickness of 5000 wavelengths and the refractive index of water, the theoretical σ-reflectance reaches magnitudes between 2% and 10% in the vicinity ($\sim 70°$) of the maxima of the lower and upper

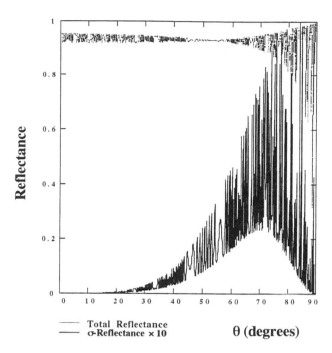

Figure 11.10. Total reflectance and σ-polarized reflectance for π-polarized light incident upon a transparent chiral sample with thickness $d = 5000\,\lambda$ and indices $n_{2\pm} = 1.33 \times (1 \pm 10^{-5})$; the ambient medium is air ($n_1 = n_3 = 1$).

enveloping curves. Phase modulation and synchronous detection techniques would hardly be necessary here. One simply turns a linear polarizer to block out the orthogonal π-component—and looks; the chiral signal is practically blinding.

11.6 GATHERINGS: PLUS ÇA CHANGE ...

When I began my studies of the optics of chiral media, I knew of very few physicists who were interested in the subject—and these exclusively in the time-honored methods of measuring optical rotation and circular dichroism by light transmission. Since then, the fascination with reflection and scattering of electromagnetic waves from chiral structures, either naturally occurring ones of molecular size or synthetically produced macroscopic objects, has grown extensively. Indeed, many disparate lines of investigation previously hidden within the traditional fields of physics, chemistry, engineering, and life sciences have become to come together to define the emerging field of "chirality."

In the domain of light optics (IR, visible, and UV), such studies have long provided deep insights into stereochemical structure and physical interactions. But the interest in chirality transcends traditional optics and forms an important part, as well, of radio and microwave electromagnetics.

A few years ago, in the small town of Savignac, France, in the region of Périgord justly renowed for its gastronomic specialities and fine wines, about one hundred specialists representing some twenty nationalities assembled to discuss the electromagnetic behavior of chiral systems. This international workshop, the first of its kind, was generously sponsored by the French Atomic Energy Agency (CEA), which, along with the mayor and other worthies of the town, spared no effort to ensure that its guests returned home with beautiful memories of France—and a copious supply of paté de foie gras. Now, chirality is an intriguing subject to be sure, but why would it interest the CEA and analogous government agencies of other countries who paid the expenses of their participants?

As one of the very few physicists in what was basically a gathering of engineers, I soon learned the reason for this curious circumstance. In stark contrast to my own objective of enhancing a chiral signal in scattered light, the engineers were examining all conceivable ways to suppress it. They were interested in the problem of "stealth"—that is, how to evade radar detection. In this regard the chiral parameter provided an additional "handle," besides the ordinary electric and magnetic material constants, that could possibly be adjusted to create new materials that reflect radar waves especially poorly. At least, that was what governments funding chiral research were hoping. Would these hopes be realized? On this point the "chirilists" had reached no consensus.

I thought about this feature of the conference, which for me was unexpected, as I drove back to Paris, stopping every so often to examine the beautiful, yet sobering, landscape. Like mushrooms in a verdant meadow, castles and fortresses dotted the green landscape of Périgord, the former Aquitaine over which French and English armies so often collided centuries ago. Time has passed, technology has advanced, and yet—as the underlying and unstated theme of the chiral workshop reminded me—the concerns of war and defense are ever with us.

APPENDIX: CHIRAL REFLECTION AND TRANSMISSION AMPLITUDES

1. External Reflection Matrix **R** for Reflection in Achiral Medium 1 at the 1-2 Interface

Circular Polarization Basis (LCP = +; RCP = −)

$$r_{++} = \frac{2q_{21}}{\Delta}[(\beta_1^2 - \beta_+\beta_-) - \beta_1(\beta_+ - \beta_-)], \tag{A1}$$

$$r_{+-} = r_{-+} = \frac{-(q_{21}^2 - 1)}{\Delta}(\beta_+ + \beta_-), \tag{A2}$$

$$r_{--} = \frac{2q_{21}}{\Delta}[(\beta_1^2 - \beta_+\beta_-) + \beta_1(\beta_+ - \beta_-)] \tag{A3}$$

where

$$\beta_1 = \cos\theta_1, \qquad \beta_\pm = \cos\theta_\pm, \tag{A4}$$

$$\Delta = (2q_{21})(\beta_1^2 + \beta_+\beta_-) + (1 + q_{21}^2)\beta_1(\beta_+ + \beta_-), \tag{A5}$$

$$q_{21} = \frac{n_2\mu_1}{n_1\mu_2}. \tag{A6}$$

Linear Polarization Basis (σ-polarization = 1; π-polarization = 2)

$$r_{11} = \frac{2}{\Delta}\left[q_{21}(\beta_1^2 - \beta_+\beta_-) - \beta_1(\beta_+ + \beta_-)\left(\frac{q_{21}^2 - 1}{2}\right)\right], \tag{A7}$$

$$r_{12} = -r_{21} = \frac{2iq_{21}}{\Delta}\beta_1(\beta_+ - \beta_-), \tag{A8}$$

$$r_{22} = \frac{2}{\Delta}\left[q_{21}(\beta_1^2 - \beta_+\beta_-) + \beta_1(\beta_+ + \beta_-)\left(\frac{q_{21}^2 - 1}{2}\right)\right]. \tag{A9}$$

In the absence of chirality, $r_{11} = r_\sigma$, $r_{22} = r_\pi$, $r_{12} = r_{21} = 0$.

2. Internal Reflection Matrix R′ for Reflection in Chiral Medium 2 at the 1-2 Interface

$$r'_{++} = \frac{[(1 + q_{12}^2)\beta_1(\beta_+ - \beta_-) - 2q_{12}(\beta_1^2 - \beta_+\beta_-)]}{\Delta'}, \tag{A10}$$

$$r'_{-+} = \frac{2(1 - q_{12}^2)\beta_1\beta_+}{\Delta'}, \tag{A11}$$

$$r'_{+-} = \frac{2(1 - q_{12}^2)\beta_1\beta_-}{\Delta'}, \tag{A12}$$

$$r'_{--} = \frac{-[(1 + q_{12}^2)\beta_1(\beta_+ - \beta_-) + 2q_{12}(\beta_1^2 - \beta_+\beta_-)]}{\Delta'} \tag{A13}$$

where

$$\Delta' = (2q_{12})(\beta_1^2 + \beta_+\beta_-) + (1 + q_{12}^2)\beta_1(\beta_+ + \beta_-) = \frac{\Delta}{q_{21}^2}, \tag{A14}$$

$$q_{12} = \frac{n_1\mu_2}{n_2\mu_1} = \frac{1}{q_{21}}. \tag{A15}$$

3. Transmission Matrix **T** for Transmission from Achiral Medium 1 to Chiral Medium 2

Circular polarization basis

$$t_{++} = \frac{(1 + q_{21})\beta_1(\beta_1 + \beta_-)}{\Delta}, \tag{A16}$$

$$t_{-+} = \frac{2(q_{21} - 1)\beta_1(\beta_1 - \beta_+)}{\Delta}, \tag{A17}$$

$$t_{+-} = \frac{2(q_{21} - 1)\beta_1(\beta_1 - \beta_-)}{\Delta}, \tag{A18}$$

$$t_{--} = \frac{(1 + q_{21})\beta_1(\beta_1 + \beta_+)}{\Delta}. \tag{A19}$$

Linear-to-circular polarization transformation

$$t_{1\pm} = \frac{2\sqrt{2}}{\Delta}\beta_1(\beta_\mp + q_{21}\beta_1), \tag{A20}$$

$$t_{2\pm} = \frac{\pm i2\sqrt{2}}{\Delta}\beta_1(q_{21}\beta_\mp + \beta_1), \tag{A21}$$

where $t_{j\alpha}$ $(j = 1, 2; \alpha = +, -)$ is the amplitude for conversion of linear polarization j to circular polarization α upon transmission from medium 1 to medium 2.

4. Transmission Matrix **T'** for Transmission from Chiral Medium 2 to Achiral Medium 1

Circular polarization basis

$$t'_{++} = \frac{(1 + q_{12})\beta_+(\beta_1 + \beta_-)}{\Delta'}, \tag{A22}$$

$$t'_{-+} = \frac{2(1 - q_{12})\beta_+(\beta_1 - \beta_-)}{\Delta'}, \tag{A23}$$

$$t'_{+-} = \frac{2(1 - q_{12})\beta_-(\beta_1 - \beta_+)}{\Delta'}, \tag{A24}$$

$$t'_{--} = \frac{2(1 + q_{12})\beta_-(\beta_1 + \beta_+)}{\Delta'}. \tag{A25}$$

Circular-to-linear polarization transformation

$$t'_{\pm1} = \frac{2\sqrt{2}}{\Delta'}\beta_\pm(\beta_1 + q_{12}\beta_\mp), \tag{A26}$$

$$t'_{\pm 2} = \frac{\mp i 2\sqrt{2}}{\Delta'} \beta_\pm (\beta_\mp + q_{12}\beta_1) \tag{A27}$$

where $t_{\alpha j}$ $(j = 1, 2; \alpha = +, -)$ is the amplitude for conversion of circular polarization α to linear polarization j upon transmission from medium 2 to medium 1.

5. Internal Reflection Matrix $\mathbf{R}''$ for Reflection in Chiral Medium 2 at the 2-3 Interface

Form the elements of $\mathbf{R}''$ from those of $\mathbf{R}'$ by making the following substitutions:

$$\beta_1 \Rightarrow \beta_3 = \cos\theta_3, \tag{A28}$$

$$q_{12} \Rightarrow q_{32} = \frac{n_3 \mu_2}{n_2 \mu_3}. \tag{A29}$$

Note that for intrinsically nonmagnetic systems, such as those examined in the text, the permeabilities $\mu_1 = \mu_2 = \mu_3 = 1$.

6. Conservation of Energy

The application of Stokes's law to reflection and refraction at a chiral interface leads to the following relationships among the amplitudes:

External reflection

Incident LCP

$$\beta_1(1 - |r_{++}|^2 - |r_{-+}|^2) - q_{21}(\beta_+|t_{++}|^2 + \beta_-|t_{-+}|^2) = 0 \tag{A30}$$

Incident RCP

$$\beta_1(1 - |r_{+-}|^2 - |r_{--}|^2) - q_{21}(\beta_+|t_{+-}|^2 + \beta_-|t_{--}|^2) = 0 \tag{A31}$$

Internal reflection

Incident LCP

$$\beta_+(1 - |r'_{++}|^2 - \beta_-|r'_{-+}|^2) - q_{12}\beta_1(|t'_{++}|^2 + |t'_{-+}|^2) = 0 \tag{A32}$$

Incident RCP

$$\beta_-(1 - |r'_{--}|^2 - \beta_+|r'_{+-}|^2) - q_{12}\beta_1(|t'_{+-}|^2 + |t'_{--}|^2) = 0 \tag{A33}$$

NOTES

1. L. Pasteur, "Researches on the Molecular Asymmetry of Natural Organic Products," in *Treasury of World Science*, ed. D. D. Runes (Patterson, N.J.: Littlefield, Adams, 1962), pp. 821–27.

2. "Chiral" from the Greek for "hand."

3. "Enantiomer" from the Greek for "opposite."

4. L signifies "levo" or left, in contrast to D for "dextro" or right.

5. I discuss the symmetries, invariancies, and underlying quantum basis of optical activity in *More Than One Mystery: Explorations of Quantum Interference* (New York: Springer-Verlag, 1995), ch. 6, "The Quantum Physics of Handedness."

6. The word "racemic" has nothing to do with mirrors or inversions; it comes from the Greek for "grapes." Racemic acid, derived from grapes, is a 50–50 mixture of the enantiomeric forms of the chiral molecule tartaric acid.

7. D. H. Deutsch, "A Mechanism for Molecular Asymmetry," *J. Mol. Evol.* 33 (1991): 295–96.

8. T. D. Lee and C. N. Yang, "Question of Parity Conservation in Weak Interactions," *Phys. Rev.* 104 (1956): 254–58.

9. For a comprehensive account of gauge field theories see E. Abers and B. Lee, "Gauge Theories," *Phys. Rep.* 9C, no. 1 (1973).

10. The Compton wavelength λ_C of a particle of mass m defines the momentum $p = h/\lambda_C$ for which the rest mass energy mc^2 and relativistic kinetic energy pc are equal; thus $\lambda_C = h/mc$.

11. M.-A. Bouchiat and L. Pottier, "Optical Experiments and Weak Interactions," *Science* 234 (1986): 1203–10.

12. R. P. Feynman, "The Value of Science," reprinted in *The Shape of This Century*, ed. D. W. Rigden and S. S. Waugh (New York: Harcourt Brace Jovanovich, 1990), pp. 352–57.

13. James Clerk Maxwell, *A Treatise on Electricity & Magnetism* (New York: Dover, 1954; unabridged republication of the 3d edition, published by the Clarendon Press in 1891).

14. *The Scientific Papers of James Clerk Maxwell*, ed. W. D. Niven (New York: Dover, 1952).

15. A. Einstein, "On the Electrodynamics of Moving Bodies," *Annalen der Physik* 17 (1905): 891–921.

16. J. D. Jackson, *Classical Electrodynamics*, 2d ed. (New York: Wiley, 1975), pp. 236–37.

17. The mémoire read by Fresnel before the French Academy in 1823 was apparently lost, subsequently found among the papers of Fourier, then published in *Annales de Chemie et de Physique* in 1831.

18. A. Fresnel, "Mémoire sur la loi des modifications que la réflexion imprime à la lumière polarisée," no. 30, in *Oeuvres complètes d'Augustin Fresnel* (Paris: Imprimerie Impériale, 1866), pp. 767–99.

19. J. C. Maxwell, Article 97, "Ether," in *The Scientific Papers*, pp. 763–75.

20. "Gyrotropic" from the Greek for "turning in a circle."

21. E. U. Condon, "Theories of Optical Rotatory Power," *Rev. Mod. Phys.* 9 (1937):

432–57.

22. M. Born, *Optik* (Heidelberg: Springer-Verlag, 1972), p. 412.

23. M. P. Silverman, "Specular Light Scattering from a Chiral Medium: Unambiguous Test of Gyrotropic Constitutive Relations," *Lettere al Nuovo Cimento* 43 (1985): 378–82; "Reflection and Refraction at the Surface of a Chiral Medium," *J. Opt. Soc. Am. A* 3 (1986): 830–37; correction of printing errors in *J. Opt. Soc. Am. A* 4 (1987): 1145. The extension to birefringent chiral media likewise requires a generalization of the symmetric set of constitutive relations; see M. P. Silverman and J. Badoz, "Light Reflection from a Naturally Optically Active Birefringent Medium," *J. Opt. Soc. Am. A* 7 (1990): 1163–73.

24. J. D. Jackson, *Classical Electrodynamics*, p. 252; C. W. Misner, K. S. Thorne, and J. A. Wheeler, *Gravitation* (San Francisco: Freeman, 1973), pp. 105–10.

25. The theorem is derived in M. Born and E. Wolf, *Principles of Optics*, 4th ed. (London: Pergamon, 1970), pp. 100–108.

26. M. P. Silverman and J. Badoz, "Large Enhancement of Chiral Asymmetry in Light Reflection near Critical Angle," *Opt. Comm.* 74 (1989): 129–33; M. P. Silverman and J. Badoz, "Multiple Reflection from Isotropic Chiral Media and the Enhancement of Chiral Asymmetry," *J. Electromagnetic Waves and Applications* 6 (1992): 587–601.

27. M. P. Silverman and T. C. Black, "Experimental Method to Detect Chiral Asymmetry in Specular Light Scattering from a Naturally Optically Active Medium," *Phys. Lett. A* 126 (1987): 171–76; M. P. Silverman, N. Ritchie, G. M. Cushman, and B. Fisher, "Experimental Configurations Using Optical Phase Modulation to Measure Chiral Asymmetries in Light," *J. Opt. Soc. Am. A* 5 (1988): 1854–62.

28. M. P. Silverman, J. Badoz, and B. Briat, "Chiral Reflection from a Naturally Optically Active Medium," *Opt. Lett.* 17 (1992): 886–88.

29. M. P. Silverman and J. Badoz, "Interferometric Enhancement of Chiral Asymmetries: Ellipsometry with an Optically Active Fabry-Perot Interferometer," *J. Opt. Soc. Am. A* 6 (1994): 1894–1917.

30. M. P. Silverman and R. B. Sohn, "Effects of Circular Birefringence on Light Propagation and Reflection," *Am. J. Phys.* 54 (1986): 69–76.

31. M. Born, *Optik*, pp. 119–26.

CHAPTER 12

Through a Glass Brightly: "Fresnel Amplification"

> It is as if tiny mechanical men, all wound up to a
> certain energy and facing along the axis of the laser
> enclosure, were successively set in motion by other
> marchers and fell into step until they became an
> immense army marching in unison row on row (the
> plane wave fronts) back and forth in the enclosure.
> After the laser light has built up in this way it
> emerges through the partly reflecting mirror at one
> end as an intense, highly directional beam.
> (*Arthur L. Schawlow, 1968*)[1]

THE REFLECTION of light at the surface of a medium with stored energy—that is, a medium with a population inversion—raises questions of both fundamental and practical import. Is it possible to have more light leave the surface than is incident upon it? By what mechanism could such a process occur? And if light amplification is theoretically possible in this way, then under what conditions can it actually be realized? For example, will any angle of incidence suffice, or is there a threshold angle beyond or below which the conversion of chemical to optical energy does not occur? Once a controversial issue beclouded by theoretical and experimental inconsistencies, the phenomenon has since been examined thoroughly and systematically. Enhanced reflection violates no laws of physics, does indeed occur, and may well be extremely useful.

12.1 THE ENIGMA OF ENHANCED REFLECTION

As the original acronym indicates, the laser amplifies light by means of stimulated emission. Schawlow's colorful anthropomorphic description of lasing reveals several features basic to the design of the most familiar lasers. First, the atoms whose energy of excitation will be transformed into light are contained in an enclosure, the optical cavity; second, the atoms are de-excited by interaction with lightwaves within the cavity; third, it is by multiple passages of a stimulating wave through the cavity that the intensity of the wave grows; and fourth, the light that is ultimately available for useful purposes "leaks" out one end of the cavity through a mirror with a high, but not perfect, reflectivity.

But suppose the atoms are *inside* the enclosure and the light *outside*—and the light merely makes a momentary reflective pass from the cavity surface. Can the atoms give up their energy "as an immense army marching in unison"? Will the outside light be amplified?

When light is reflected at the interface between two transparent regions, the law of energy conservation requires that the reflectance always be less than or equal to unity for all angles of incidence. A medium in which energy is stored is not transparent, however, but may be thought to have a complex refractive index ($\tilde{n} = n - i|\kappa|$) with intrinsically negative absorption coefficient κ. A negative κ signifies that the amplitude $A \sim \exp(i\tilde{n}kz)\exp(-i\omega t)$ of a wave traveling through the medium will grow exponentially with penetration z. It is by no means obvious, however, what effect such a medium will have on an externally reflected wave.

The expressions Fresnel derived for the amplitude of light reflected from a transparent homogeneous dielectric medium are equally valid for an absorbing medium in which case the imaginary part of the refractive index $\tilde{n} = n + i\kappa$ is a positive number, signifying diminution in intensity with penetration. These amplitudes are frequently used, for example, to analyze the reflectance from metals (where κ depends on the conductivity σ). Although it may seem like the most obvious thing to do, substituting a negative κ into the Fresnel amplitudes for a nontransparent medium with uniform gain does not resolve the question of light amplification; rather, the results present a curious ambiguity.

Consider the geometric configuration of figure 11.2 with a σ-polarized wave incident at angle θ to the normal. The light originates in a transparent medium with real-valued index n_1 and reflects from the surface of an excited medium with complex index $\tilde{n}_2 = n_2(1 - i\gamma)$; $\tilde{n}_2$ is so defined here that γ is a positive gain parameter. The Fresnel reflection amplitude r_σ has precisely the same form as Eq. (11.14a),

$$r_\sigma = \frac{n_1 \cos\theta - \tilde{n}_2 \cos\phi}{n_1 \cos\theta + \tilde{n}_2 \cos\phi}, \tag{1}$$

except that now the material parameters of the second medium are complex-valued. The complex angle ϕ, as explained previously, does not represent the direction of propagation in the second medium. Indeed, it has no individual physical significance; rather, the quantities $\tilde{n}_2 \sin\phi$ and $\tilde{n}_2 \cos\phi$ are the components, respectively parallel and perpendicular to the interface, of the propagation vector $\mathbf{k}_2/k_0$ in medium 2. (Recall that $k_0 \equiv \omega/c = 2\pi/\lambda$ is the vacuum wavenumber.) The first component is readily deduced from Snel's law:

$$\tilde{n}_2 \sin\phi = n_1 \sin\theta. \tag{2}$$

The second component, obtained from the trignometric identity $\cos^2\phi = 1 - \sin^2\phi$ with substitution of Snel's law, may be written in the form

$$\tilde{n}_2 \cos\phi = \sqrt{\tilde{n}^2 - (n_1 \sin\theta)^2} \equiv m' + im''. \tag{3}$$

Substituting Eq. (3) into the Fresnel amplitude (1) leads to the reflectance

$$R_\sigma = |r_\sigma|^2 = \frac{(n_1 \cos\theta - m')^2 + (m'')^2}{(n_1 \cos\theta + m')^2 + (m'')^2} \quad \left\} \begin{array}{l} m' > 0 \Rightarrow R_\sigma \leq 1 \\ m' < 0 \Rightarrow R_\sigma \geq 1 \end{array} \right. \quad (4)$$

from which it is seen that, depending on the sign of the real part of $\tilde{n}_2 \cos\phi$, the light can either be amplified or not. Since $\tilde{n}_2 \cos\phi$ is deduced from a quadratic expression, two roots are possible that occur with opposite signs,

$$(A) \quad m' > 0, \quad m'' < 0$$

and

$$(B) \quad m' < 0. \quad m'' > 0,$$

and have the following interpretations: Root (A) characterizes an amplified wave $[e^{+k_0|m''|z}e^{ik_0|m'|z}]$ moving through the medium away from the interface, whereas root (B) represents a decaying wave $[e^{-k_0|m''|z}e^{-ik_0|m'|z}]$ approaching the boundary from within the medium.

At first glance, it may seem reasonable to select root A—whereupon the reflected wave is never amplified—for under the circumstances of illumination from upper medium 1 how could there be a wave approaching the boundary from the infinite depths of lower medium 2? In the case $n_2/n_1 < 1$, however, all light ought to be totally reflected for incident angles beyond critical angle—that is, for $\theta \geq \theta_c = \sin^{-1}(n_2/n_1)$—in which case it would be difficult to justify an amplified wave penetrating medium 2. Here, perhaps, one should select root B with the consequence that the reflected wave is amplified. Apart from the speculative (rather than definitive) selection of roots A or B, this solution unsatisfactorily leads to a discontinuity in reflectance precisely at θ_c.

We are back at square one. Is the light amplified or not—and, if so, under what conditions?

12.2 WAVES IN AN EXCITED MEDIUM

The puzzling issue of enhanced reflection was one of the first problems in optics that I took up with a graduate student colleague (Raymond Cybulski Jr.). Together we resolved the problem theoretically and confirmed experimentally[2] that the enhanced reflection of light does indeed occur—and that it occurs principally in the vicinity of critical angle under conditions of total reflection. The ambiguity of the simple uniform gain model outlined in the preceding section arises from its infinite extent, which brings into play amplified waves at both "ends" of the medium. Realistically, such an infinitely extended population inversion does not exist; rather, the gain is established by an external perturbation—as, for example, laser pumping—and diminishes with depth in accordance with some prescribed law.

For tractability of analysis, as well as reasonable conformity with experiment, let us consider an amplifying medium in which the gain decreases exponentially with distance z from the interface (taken to be $z = 0$). The refractive index of the medium can be written as

$$\tilde{n}_2 = n_2(1 - i\gamma e^{-z/d}) \tag{5}$$

where d is a characteristic depth or penetration parameter, and the gain parameter γ is again real-valued and positive. Optical excitation is a common method for producing a population inversion and is the method employed in the experiments I will describe shortly. Under conditions in which the active medium does not saturate—that is, become incapable of further absorption—the intensity $I(z)$ of the pumping light beam, and therefore the gain established within the absorbing medium, diminish exponentially with depth in accordance with Beer's law:

$$I(z) = I_0 \exp(-\sigma_a N z) \equiv I_0 \exp\left(-\frac{z}{d}\right). \tag{6}$$

Here σ_a is the absorption cross section at the frequency of the pump beam, N is the concentration (number/volume) of the absorbing molecules, and from the two forms of Eq. (6) it is clear that the depth parameter is

$$d = (\sigma_a N)^{-1}. \tag{7}$$

Physically, d is the distance at which the imaginary part of $\tilde{n}_2$ falls to e^{-1}, its value at the interface. For distances $z \gg d$ within the medium, the region is essentially transparent.

The reflecting waves (the probe beam—to be distinguished from the pump beam) originate in a homogeneous transparent medium of refractive index n_1. Under the conditions that lead to amplified reflection, $n_{12} \equiv n_1/n_2 > 1$, one can define a critical angle with respect to the far-field region ($z \gg d$) by

$$\sin \theta_c = \frac{1}{n_{12}}. \tag{8}$$

The incident and reflected waves in medium 1 are homogeneous plane waves. By contrast, the transmitted wave propagates through an optically inhomogeneous medium and has a more complicated form. Nevertheless, irrespective of the nature of the various waves, phase matching at the interface and at any plane $z =$ constant lead, respectively, to the equality of the angles of incidence and reflection, and to a generalization of Snel's law (Eq. [2]), which in the present case may be written as

$$\sin \phi(z) = \frac{n_{12} \sin \theta}{1 - i\gamma e^{-z/d}}. \tag{9}$$

To deduce the wave which propagates through the amplifying medium one must solve Maxwell's wave equation

$$\left(\nabla^2 - \frac{\varepsilon\mu}{c^2}\frac{\partial^2}{\partial t^2}\right)\mathbf{E}(\mathbf{r}, t) = 0 \tag{10}$$

in which for all practical purposes the magnetic permeability μ is unity and the dielectric constant $\varepsilon(z)$ takes the form

$$\varepsilon(z) = \tilde{n}_2^2 \sim n_2^2(1 - 2i\gamma e^{-z/d}). \tag{11}$$

The second (approximate) relation in Eq. (11) is based on the assumption that the gain parameter is much smaller than one; for example, in the experiments to be discussed, $\gamma \sim 10^{-5}$. Since ε is a function only of the perpendicular distance from the interface, plane wave solutions may be chosen for the time dependence of the wave and for the spatial variation parallel to the interface. Thus, it is only the z-dependence of the wave that is of interest.

As is customarily the case, let us examine the two basic orientations of the electric field: normal (σ-polarization) and parallel (π-polarization) to the plane of incidence.

σ-Polarization

Without any loss of generality, one can represent the electric field by

$$\mathbf{E}_\sigma = \mathbf{y}F(z)\exp[i(k_x x - \omega t)] \tag{12}$$

where

$$k_x = n_2 k_0 \sin\theta \tag{13}$$

is the lateral component of the propagation vector. Substitution of Eq. (12) into the wave equation (10) leads to the one-dimensional equation

$$\frac{d^2 F}{dz^2} + [\varepsilon(z)k_0^2 - k_x^2]F = 0 \tag{14}$$

which, though complicated by the exponential variation of the dielectric constant, can actually be cast into a (more-or-less) familiar expression by the following change of variable:

$$\xi = 2k_0 d(-2i\gamma)^{1/2}e^{-z/d}. \tag{15}$$

This yields a form of Bessel's equation

$$\frac{d^2 F}{d\xi^2} + \frac{1}{\xi}\frac{dF}{d\xi} + \left[\frac{(2k_0 d \cos\phi_\infty)^2}{\xi^2} + 1\right]F = 0 \tag{16}$$

with the solution

$$F(\xi) = A J_\nu(\xi) \tag{17}$$

where A is a complex-valued constant and

$$\nu = -i2k_0d \cos \phi_\infty \tag{18}$$

is the complex-valued order of the Bessel function. The angle ϕ_∞ is the angle of refraction deduced from Eq. (9) in the limit that $z \to \infty$; since the medium is transparent in the far-field limit, ϕ_∞ is a real angle that represents the actual direction of wave propagation at infinity.

Since the wave equation is of second order, there is a second solution. This solution, however, diverges at the interface and must be discarded on physical grounds.

π-Polarization

Because the electric field now lies in the plane of incidence, it is actually more convenient to solve for the magnetic field which is normal to the plane of incidence. The procedure is similar in principle to that followed in the preceding case. Represent the magnetic field by the expression

$$\mathbf{H}_\pi = \mathbf{y}G(z) \exp[i(k_x x - \omega t)] \tag{19}$$

which is then substituted into Eq. (10) to produce the one-dimensional wave equation

$$\frac{d^2G}{dz^2} + \frac{1}{d}\left[1 - \frac{1}{\varepsilon(z)}\right]\frac{dG}{dz} + [\varepsilon(z)k_0^2 - k_x^2]G = 0. \tag{20}$$

Upon implementing the same change of variable as before and making the approximation $\varepsilon(\xi)^{-1} \approx 1 - (\xi/2k_0d)^2$, one obtains an equation

$$\frac{d^2G}{d\xi^2} + \left[\frac{1}{\xi} - \frac{\xi}{2k_0^2d^2}\right]\frac{dG}{d\xi} + \left[\frac{(2k_0d \cos \phi_\infty)^2}{\xi^2} + 1\right]G = 0 \tag{21}$$

that regrettably does not correspond, as far as I know, to any "named" equation of mathematical physics. It is the reciprocal of the dielectric constant in Eq. (20), or the term deriving from it in Eq. (21), that complicates matters; otherwise we would have again arrived at a form of Bessel's equation. Although Bessel functions of complex argument and complex order are not trivial to work with, at least there are known theorems and expansions that can be useful. In the present case we must generate our own solutions.

A standard procedure for solving differential equations is to express the solution in a so-called Frobenius series

$$G(\xi) = \sum_{j=0}^{\infty} a_j \xi^{\nu+2j} \tag{22}$$

where the index ν of the new function is given by Eq. (18), and the expansion coefficients a_j satisfy the following recursion relation:

$$a_{j+1} = a_j \frac{(\nu + 2j)/(2k_0^2 d^2) - 1}{(\nu + 2j + 2)^2 - \nu^2}. \qquad (23)$$

For want of another term, I shall refer to the series (22)–(23) as the Cybulski-Silverman or "Bussil" function, to be distinguished from the Bessel function. Again, there is a second solution which diverges at the interface and must be discarded.

Far from the interface, where medium 2 is essentially transparent, the solutions for both polarizations should approach an outgoing plane wave for angles of incidence less than the critical angle and an evanescent wave for angles greater than the critical angle. That this is indeed the case may be seen by examining the series (22) and the comparable series representation of the Bessel function (17), each term of which contains ξ^ν. Substitution of the defining expression (18) for ν leads to

$$\xi^\nu \propto \left[\exp\left(\frac{-z}{2d} \right) \right]^\nu = \exp(i k_0 z \cos \phi_\infty). \qquad (24)$$

Since $\cos \phi_\infty$ is real for $\theta < \theta_c$ and positive imaginary for $\theta > \theta_c$, the solutions behave as expected.

12.3 BEER'S LAW AND FRESNEL'S AMPLITUDES

With the waveforms in the inhomogeneous amplifying medium now known, the imposition of Maxwellian boundary conditions at the interface uniquely determines the amplitudes of the reflected and transmitted waves relative to the amplitude of the incident wave. From the former follow the Fresnel reflection coefficients for each basic polarization

$$r_\sigma = \frac{n_{12} \cos \theta - i \sqrt{-2i\gamma} \left[J_\nu'(\xi_0)/J_\nu(\xi_0) \right]}{n_{12} \cos \theta + i \sqrt{-2i\gamma} \left[J_\nu'(\xi_0)/J_\nu(\xi_0) \right]}, \qquad (25a)$$

$$r_\pi = \frac{\cos \theta - i n_{12} \frac{\sqrt{-2i\gamma}}{1-2i\gamma} [G_\nu'(\xi_0)/G_\nu(\xi_0)]}{\cos \theta + i n_{12} \frac{\sqrt{-2i\gamma}}{1-2i\gamma} [G_\nu'(\xi_0)/G_\nu(\xi_0)]} \qquad (25b)$$

in which $J_\nu'(\xi_0)$ and $G_\nu'(\xi_0)$ are, respectively, the derivatives of the Bessel and Bussil functions with respect to ξ evaluated at the interface $z = 0$, or $\xi_0 \equiv 2k_0 d \sqrt{-2i\gamma}$. In the absence of gain (i.e., either $\gamma = 0$ or $d = 0$), relations (25a, b) reduce to the usual Fresnel formulas

$$r_\sigma = \frac{n_{12} \cos \theta - \cos \phi_\infty}{n_{12} \cos \theta + \cos \phi_\infty}, \qquad (26a)$$

$$r_\pi = \frac{\cos\theta - n_{12}\cos\phi_\infty}{\cos\theta + n_{12}\cos\phi_\infty} \qquad (26b)$$

for a uniform transparent medium.

Figure 12.1 illustrates the theoretical variation of the reflectances $R_\sigma = |r_\sigma|^2$ and $R_\pi = |r_\pi|^2$ as a function of incident angle over the "interesting" part of the angular range, that is, the portion from the vicinity of critical angle, which is 89° in the figure, to grazing incidence where amplification is predicted. From normal incidence to within less than a degree of critical angle, the reflectance curves (not shown) are effectively the same as those deduced from the Fresnel amplitudes for a homogeneous transparent medium; R_σ and R_π differ primarily in the vicinity of Brewster's angle and are essentially indistinguishable in the vicinity of critical angle; in neither case is there enhancement. As the figure shows, however, the reflectance from a medium with an exponential gain profile "overshoots" unity near the critical angle and can become fairly large depending on the value of the gain parameter. The reflectance is sharply peaked about θ_c but remains greater than unity for all angles between θ_c and 90°. Actually, theory shows that the angle at which maximum reflectance occurs decreases slightly from θ_c as γ or d becomes larger.

To determine whether the predicted light amplification by reflection actually occurs and if the Beer's law model accounts for its features, my student and I performed the experiment shown schematically in figure 12.2. An amplifying medium with exponential gain was prepared from a solution of the dye rhodamine B in ethyl and benzyl alcohols by excitation with a pulsed dye laser operating at 540 nm, close to the center of the absorption curve of the dye. At the maximum pump power employed ($\sim$80 kW), the photon density was some five orders of magnitude lower than the density of dye molecules in the experiment, and consequently an exponential attenuation with penetration of the pump beam into the dye solution could be expected.

The dye, which was contained in a pocket milled out of an aluminum block, was covered with a transparent fused-quartz window with refractive index $n_1 = 1.4570$ at the wavelength $\lambda = 633$ nm of the probe beam obtained from a continuous helium-neon laser. Since the refractive index of ethyl alcohol is less than that of quartz, whereas the index of benzyl alcohol is greater, one could set the refractive index n_2 of the gain medium, and hence the critical angle θ_c, approximately by adjusting the composition of the alcohol solution and then precisely by controlling the temperature. The closer n_2 matched n_1, the nearer θ_c approached grazing incidence, and the greater was the predicted light amplification.

The choice of the He-Ne red line for the probe was particularly convenient, for it fell well outside the absorption curve of rhodamine B, so that the unexcited dye was effectively transparent; however, it lay almost at the center of the dye's stimulated emission spectrum. Thus, to construct the reflectance curve $R(\theta)$ for a particular set of experimental conditions, measurements were made as follows.

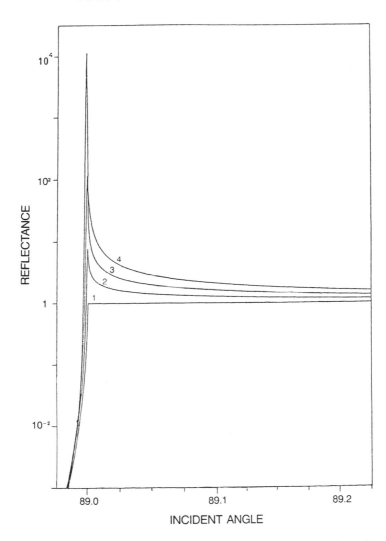

Figure 12.1. Reflectance as a function of incident angle for increasing values of the gain parameter: (1) $\gamma = 0.0$, (2) $\gamma = 1.0 \times 10^{-6}$, (3) $\gamma = 2.0 \times 10^{-6}$, and (4) $\gamma = 3.0 \times 10^{-6}$. The critical angle is $\theta_c = 89°$, and the depth parameter is $d = 96$ wavelengths. R_σ and R_π are indistinguishable over the angular range shown.

First, the dc signal from the continuous-wave probe beam was recorded at a photodetector (P-I-N photodiode) with the dye unexcited. Next, the dye was excited by the pump beam, which led to pulses of 10 ns width superimposed on the dc signal. In principal, the ratio of the pulse height to the dc background (normalized so that the dc signal was unity beyond the critical angle) should have

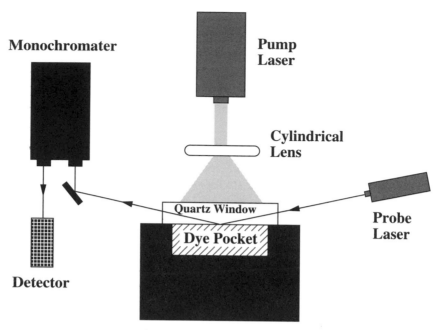

Figure 12.2. Diagram of the enhanced reflectance experiment. Pump beam is 540 nm; probe beam is 633 nm.

given the reflectance. In practice, however, there was fluorescence from the dye molecules that had to be excluded. Most of this fluoresence was removed by passing the reflected light beam through a monochromater set to 633 nm before detecting it, but a weak residual fluorescence at the probe beam wavelength was measured with the probe beam extinguished and then substracted from the 10 ns probe signal.

In the initial experiments a polarizing prism was placed in the path of the probe beam between the helium-neon laser and the dye cell. This permitted separate measurement of R_σ and R_π. However, as the reflectance was found to be independent of polarization for large incident angles (as predicted by the theory), the polarizer was subsequently removed.

Figures 12.3 and 12.4 show examples of the reflection curves that resulted, respectively, when the dye was not excited and when it was pumped under conditions with $\theta_c = 88.8°$ and $d = 58\ \lambda$. In the first case the dye was transparent and no amplification occurred. In the second case the level of pumping engendered a negative absorption parameter ($\gamma \sim 2.6 \times 10^{-5}$) at the probe wavelength with resulting peak reflectance of approximately $R = 2$. Theory and experiment agree well in all respects: line shape, peak location, and peak height. Indeed, many such reflectance curves were measured for various critical angles and depth parameters, and in all cases the theory accounted for the

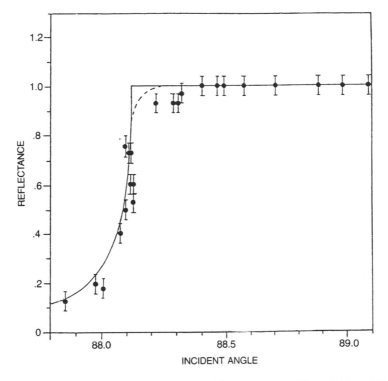

Figure 12.3. Reflection from a dye region with no pumping. The solid line follows from the Fresnel reflectance formulas for a transparent medium. The dashed line is the Fresnel reflectance averaged over beam divergence. Critical angle $\theta_c = 88.12°$.

observed results quite satisfactorily. The full series of measurements verified the increased amplification as θ_c approached grazing incidence.

It is to be noted that the region of enhancement in figure 12.4 is not a sharp spike, as portrayed in figure 12.1, but a more rounded peak. Although the details of analysis must be left to the cited references, the primary reason for this difference in shape is the divergence of the probe beam. To account for the finite divergence of the probe—whose TEM_{00} mode had a Gaussian profile of half-angle 0.045° across the beam waist—the theoretical reflectance curves were averaged over an incident angle with a Gaussian distribution function. This averaging, which represented the experimental conditions more closely, "rounded" the peak, but had little effect elsewhere on the reflectance curve.

Despite the smoothing effect of beam divergence, the sharpness of the resonance-like maximum can nevertheless be gauged by the horizontal scale: Critical and incident angles were measured to a precision of ±0.01°. This uncertainty is indicated by the width of the points in the figures. Like Newton's

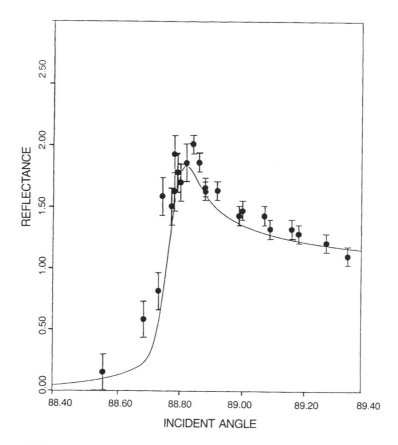

Figure 12.4. Reflection from a laser-pumped dye with thickness parameter $d = 58\,\lambda$ and gain parameter $\gamma \sim 2.6 \times 10^{-5}$. The solid line is the theoretical reflectance curve averaged over beam divergence.

measurement of the diameter of a very small aperture (see chapter 5), the attainment of this level of precision was actually not difficult. Experimentally, the probe beam was detected at a distance of 5 m from the dye cell, which permitted minuscule variations in angle to be converted to measurable variations in length. The uncertainty in reflectance (vertical error bars) at angles greater than about 0.05° past θ_c was determined by the resolution of the oscilloscope scale. For smaller angles, pulse-to-pulse fluctuations in the pump power led to a greater variation in the signal.

At the time these experiments were undertaken, the existence of enhanced reflection was a confused and controversial issue. Theoretical analyses disagreed with one another and, in any event, were incapable of accounting for reported observations, which were orders of magnitude too large, nonreproducible, and attributable to other phenomena such as photoinduced or thermally induced refractive index gradients. The investigations recounted here showed, among other things, that under well-defined experimental conditions, light amplification by reflection did indeed exist and was amenable to treatment within the framework of classical electrodynamics. In no case was there a need to invoke additional mechanisms to account for anomalously high levels of amplification.

Although this theoretical study has led to unambiguous reflectance formulas in agreement with experiment, it is still instructive to inquire why, physically, amplification occurs. In interesting contrast to the laser, we have here a mechanism of stimulated emission by an evanescent wave, for, though the probe beam propagates outside the excited medium, the exponential "tail" of the reflected wave penetrates to an extent marked by the depth parameter d.

In addition, although the reflected wave serves as the "trigger" to release the energy stored in the pumped medium, it is the transmitted wave that transports that energy to the surface. A close examination of the series representation for either σ- or π-polarized light in the excited medium shows that it consists of a linear superposition of two series representing waves that propagate respectively away from, and toward, the interface. There is no violation of physical law, for, as Eq. (24) demonstrates, only an outgoing wave is to be found in the far-field limit where medium 2 is effectively transparent. Nevertheless, for enhanced reflection to take place, both types of waves must be present simultaneously in the region with gain, their relative amplitudes and phases being specified by the theory with no arbitrariness. In the Beer's law gain model there is an angular region, including normal incidence, for which only the transmitted wave traveling away from the interface exists; thus, for a given set of experimental parameters, the theory predicts a lower limit to the angle of incidence at which the reflected wave will be amplified. No such lower limit occurs in the simple extension of the Fresnel amplitudes to a medium with uniform gain.

One of the most significant differences between ordinary reflection—that is, from an interface between two transparent media—and enhanced reflection concerns the energy transport by the reflected and transmitted waves. Since the component of the transmitted wave traveling toward the interface enhances the reflected wave, the transmitted wave must be comparable in magnitude to, or larger than, the reflected wave if amplification is to occur. Thus, the reflectance R is large when the transmittance T is large in marked contrast to the case of ordinary reflection in which the Fresnel formulas lead to the relation $R + T = 1$ in conformity with energy conservation. In enhanced reflection, the reflecting medium itself is a source of energy.

The power flux of the transmitted wave in the exponentially nonuniform gain region is given by the time-averaged Poynting vector

$$\mathbf{S}_t = \frac{c}{8\pi} \mathrm{Re}\{\mathbf{E}_t \times \mathbf{H}_t^*\}. \tag{27}$$

Of particular interest is the transmittance

$$T = \frac{\mathbf{S}_t \cdot \mathbf{z}}{\mathbf{S}_i \cdot \mathbf{z}} \tag{28}$$

(where $\mathbf{S}_i$ is the Poynting vector of the incident wave) in the far-field region ($z \gg d$). For illustrative purposes, let us consider the case of σ-polarization; the electric field $\mathbf{E}_\sigma$ is given by Eq. (12), and the associated magnetic field is $\mathbf{H}_\sigma = (\mathbf{k}_2/k_0) \times \mathbf{E}_\sigma$ as deduced from Maxwell's equations (see Eq. [11.11]). The analytical details are straightforward, but cumbersome, and lead in the limit $z \gg d$ to the expression

$$T = \left| \frac{2n_{12}(\xi_0/2)^2 \cos\theta / \Gamma(\nu+1)}{(n_{12}\cos\theta + \cos\phi_0)J_\nu(\xi_0) - i(\xi_0/2k_0d)J_{\nu+1}(\xi_0)} \right|^2 \frac{\cos\phi_0}{\cos\theta} \tag{29}$$

(where Γ is the gamma function) for incident angles less than critical angle; for $\theta \geq \theta_c$, the transmittance is null since $\mathbf{S}_t$ vanishes identically. Figure 12.5 shows a comparison of the far-field transmittance and reflectance as a function of incident angle; both curves peak at values greater than unity at virtually the same angle. The transmittance, however, attains values approximately two orders of magnitude greater than the reflectance over the angular domain below critical angle.

12.4 SINGULARITIES—OR IS THERE AN UPPER LIMIT?

With a successful theory of enhanced reflection at hand, it is natural to ask what experimental conditions—critical angle, gain, penetration parameter— will produce the largest amplifications. Indeed, what might be a practical upper limit to enhanced reflectance?

Unfortunately, it is difficult to answer these questions analytically from the rather complicated expressions for the Fresnel amplitudes r_σ and r_π. Computer simulations, however, have led to large values of reflectance—in excess of 10^4, for example—at certain gain parameters and incident angles. It would appear, in fact, that the calculated reflectance actually becomes infinite under these special conditions. Singularities in the reflectance violate energy conservation and therefore signify that the limits of validity of one or more assumptions underlying the theory have been surpassed.

Such violations should perhaps not be too surprising, given the assumption that the gain parameter γ is independent of the energy loss from the amplifying medium. Consider, for example, a beam propagating (along the z-axis) through a uniform amplifying medium. The rate at which the power per unit area (S_t) of this beam increases with penetration is described by an equation of the form

$$\frac{dS_t}{dz} = \mathcal{G}S_t \tag{30a}$$

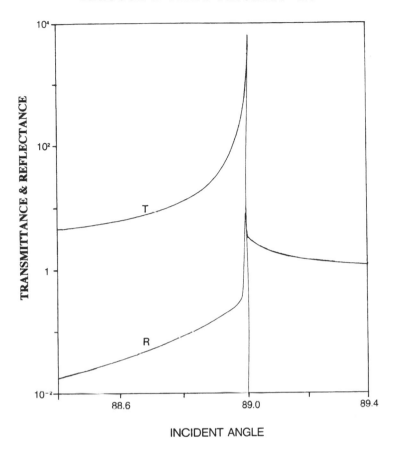

Figure 12.5. Far-field transmittance T and reflectance R as a function of incident angle for $\theta_c = 89°$, $\gamma = 2.6 \times 10^{-5}$, and $d = 96\ \lambda$. T is strictly 0 beyond critical angle.

in which

$$\mathcal{G} = 2n\gamma k_0 \tag{30b}$$

is the gain per unit of length.[3] Equation (30a) leads to exponential growth if $\mathcal{G}$ is independent of S_t, but this cannot be the case generally.

Let us examine a simple two-level model of an optical medium with ground and excited-state population densities n_g and n_e. The law of molecular conservation requires that $n_g + n_e = N = $ constant, where N is the total dye molecule concentration. If σ_e and σ_a are, respectively, the cross sections for stimulated emission and absorption at the frequency ν_t of the transmitted beam (same frequency as the probe beam), then the gain $\mathcal{G}$ is expressible in terms of molecular properties by the relation

$$\mathcal{G} = \sigma_e n_e - \sigma_a n_g. \tag{31}$$

For the purpose of illustrating the basic dependence of $\mathcal{G}$ on the intensities of the pump and transmitted beams, it is adequate to assume singlet ground and excited states and to neglect any nonradiative energy-loss mechanisms.

Under steady-state conditions, the rate of molecular excitation within the dye medium, pumped by a beam of power flux S_p and supporting a transmitted flux S_t (produced by reflection of the probe beam), is attributable to four processes as described by the following equation:

$$\frac{dn_e}{dt} = \sigma_{ap}n_g\frac{S_p}{h\nu_p} - \frac{n_e}{\tau} - \sigma_e n_e\frac{S_t}{h\nu_t} + \sigma_a n_g\frac{S_t}{h\nu_t} = 0. \tag{34}$$

The first term, in which σ_{ap} is the absorption cross section at the pump frequency ν_p, represents the increase in excited-state population by absorption from the pump beam. The second term, in which τ is the mean excited-state lifetime, signifies loss of excited states by spontaneous emission (fluorescence). The third term represents a further loss by stimulated emission at the frequency ν_t. The last term characterizes the repopulation of the excited state by absorption of photons emitted by molecules of the medium.

Combining relations (31) and (34) leads to a gain

$$\mathcal{G} = \frac{N(\sigma_e\sigma_{ap}S_p/h\nu_p - \sigma_a/\tau)}{\sigma_{ap}S_p/h\nu_p + (\sigma_e + \sigma_a)S_t/h\nu_t + 1/\tau} \tag{35}$$

that clearly depends on the intensity S_t of the transmitted wave.

In the vicinity of a reflectance singularity, the term in the denominator containing S_t dominates the expression, for it was demonstrated in the previous section that under the conditions of enhanced reflection a large reflectance is accompanied by a large transmittance. Thus, as S_t becomes increasingly large, $\mathcal{G}$ approaches zero. This leads to an inconsistency, however, since a transparent region cannot produce a transmitted wave with power flux greater than that of the incident wave. The gain profile of an optically pumped medium, therefore, cannot have a simple exponential dependence when the transmitted wave is sufficiently large.

Equation (35) allows one to determine the conditions under which the Beer's law gain model is expected to be valid. If we assume—as was experimentally the case—S_p sufficiently low that depletion of the population inversion by the transmitted wave is negligible, yet high enough that the term σ_a/τ can be omitted from the numerator, then the expression for gain reduces to

$$\mathcal{G} = \frac{N\sigma_e\sigma_{ap}(S_p/h\nu_p)}{\sigma_{ap}S_p/h\nu_p + 1/\tau}. \tag{36}$$

The condition is not difficult to meet since σ_a is ordinarily much smaller than σ_e. Let us further suppose that the effective ground-state lifetime—that is, the reciprocal of the pumping rate—is much longer than the excited-state lifetime[4]

$$\frac{h\nu_p}{\sigma_{ap}S_p} \gg \tau. \tag{37}$$

In this case G becomes

$$G = \left(\frac{N \tau \sigma_e \sigma_{ap}}{h \nu_p} \right) S_p \tag{38}$$

and is directly proportional to S_p. The gain therefore has the same spatial dependence as the pump beam, and an exponentially decreasing gain is realized in accordance with Beer's law.

Upon increasing S_p further so that the inequality (37) is reversed and the ground-state lifetime becomes shorter than the excited-state lifetime, depopulation occurs primarily by stimulated emission, and G is simply a constant

$$G = N \sigma_e. \tag{39}$$

The gain is then spatially uniform until enough energy has been removed from the pump beam so that the term containing S_p no longer dominates the denominator of Eq. (36).

Although no physical model depicts nature exactly, the Beer's law gain model is nevertheless valid over a wide range of experimental conditions. For one thing, the occurrence of reflectance singularities was found to be rare for the range of experimental conditions surveyed. Also, computer calculations showed that the transmittance could be about two orders of magnitude larger than the reflectance for incident angles below critical angle, and yet contribute less than 10% to the denominator of Eq. (35) for a reflectance as great as 100. Roughly, therefore, the applicability of the model described here becomes questionable only over the extremely narrow range of incident angles for which the peaks of the reflectance curves may well exceed 100. Finally, even if the model is not strictly valid at, and in the immediate neighborhood of, a singularity, it may still be expected to work reasonably well when account is taken of the small divergence of the probe beam. One must then average the theoretical reflectance at each incident angle over the angular width of the beam. The angular width of the peak in the reflectance curve as a singularity is approached is extremely narrow (less than 0.01°) and therefore makes virtually no contribution to the averaged theoretical signal.

To go beyond a model based on Beer's law and contruct a comprehensive theory of enhanced reflection, presumably free of singularities, requires that the dependence of gain on the transmitted power flux be part of the solution of Maxwell's equations for the transmitted wave. This would lead to a highly nonlinear wave equation for which an analytically tractable result would almost assuredly exceed the mathematical virtuosity of even Maxwell himself.

12.5 PRACTICAL THOUGHTS

Apart from pioneering experiments of principle, of what use is the amplification of light by reflection? The laser, found everywhere from the eye surgeon's office to the grocery store checkout counter and myriad points in between, is perhaps as familiar to scientist and layman alike as the telephone. By contrast, I have found that enhanced reflection (ER) is an obscure phenomenon even to most scientists I know. And yet someday it, too, may be as familiar as the telephone; indeed, when future electromagnetic communications are effected in an all-optical system, ER may well become an indispensable part of the telephone system.

There are advantages to ER that make it particularly suitable for applications in telecommunications. For one thing, the reflecting wave interacts with the active medium within a fairly narrow region—typically less than 100 wavelengths—close to the interface. Bulk inhomogeneities in the active medium, therefore, do not seriously distort or degrade the waveform. Moreover—and this is a significant point—small quantities of lasing materials are required, and only small regions need to be pumped.

In conventional fiber optics, light is carried in a filament of glass or quartz surrounded by a material (the cladding) of lower refractive index. Although the characteristics of the transmitted waves are most fully elucidated by treating the fiber as a cylindrical waveguide and solving Maxwell's equations for the allowed modes, certain key features can be understood in terms of simple reflection and refraction of rays. Consider a cone of radiation entering the end of the core of a fiber. Since the light originates in the medium of higher refractive index, there is a range of incident angles down to some threshold critical angle for which all rays are totally reflected at the boundary between the core and cladding. One may think of these rays, therefore, as propagating through the fiber by multiple total internal reflections.

For long-distance transmission to be practical, optical losses in the fiber need to be kept low. For example, in a fiber with attenuation coefficient of 1 dB/km, the transmittance through a 100 km fiber would be 1%.[5] Because an impurity level of only a few parts in 10^9 of certain elements can produce an attenuation of 1 db/km, the requirements on fiber purity are stringent. However, if it were possible to amplify at regular intervals the light propagating through a long fiber, the prospects of attaining an all-optical telecommunication system would be so much the better. Suppose, therefore, that the cladding surrounding the fiber core is not a passive substance, but a material that can be excited by optical pumping. Under the conditions delineated above, total internal reflection would give rise to enhanced reflection. It was precisely by such means that evidence for ER was first adduced in a fiber with passive glass core and neodymium-doped glass cladding.[6]

Incorporated into a fiber optic communication system, ER can be used to amplify a light signal without the need for conversion into an electronic signal followed by subsequent electronic amplification and optical reconversion. In addition, ER-based devices can be employed as high-speed optical switches; by appropriate design, a signal could be made to propagate only when the cladding is excited.

In Fresnel's time, the study of reflection helped elucidate the nature of light. During the twenty-first century, the same process will help carry that light around the world.

NOTES

1. A. Schawlow, "Laser Light," *Scientific American* (September 1968), reprinted in *Lasers and Light: Readings from Scientific American* (San Francisco: W. H. Freeman, 1969), pp. 282–90. Quotation on p. 282.

2. R. F. Cybulski Jr. and M. P. Silverman, "Investigation of Light Amplification by Enhanced Internal Reflection. Part I: Theoretical Reflectance and Transmittance of an Exponentially Nonuniform Gain Region," *J. Opt. Soc. Am.* 73 (1983): 1732–38; M. P. Silverman and R. F. Cybulski Jr., "Investigation of Light Amplification by Enhanced Internal Reflection. Part II: Experimental Determination of the Single-Pass Reflectance of an Optically Pumped Gain Region," *J. Opt. Soc. Am.* 73 (1983): 1739–43.

3. Eq. (30b) follows from the form of the refractive index, Eq. (5), with $d = \infty$. The transmitted wave is then proportional to $e^{ink_0 z} e^{n\gamma k_0 z}$, and the relative intensity is $e^{2n\gamma k_0 z} = e^{gz}$.

4. Under conditions of strong pumping, when the lifetime of the excited state is longer than that of the ground state, an atom or molecule can be driven by absorption and stimulated emission back and forth between ground and excited states multiple times during the passage of a single light pulse. This leads to a variety of interesting nonlinear optical effects such as the restitution of quantum beats in radiative decay. Such phenomena are discussed in the chapter "Quantum Boosts and Quantum Beats" of my book, *More Than One Mystery: Explorations of Quantum Interference* (New York: Springer Verlag, 1995).

5. The attenuation α of a fiber of length L and transmittance T (i.e., the ratio of output to input power) is given by $\alpha L = -(10 \text{ db}) \log_{10} T$.

6. C. J. Koester, "Laser Action by Enhanced Total Internal Reflection," *IEEE J. Quantum Electronics* QE 2 (1966): 580–84.

CHAPTER 13

A Penetrating Look at Scattered Light

> I've learned that you can't hide a piece of broccoli in
> a glass of milk.
> *(Anonymous 7-year-old)*[1]

VIRTUALLY EVERY aspect of physical optics is an example of light scattering—even the undeviated transmission of light. Because of light scattering, the purest dust-free atmosphere is still turbid. When light is multiply scattered, a sample may become so turbid as to appear opaque. But, in fact, it is not. Selective detection of polarized light can reveal properties of both the ambient medium and the particles suspended in it. More remarkable still, using a photoelastic modulator, one can actually look into a turbid medium and see embedded objects otherwise invisible to the unaided eye.

13.1 HARMONY OF A SPHERE

There is an old adage that to the worker with only a hammer, every problem looks like a nail. I think of that sometimes when I consider the evolution of light scattering: every scatterer, more or less, looked like a sphere. This is of course an overstatement, but not without truth. Spheres are useful; it is possible to obtain exact analytical expressions for the scattering of waves from isotropic systems and very few others (e.g., ellipsoids and cylinders). Moreover, despite the high degree of symmetry and seeming simplicity of a spherical geometry, the resulting unapproximated solutions, derived by G. Mie in 1908,[2] are not at all simple, but consist of complicated infinite series of Legendre functions and two variations of spherical Bessel functions (Ricatti-Bessel and Ricatti-Hankel).

Light scattering as a distinct subject did not begin with the full complexity of Mie's theory. Rather, it began with a question that must have perplexed curious people for millennia: Why is the sky blue? Later, following Arago's chance observation (~1809) of skylight through a Nicol prism, there arose a second question equally puzzling: Why is skylight polarized (in fact, nearly 100% when received from the zenith, the Sun lying close to the horizon)?

The subtleties of these questions are perhaps not fully appreciated by those of us who have learned the "right" answer early in our student careers and cannot imagine the result being otherwise. Nevertheless, the subject remained a contentious scientific issue for a long time, and even sharp-minded thinkers provided answers wide of the mark. Newton, for example, thinking of the interference colors of thin films, attributed the blueness to interference of light reflected from tiny water droplets in the air. Although this model would be criticized and modified by others, the confusion of reflection and light scattering was to persist well into the nineteenth century.

Among the inconsistencies of Newton's explanation was the fact that, if the air of a cloudless sunny day actually contained all these drops, then the Sun, Moon, and stars would be permanently surrounded by diffraction coronas, which is not the case. Furthermore, a clear blue sky is not accompanied by permanent rainbows as one might expect to find in an atmosphere filled with water drops. Reluctant to dismiss Newton's model entirely, however, German physicist Rudolph Clausius—known best for his contributions to thermodynamics—replaced the droplets with thin-walled bubbles for which the reflection-interference hypothesis would seem more appropriate. But an atmosphere filled with hollow bubbles was hardly more plausible than solid globules. Apart from the troubling question of how such bubbles might form, the objection was also raised that the sky is of a much purer blue than the color which selective reflection of white sunlight could produce.

It was the experiments of John Tyndall (~1869) that ultimately provided the first data critical to unraveling the mystery of skylight. Simulating smog by means of photochemically generated aerosols, Tyndall noted the azure blueness of the light scattered from droplets believed to be much smaller in size than visible wavelengths. More auspicious still, his painstaking measurements of the polarization of scattered light revealed a maximum degree of polarization at a scattering angle of 90°. As the particles comprising the aerosols increased in size, the color of the scattered light whitened and the degree of polarization diminished. Thus did Tyndall's model system reproduce the salient features of skylight, but the nature of the underlying process eluded him. Thinking still in terms of reflection, he wrote, "the polarization of the beam by the incipient cloud has thus far proved itself to be *absolutely independent of the polarizing-angle*. The law of Brewster does not apply to matter in this condition; and it rests with the undulatory theory to explain why. Whenever the precipitated particles are sufficiently fine, no matter what the substance forming the particles may be, the direction of maximum polarization is at right angles to the illuminaing beam. . . . This I consider to be a point of capital importance."

Lord Rayleigh (then still John Strutt), cited Tyndall's words a few years later in his groundbreaking paper (1871) on light scattering, and replied: "As to the importance there will not be two opinions; but I venture to think that the difficulty is imaginary, and is caused mainly by misuse of the word reflection. Of course there is nothing in the etymology of reflection or refraction to forbid their application in this sense; but the words have acquired technical meanings, and become associated with certain well-known laws called after them."[3] These well-known laws like Snel's law and the Fresnel relations, which Rayleigh had no need to enumerate, were applicable only if the surface of the disturbing body was much larger than many square wavelengths, a condition clearly violated in the case of Tyndall's finest aerosols.

Working within the framework of the elastic-solid model of light—for at the time Maxwell's theory of electromagnetic waves was still untested and by no means widely understood or accepted—Rayleigh first deduced the wavelength dependence of his eponymous process by a simple and incisive use of dimensional analysis. The sought-for ratio of scattered and incident light amplitudes is a dimensionless number, and from the nature of the problem the spatial variables upon which it could depend were limited to the particle volume V, distance r from the scatterer to the observation point, and the wavelength λ. Thus $V/r\lambda^2$ was the only dimensionless combination consistent with the dynamics in question (the far-field amplitude of a radiated wave must diminish as $1/r$) and it then followed that the corresponding ratio of light intensities must vary inversely with λ^4. Furthermore, this variation was decisive in distinguishing between scattering from small spheres, as Rayleigh proposed, and reflection from thin plates, the process envisioned since Newton. In the latter case, apart from the wavelength, there is only the plate thickness d, and the intensity ratio goes as $(d/\lambda)^2$.

The model of a scalar wave in an isotropic elastic ether does not accommodate the property of light polarization in a consistent way, but Rayleigh, like Fresnel, assumed transverse vibrations at the outset and arrived at the correct $\cos^2 \theta$ angular distribution for the flux of π-polarized light scattered at an angle θ to the incident direction.[4] For σ-polarized light, the angular distribution was clearly isotropic, since the vibrations of the incident and outgoing waves remained parallel irrespective of the direction of scattering. Thus, Rayleigh's analysis yielded the characteristic linear intensity difference (Eq. [9.38]) for small-particle scattering

$$\delta_{\mathrm{L}} = \frac{I_\sigma - I_\pi}{I_\sigma + I_\pi} = \frac{\sin^2 \theta}{1 + \cos^2 \theta} \tag{1}$$

although he did not explicitly record this relation. In the above equation I_σ and I_π are the scattered light intensities of designated polarization, and δ_{L} attains its maximum value of 1 for $\theta = 90°$.

Although Rayleigh's analysis accounted well for the characteristic features of skylight, there remained an unsettled matter of great importance. Whether it would have in good time occurred to Rayleigh on his own I cannot say (although I do believe it likely), but the issue was in any event brought to his attention by Maxwell in a letter dated two years later (1873). In brief, Maxwell wanted to know the size of the molecules of air. Writing from his estate in Scotland, he could not calculate the desired figure for himself, as he had left Rayleigh's papers containing the appropriate equations in Cambridge, and therefore requested that

Strutt "stick the data into your formula and send me the result." Maxwell was kindly, but direct.

The substitution must not have seemed so simple to Rayleigh, for the question remained in his mind over twenty-five years. And when he wrote his definitive paper (1899) on the transmission of light through the atmosphere,[5] he had at last recognized the essential contribution of *molecular* scattering. "It will appear," he concluded, "that even in the absence of foreign particles we should still have a blue sky." At this point, on the threshold of the new century, I would say that the two fundamental questions of skylight were finally answered.

To the extent that the ultimate objective of physics is to account for the broadest range of phenomena with the narrowest conceptual base, the long-standing confusion between reflection and scattering is not merely an irrelevant historical detail. On the contrary, it illuminates fundamental and far-reaching connections that even today few physicists appreciate. Although all scattering may not be equated with reflection, it is completely accurate to regard reflection (and refraction) as legitimate examples of light scattering by electric and magnetic dipoles. Indeed, starting from this premise one can arrive at precisely the Fresnel formulas, albeit with a mathematical effort greater than that ordinarily expended in deducing these relations directly from Maxwell's equations with no explicit attention to the molecular constitution of matter. I have rarely seen this link to scattering stressed except, perhaps, in addressing the question of Brewster's law.

Why is there no reflected wave when π-polarized light illuminates the plane surface of a nonconducting material at Brewster's angle? The familiar answer is that the incident wave induces oscillating electric dipole moments in the penetrated medium which radiate secondary wavelets whose superposition constitutes the reflected wave. At Brewster's angle, however, the axis of each dipole coincides with the propagation direction of the reflected wave (if there were one), but since a dipole does not radiate along its axis, there is consequently no reflected wave.

Although not actually incorrect, the preceding explanation is incomplete and problematic. First, one might infer by such reasoning that σ-polarized light is always reflected at every angle of incidence, for in no case would the axes of the radiating dipoles and the propagation direction of the reflected wave coincide. This is not true. There is a Brewster angle for reflection of σ-polarized light from magnetic media (permeability $\mu \neq 1$). Second, there is also a Brewster's angle for internal reflection in which case the penetrated "medium" may be the vacuum devoid of all material dipoles. What, then, generates the reflected wave? There is a certain irony in the fact that, applied uncritically, the scattering theory of reflection falters over precisely the issue that troubled the reflection theory of scattering, namely, Brewster's law.

Applied carefully, however, scattering theory leads to a complete and consistent derivation of the Fresnel relations and a physical explanation of all the phenomena (e.g., Brewster's law) contained therein. The analysis is, in fact, an application of the Ewald-Oseen extinction theorem of which I have already spoken.[6] In contrast to the treatment of reflection and refraction based on differential equations as embodied in the familiar formulation of Maxwell's theory, the extinction theorem is formulated in terms of integral equations. The volume integral containing the dipole-generated wave in the refractive medium can be reduced to a surface integral by means of Green's theorem (the same mathematical basis underlying the treatment of diffraction). Thus, despite the ultimate emergence of surface waves, all the induced dipoles of a material medium— and not just those near the surface—contribute to the reflected and transmitted waves for both ordinary and internal reflection.

Examined from the perspective of scattering, the mathematical form of the Fresnel relations is amenable to a most interesting interpretation.[7] Each reflection and transmission amplitude is expressible as a product of two factors: (1) a dynamical scattering function D characteristic of the isolated dipoles and dependent upon the polarization of the incident wave, and (2) a kinematical scattering function A, dependent only on the angles of incidence (θ) and transmission (ϕ), which characterizes the coherently excited array of dipoles within the medium. For example, for a magnetic medium within which electric ($\mathbf{E}$) and magnetic ($\mathbf{H}$) fields excite electric ($\boldsymbol{\mu}_E$) and magnetic ($\boldsymbol{\mu}_M$) dipoles of volume V, the dynamical (or dipole) function D is

$$D = \frac{4\pi}{V} |\hat{\mathbf{k}} \times \hat{\mathbf{k}} \times \boldsymbol{\mu}_E + \hat{\mathbf{k}} \times \boldsymbol{\mu}_M| \tag{2a}$$

in which $\hat{\mathbf{k}}$ is a unit vector in the direction of propagation of the scattered wave; the induced dipoles are related to the inducing fields by the companion expressions

$$\boldsymbol{\mu}_E = \frac{(\varepsilon - 1)}{4\pi} V \mathbf{E}, \tag{2b}$$

$$\boldsymbol{\mu}_M = \frac{(\mu - 1)}{4\pi} V \mathbf{H}. \tag{2c}$$

It is assumed that the transmitted fields $\mathbf{E}$ and $\mathbf{H}$ arise from an electromagnetic wave incident in vacuum (medium 1). If, however, medium 1 is itself material, then the symbols ε and μ represent, respectively, the relative permittivity ($\varepsilon_2/\varepsilon_1$) and permeability ($\mu_2/\mu_1$) of the two media.

The Fresnel reflection coefficients for the two basic states of linear polarization may then be written as

$$
\begin{aligned}
r_\sigma &= D_\sigma A \\
&= \left(-\frac{(\varepsilon - 1)\sqrt{\mu} - (\mu - 1)\sqrt{\varepsilon} \cos(\theta + \phi)}{(\varepsilon - 1)\sqrt{\mu} + (\mu - 1)\sqrt{\varepsilon} \cos(\theta - \phi)} \right) \times \left(\frac{\sin(\theta - \phi)}{\sin(\theta + \phi)} \right),
\end{aligned} \tag{3a}
$$

$$r_\pi = D_\pi A$$

$$= \left(-\frac{(\mu - 1)\sqrt{\varepsilon} - (\varepsilon - 1)\sqrt{\mu}\cos(\theta + \phi)}{(\mu - 1)\sqrt{\varepsilon} + (\varepsilon - 1)\sqrt{\mu}\cos(\theta - \phi)} \right) \times \left(\frac{\sin(\theta - \phi)}{\sin(\theta + \phi)} \right), \quad (3b)$$

where the first factor (D_σ, D_π) is the dipole function appropriate to the specified polarization, and the second factor (A) is the array function common to all states of polarization. In the case of nonmagnetic media $(\mu = 1)$, the dipole functions reduce to

$$D_\sigma^{\mu=1} = 1, \qquad (4a)$$

$$D_\pi^{\mu=1} = \frac{\cos(\theta + \phi)}{\cos(\theta - \phi)}, \qquad (4b)$$

and Eqs. (3a, b) become identical to the Fresnel coefficients (Eqs. [11.14a, b]) discussed in chapter 11.

The existence of a Brewster's angle is determined by the vanishing of D, since there is no angle (for $\varepsilon \neq 1$) at which A is null. Nonmagnetic media Eqs. (4a, b) reproduce the well-known facts that only π-polarized light can be 100% suppressed by reflection, and that the polarizing angle θ_B^π satisfies $\theta_B^\pi + \phi = \pi/2$, or, from Snel's law, $\tan\theta_B^\pi = \sqrt{\varepsilon} = n$. In the more general case of a homogeneous dielectric medium with arbitrary ε and μ, Eqs. (3a, b) lead to polarizing angles for both σ- and π-polarized light

$$\tan\theta_B^\sigma = \sqrt{\frac{\mu(\mu - \varepsilon)}{\varepsilon\mu - 1}}, \qquad (5a)$$

$$\tan\theta_B^\pi = \sqrt{\frac{\varepsilon(\varepsilon - \mu)}{\varepsilon\mu - 1}}. \qquad (5b)$$

The factorization of the Fresnel amplitudes into dipole and array functions has a direct counterpart in the theory of diffraction and interference. This is not a coincidence, but another example of the conceptual ties between ostensibly unrelated optical processes—all of which are manifestations of light scattering. The array theorem, as this counterpart is known, addresses the following question. What is the far-field (Fraunhofer) diffraction pattern of an object consisting of an arbitary number of identically shaped apertures? This is a problem of notable importance to both image processing and molecular structure analysis. The X-ray diffraction pattern of a molecular crystal, for example, is formed from the superposition of electromagnetic waves scattered by all the (identical) molecules comprising the crystal lattice. How this pattern relates to the diffraction pattern of a single molecule whose structure is being sought lies at the heart of many a crystallographic study.

The reader will recall that the Fraunhofer amplitude $\Psi(\mathbf{x})$ of a wave diffracted by an object and received at a distant observation point $\mathbf{x}$ is the Fourier transform of the amplitude distribution over the diffraction plane. It is not difficult to show that, for a composite object, this distribution is the convolution (Eq. [7.32]) of the mathematical functions representing (1) the wave illuminating an individual aperture and (2) the spatial distribution of the apertures in the array. According to the array theorem,[8] the total diffraction amplitude $\Psi(\mathbf{x})$ is then simply the product

$$\Psi(\mathbf{x}) = \psi(\mathbf{x})A(\mathbf{x}) \tag{6}$$

of the wave $\psi(\mathbf{x})$ diffracted by a single aperture and the Fourier transform $A(\mathbf{x})$ of the geometric array.

Like Eqs. (3a,b), Eq. (6) provides an insightful interpretation of a scattering pattern otherwise deducible by uninspired evaluation of a complex diffraction integral. For example, one can look upon the two-slit interference experiment (chapter 6) as the simplest application of the array theorem. Consider the one-dimensional case of two infinite slits of width a and center-to-center separation d, illuminated by light of wavelength λ received at an observation screen a distance $z \gg d^2/\lambda$ from the diffraction plane. As derived in any optics text, the single-slit diffraction amplitude is the familiar "sinc" function

$$\psi(x) = \frac{\sin \alpha}{\alpha} \tag{7a}$$

with argument

$$\alpha = \frac{\pi a x}{\lambda z} \tag{7b}$$

where x is the distance of the image point from the intersection of the optic axis (z) with the image plane. The array function in this case—a function of the lateral coordinate ξ in the diffraction plane—is simply the sum of two Dirac delta functions

$$A(\xi) = \delta(\xi - \tfrac{1}{2}d) + \delta(\xi + \tfrac{1}{2}d), \tag{8}$$

the Fourier transform of which readily yields

$$A(x) = 2\cos\beta \tag{9a}$$

with associated argument

$$\beta = \frac{\pi d x}{\lambda z}. \tag{9b}$$

The product of Eqs. (7a) and (9a) gives the composite diffraction amplitude $\Psi(x)$. Although one commonly describes the resulting light intensity pattern $|\Psi(x)|^2$ as a modulation of two-slit interference by the single-slit diffraction "envelope," the array theorem makes the mathematical basis of this interpretation explicit. Its utility becomes all the more apparent in more complex situations involving diffraction from two-dimensional objects with larger number of apertures.

There is yet another way in which the outwardly dissimilar processes of reflection/refraction and diffraction/interference—both perceived as forms of light scattering—show unifying connections. I have discussed earlier in conjunction with the Newton two-knife experiment (chapter 5) that the treatment of diffraction in terms of geometric and edge waves provides an alternative to the standard Fresnel-Kirchhoff formulation which superposes radiation from fictitious point sources over the "empty space" of an aperture. Although, in a certain sense, the extinction theorem does something similar for reflection by converting radiation from volume sources into surface waves, the parallel is even more directly seen from the unusual perspective of Oliver Heaviside.

According to Heaviside, the incident wave creates a boundary source that travels along the interface between two refractive media at a speed $v = \frac{c/n_1}{\sin\theta}$ where θ is again the angle of incidence.[9] If v exceeds the phase velocity of light ($v_1 = c/n_1$ in incident medium 1 and $v_2 = c/n_2$ in medium 2), the boundary source radiates a wedgelike bow wave similar to that of an optical sonic boom. Heaviside demonstrated that superposition of the contributions from all source points on the boundary leads to the reflected and transmitted waves. The process is in fact an analog for bound sources of the Cerenkov radiation produced by free relativistic charged particles; Cerenkov radiation, however, was not reported until some three decades later.

If the awesome splendor of Maxwell's kingdom lies in its inclusive embrace of all electrical, magnetic, and optical phenomena of a nonquantum nature, then, within the extensive subdomain of optics alone, it is the concept of light scattering that ties all phenomena together. And there is perhaps no more breathtaking illustration of the full sweep of this broad synthesis than—to return to our starting point—Mie's theory of scattering from spheres. Out of this theory one sees explicitly and quantitatively, not merely enticing analogs and similarities, but the rigorous emergence of the distinct laws of geometrical and physical optics from a continuum of special cases. A comprehensive treatment of Mie theory lies well beyond the envisioned scope of this book and is, in any event, handled expertly in van de Hulst's classic monograph.[10] For the reader, however, who has never wandered through this esoteric province of physics, the following synopsis succinctly lays bare its basic structure and unifying connections.

Mie theory provides a complete vector-field solution to Maxwell's equations applied to the scattering of electromagnetic plane waves from a homogeneous sphere of arbitrary size and refractive index. The theory reminds me very much in some ways of Einstein's general relativity, the exact solutions of which are few in number, arduous to obtain, and in general difficult to interpret, but whose basic formulation rests upon equations of exquisite beauty and compelling simplicity. Electromagnetic waves of frequency ω scattered by a homogeneous object of any symmetry, spherical or otherwise, must satisfy the vector wave equation

$$\nabla^2 \mathbf{F} + k^2 \mathbf{F} = 0 \tag{10}$$

where $k = n\omega/c = nk_0$ is the wavenumber in a medium of refractive index $n = \sqrt{\varepsilon\mu}$, and **F** represents either **E** or **H**. If the scattering system, which includes both the scatterer and the ambient environment (a consideration especially relevant to the next section), is isotropic, then there are elementary solutions to Eq. (10) that are related to one another in a particularly gratifying way.

Suppose ψ to be a scalar solution of the wave equation (10) and **M** a vector field defined by the curl

$$\mathbf{M} \equiv \nabla \times (\mathbf{r}\psi) \tag{11}$$

with **r** the coordinate vector from the center of symmetry to the observation point. In a spherical coordinate system, the point designated by **r** is labeled by the magnitude r, azimuth φ, and polar angle θ, which is identified with the scattering angle. From the form of expression (11), it is clear that the divergence of **M** vanishes. Defining a second (divergence-free) vector field by

$$\mathbf{N} \equiv \frac{1}{k}\nabla \times \mathbf{M}, \tag{12a}$$

one finds from straightforward vector calculus that the first vector field satisfies a relation of identical form

$$\mathbf{M} = \frac{1}{k}\nabla \times \mathbf{N}, \tag{12b}$$

and that both **M** and **N** are solutions to the wave equation.

Since the coupled equations (12a, b) do not have the same relative sign as the pair of Maxwell equations corresponding to Faraday's law and Ampère's law (with displacement current), **M** and **N** are not directly proportional to **E** and **H**. However, the correct symmetry is attained by the linear combinations

$$\mathbf{E} = \mathbf{M}_1 + i\mathbf{N}_2, \tag{13a}$$

$$\mathbf{H} = \frac{k}{\mu k_0}(\mathbf{M}_2 - i\mathbf{N}_1) = \sqrt{\frac{\varepsilon}{\mu}}(\mathbf{M}_2 - i\mathbf{N}_1). \tag{13b}$$

In Eqs. (13a, b), the subscripts refer to two independent scalar solutions (ψ_1, ψ_2) of the wave equation which differ only in their dependence on the azimuthal coordinate φ. The exact form of ψ need not be specified here; it suffices to say that ψ is a product of $\sin\varphi$ or $\cos\varphi$, an associated Legendre polynomial of the scattering angle θ, and a radial function of kr which (depending on boundary conditions) may be any of the spherical Bessel functions. Since it is only the far-field properties of the solutions that are of interest in a scattering problem, the radial functions will ultimately be replaced by their asymptotic expansions.

There remains, of course, the nontrivial task of applying the Maxwellian boundary conditions to the incident, internal, and scattered fields at the surface of the sphere. The eventual outcome of this lengthy derivation is a pair of scattering amplitudes $S_j(\theta)$ $(j = 1, 2)$ relating the electric fields of the incident and scattered waves of polarization σ $(j = 1)$ or π $(j = 2)$ as follows:

$$E_j = S_j(\theta)\frac{e^{ikr}}{ikr}E_j^{(0)}. \tag{14}$$

Each amplitude is expressible in an infinite series expansion

$$S_1(\theta) = \sum_{m=1}^{\infty}\frac{2m+1}{m(m+1)}\{a_m f_m(\cos\theta) + b_m g_m(\cos\theta)\}, \tag{15a}$$

$$S_2(\theta) = \sum_{m=1}^{\infty}\frac{2m+1}{m(m+1)}\{b_m f_m(\cos\theta) + a_m g_m(\cos\theta)\} \tag{15b}$$

of functions (f_m, g_m) constructed from the associated Legendre polynomials $P_m^1(\cos\theta)$ and their derivatives. The expansion coefficients (a_m, b_m), constructed from spherical Bessel and Hankel functions and their derivatives, depend on the relative refractive index $n = n_2/n_1$ and dimensionless size parameter $\sigma \equiv k_0 a = 2\pi a/\lambda$, with a the radius of the sphere. (There should be no confusion between σ as a polarization label and as the size parameter; context will clearly distinguish the two usages.) Note that incident σ- or π-polarized light scattered under the conditions investigated by Mie—that is, from homogeneous, optically inactive, transparent or absorbing spheres—preserves its state of polarization. It may seen plausible that such polarization conservation is an inevitable outcome of the foregoing conditions, but in fact—to anticipate the contrary result of the next section—this is not necessarily so.

As I have already implied, a glance at Mie's theory, embodied in Eqs. (15a, b), is not particularly revealing. What is one to make of it? What does it contain? Where does it lead? These are questions to which entire treatises have been devoted. To survey its content qualitatively, it is useful to make a tour of the "parameter square" of figure 13.1 defined by the values of $k_0 a$ and n (for a nonmagnetic sphere). The square is divided—roughly; the boundaries are not sharp—into two vertical strips A and C, two horizontal strips B and D, and the interior region E. The dark gray corner patches are the regions of overlap. If we imagine the wavelength to be fixed, then the variation of $k_0 a$ from 0 to ∞ represents spheres of increasing size. As we will restrict our attention to nonabsorbing spheres in an optically less dense medium—a convenience, not a necessity—n spans the range from 1 to ∞. For the purposes of this discussion, the wavelength dependence of n is not an issue and will be ignored; then, varying $k_0 a$ can equally well represent different values of λ for spheres of fixed a and n.

In addition to $k_0 a$ and n, there is a separate physical significance to the product $2k_0 a(n-1)$. This is the retardation of a ray passing through the sphere along a diameter. To speak of rays at all requires that the object through which the light passes be large in comparison to the wavelength. More precisely,

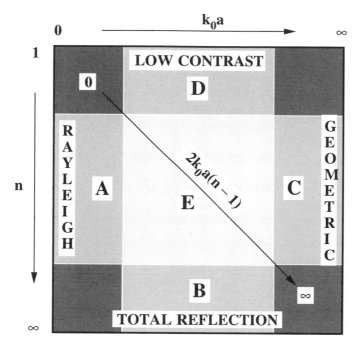

Figure 13.1. Parameter square for processes subsumed by the theory of Mie scattering.

one can infer from Fresnel diffraction theory that a "pencil" of light of length ℓ is distinguishable as a meaningful physical entity only if its width is large compared to $\sqrt{\ell\lambda}$.

Sector A of the parameter square is characterized by the condition $k_0 a \ll 1$, which implies a sphere sufficiently small relative to the incident wavelength that, practically speaking, the object is immersed in a homogeneous external field. If the sphere is also small compared with the internal wavelength λ/n—or equivalently $ka \ll 1$—then the most significant term in the series (15a, b) is the one with coefficient a_1, corresponding to electric dipole scattering. Under these conditions the Mie equations reduce to the amplitudes for Rayleigh scattering. The physical significance of the second condition is that the penetration of the incident field within the volume of the sphere occurs quickly compared to the period of oscillation. The process is nonrelativistic, and retarded times play no role. Depending on the magnitude of n, it is conceivable that the first condition, but not the second, may be fulfilled. In that case the slow penetration of the incident field gives rise to standing waves, the internal and external fields are not in phase, and the scattering process entails higher multipole contributions, each of which is resonantly amplified for certain values of a/λ. Then, even though a/λ is small, Rayleigh's theory is no longer applicable, and one must resort to the exact Mie formulas for an adequate treatment.

Under the (Rayleigh) conditions $k_0 \ll 1$, $ka \ll 1$, the concept of ray is inapplicable, and, as Tyndall and his predecessors failed to realize, the light interaction is distinct from reflection. As the size of the sphere becomes progressively larger than the wavelength, however, the Mie formulas characterize a scattering pattern, represented in sector C of the parameter square, that increasingly resembles the combined effects of reflection/refraction and Fraunhofer diffraction. The two contributions are distinguishable by their angular distribution and dependence on system parameters. For large spheres, the diffraction pattern is concentrated in the forward direction, its features determined principally by the size of the sphere. By contrast, the reflection and refraction of waves at the surface of the sphere depend on the refractive index and produce a pattern of broad angular divergence. In the limit of very large spheres, one enters the domain of geometric optics. The diffracted light is but a bright spot on the optic axis ($\theta = 0$), and it becomes meaningful to speak of distinct rays of light reflecting from the surface or undergoing multiple internal reflections before emerging at angles prescribed by Snel's law and with amplitudes given by the Fresnel relations.

Despite the fact that the laws of geometric optics emerge from Mie's theory in the domain defined by the vanishing ratio λ/a, these laws alone do not fully characterize the scattering of light no matter how small the preceding ratio becomes. For one thing, one must know the relative phases of different rays emerging in the same direction in order to deduce correctly the effects of their interference. Moreover, there are special angles—those that determine rainbows and glories—at which the geometrical optics approximation fails. And finally, with respect to the flux of radiant energy, diffraction can never be neglected; although confined essentially to the incident direction, the Fraunhofer pattern nevertheless contains one-half the total energy, the other half being borne away by reflected and refracted waves. These various complications are all intrinsically accounted for in the unapproximated formulas (15a, b) of Mie theory.

A graphic demonstration of the difference between Rayleigh scattering and general Mie scattering is provided by the following simple experiment. Direct the red beam ($\lambda = 633$ nm) of a helium-neon laser successively through two cuvettes, the first containing pure water and the second an aqueous suspension of small scattering particles (for example, latex "microspheres") of size greater than λ—as illustrated in figure 13.2—and examine the laterally scattered light. When the incident beam is polarized vertically (σ-polarization), one sees a trace of red light across both cuvettes; viewed through an analyzer, the scattered light is found to be polarized vertically. If, however, the incident beam is polarized horizontally (π-polarization), then only the cuvette with suspended particles shows a trace of light, which is also polarized in the scattering plane. The sample of pure water, whose scatterers are molecular electric dipoles of size very much less than λ, shows no trace of light for there is no radiation parallel to the dipole axis.

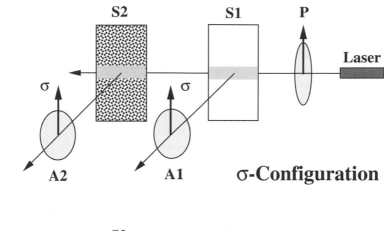

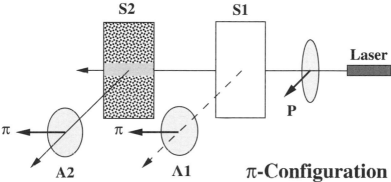

Figure 13.2. Simple experiment differentiating molecular scattering (sample S1) from large-particle scattering (sample 2). Both samples show a trace of σ-polarized light, but only sample 2 shows a trace of π-polarized light.

The preceding observations, made while I was "playing" with light in preparation for investigations of scattering from optically active media, turned out to be quite useful. They confirmed that Mie scattering (like reflection and diffraction) from achiral suspensions preferentially retained the state of incident σ- or π-polarization which could then serve as a reference for measurement of subsequent optical rotation.

Sector B of the parameter square, wherein the refractive index is very large, represents a region of total reflection, as one may readily verify in the special case of the Fresnel formulas for nonmagnetic media (Eqs. [11.14a, b]). Irrespective of polarization or incident angle, the reflectance $R \to 1$ as $n \to \infty$. In this sector, the fields within the sphere become negligible, and the equations of Mie theory again simplify.

At the opposite end of the square where the (relative) refractive index remains very close to unity, sector D characterizes what may be called the region of low contrast. Low contrast implies weak scattering irrespective of sphere size, for, if n were precisely 1, there would be no optical differentiation between object and environment. Two limiting cases for which the Mie equations reduce to distinct, recognizable processes comprise the regions where the sphere is large, $k_0a \gg 1$, and where the phase shift is small, $2k_0a(n-1) \ll 1$. The first region represents the diffraction and geometric optics of large spheres of low contrast, phenomena of particular importance in the microscopy of biological specimens.

In the second region, one has returned to the domain of Rayleigh scattering—actually to an extension of Rayleigh's theory usually associated with the names Rayleigh-Gans (RG). Here, the constraint on phase requires that the particle size satisfy $a \ll \lambda/(n-1)$, rather than the more restrictive condition $a \ll \lambda$. Since k_0a need not be small, there is an optical pathlength difference, and hence relative phase, between waves scattered from different elemental volumes within the sphere. The superposition of all such waves leads to a scattered intensity

$$I_{RG} = G(p)I_R \tag{16a}$$

that differs from Rayleigh's expression for small spheres (illuminated by unpolarized light of intensity I_0)

$$I_R = \frac{1 + \cos^2\theta}{2} \frac{k_0^4 V^2}{r^2} \left(\frac{n-1}{2\pi}\right)^2 I_0 \tag{16b}$$

by a geometrical factor

$$G(p) = \frac{3}{p^3}(\sin p - p\cos p) \tag{17}$$

with argument

$$p = 2k_0a\sin\left(\frac{\theta}{2}\right). \tag{18}$$

Finally, there is the interior sector E of arbitrary n and a. This is "pure" Mie scattering; in general, nothing short of the exact scattering formulas describes this region satisfactorily.

With complex values of the refractive index allowed, that is, $\tilde{n} = n + i\kappa$, the parameter square becomes a parameter cube, the variation in the absorption parameter κ spanning 0 to ∞ along a third perpendicular axis. All the points of this cube also fall within the purview of Mie's theory.

In his major work *De Revolutionibus Orbium Celestium*, the product of thirty-six years of labor, Copenicus wrote of spheres, "we assert that the universe is spherical; partly because this form ... is the most perfect of all; partly because it ... is best suited to contain and retain all things."[11] He wrote, of course, of heavenly bodies, not light. Nevertheless, there is a certain aptness. Through Mie's theory of light scattering, the harmony of all physical optics may be seen in a sphere.

13.2 SCATTERING IN A CHIRAL SKY

Of all the extraordinary things that the *Star Trek* crews could do, the one I found most intriguing was their capacity to detect life forms on an alien planet by remotely scanning the surface with some sort of device. I very rarely watch television, my wife and I having resolved years ago to eliminate it entirely from our household shortly after our first child was born. Nevertheless, I confess to finding *Star Trek* thought-provoking as well as entertaining.

That the life scanner fascinated me more than, let us say, "warp drive," "tele-portation," or "Bones's" five-minute brain-replacement surgery is its potential feasibility. No one, I suspect, will ever bend space-time at will in order to achieve superluminal speed. Never will a single physician install a human brain like a mechanic installs an engine. And the idea of disassembling a per-son into atoms or infomation bits, and reassembling the units elsewhere is as improbable to me as the spontaneous diffusion of all the air in my room out an open window. Such things either violate physical law or come close enough to dissuade me from wondering how to realize them.

But not the life scanner. If alien life (and not necessarily the bipedal vari-eties on television who, despite their distant cosmic abodes, all seem to speak English) bears any similarities at all to life on earth, it will be, I believe, in the chiral asymmetry of their molecular constitution. The rest, then, is a question of optics: Can one detect chiral asymmetry by light scattering?

The answer, at least in some special cases, is clearly "yes." I have already described in chapter 11 experiments culminating in the observation of a chi-ral signature in the specular reflection of light from homogeneous optically active fluids.[12] The experimental method, based on phase modulation and synchronous detection, is highly sensitive, capable of distinguishing the in-equivalent effects on light of molecules whose refractive indices for left (LCP) and right (RCP) circular polarizations differ by at most a few parts in 10^7.

Not everything, however, reflects light specularly. Indeed, the surfaces of nat-urally occurring objects or the localized inhomogeneities distributed throughout the volume of a composite system are more likely to scatter light in the sense which this word ordinarily conveys, that is, diffusively. Can chiral asymmetry in diffusively scattered light be detected? Again, the answer is affirmative, and in the next section I will discuss experiments in which this has been accom-plished under quite remarkable conditions where one might have expected all trace of a chiral signal to be destroyed.[13]

For both practical and conceptual reasons, optically active media with nonchi-ral particulate inclusions—what I would call, in thinking of Rayleigh's researches on light propagation through the atmosphere, a "chiral sky" (although it could be designated equally well a "chiral sea")—are among the most consequential systems to study.

In the domain of planetary science, the *Star Trek* mission "to explore strange new worlds, to seek out new life and new civilizations" is perhaps not so far-fetched even now. Deep channels on Mars, thought to be cut by once-flowing water, holds out the enticing prospect that life may have arisen there. In perhaps one of the most startling announcements ever to come out of a U.S. government agency, NASA scientists, examining what they believed to be Martian meteorite debris discovered in Antarctica, recently claimed to have found indirect fossil evidence of nonterrestrial microbial life forms.[14] Time will tell. Besides Mars, the Jovian satellite Europa, whose fractured surface resembles ice floes in an ocean, is also suspected of having had—and quite possibly of having presently—large amounts of water. Moreover, it is not just water that portends the prospect of life and a chiral environment. Planetary atmospheres with high molecular hydrocarbon content, such as expected on the Saturnian moon Titan, might also be an example of a turbid chiral fluid. A search for circular birefringence of the Titanian atmosphere, which my French colleague Jacques Badoz and I discussed years ago, would make a fascinating and revealing experiment.

It is not only on distant worlds where chiral light scattering conveys critical information. Closer to home, one of the outstanding problems of clinical medicine is the development of noninvasive optical methods for monitoring the concentration of blood glucose (a chiral solution) in the presence of accompanying particulate scatterers. Blood is a turbid chiral sea, and the measurement of blood sugar an activity vital to the large number of people afflicted with diabetes. Indeed, it has been estimated that glucose is the most frequently determined analyte in the hospital, comprising more than 10% of all assays in an average clinical laboratory. Optical rotation has for decades served as a gauge to ascertain the identity and concentration of largely transparent sugar solutions, but is it possible to make such measurements in the presence of incoherently scattering particles? Once more, although with optimistic reserve, I believe this can be done.[15]

Applications like those above go beyond the bounds of Mie theory, which does not apply to chiral media. Although there have been many theoretical studies of the scattering of electromagnetic waves (particularly in the radio- and microwave domains) from various shaped chiral objects in a nonchiral environment, the converse problem—the chiral sky—surprisingly had not been solved until recently when my Finnish colleague I. V. Lindell and I set our minds to it.[16] We had hoped initially that there would be a mathematical transformation to convert the problem of a nonchiral scatterer in a chiral environment into one of the previously solved problems (such as the problem of scattering from a chiral sphere),[17] but this did not seem to be the case; in any event, if such a transformation exists, we were not able to find it.

The approach to the problem described in this section is different from that of Mie theory, which, like the conventional derivation of the Fresnel amplitudes, entails the solution of differential equations with specified boundary conditions. There are at times advantages—particularly when approximations are to be made—to framing a problem in terms of integral equations. This is what is done here. The underlying idea is quite straightforward and close in spirit to the derivation of the Ewald-Oseen extinction formula. Incident waves, originating within the chiral medium, illuminate a nonchiral object of arbitrary shape, penetrate, and set up oscillating internal currents that generate the scattered wave in the far-field or radiation zone. The fields in the radiation zone, therefore, can be expressed exactly in terms of the electromagnetic fields within the object. Although the latter are not known at the outset—for they are produced by both the incident radiation and the polarization and magnetization of matter—they can be approximated iteratively to any desired degree of accuracy.

The constitutive relations of the homogeneous optically active environment

$$\mathbf{D} = \varepsilon\mathbf{E} + inf\mathbf{H},$$
$$\mathbf{B} = \mu\mathbf{H} - inf\mathbf{E} \tag{19}$$

are exactly those introduced earlier (Eq. [11.27]), but now expressed explicitly for waves with temporal dependence $e^{-i\omega t}$; f is the chiral parameter introduced in Eq. (11.28), and $n = \sqrt{\varepsilon\mu}$ is the mean refractive index. For most naturally occurring chiral materials f is very small, often five or more orders of magnitude smaller than 1. In principle, however, the magnitude of the chiral parameter is limited to the full range $-1 < f < +1$ by the requirement that the energy function of a lossless medium be positive definite. Thus, it is theoretically possible to fabricate materials with relatively large chiral parameters.

The constitutive relations of the nonchiral scatterer take the familiar forms $\mathbf{D} = \varepsilon_s\mathbf{E}$ and $\mathbf{B} = \mu_s\mathbf{H}$. It is useful to express the electromagnetic parameters in terms of ε and μ

$$\varepsilon_s = e_s\varepsilon,$$
$$\mu_s = m_s\mu \tag{20}$$

with numerical factors e_s, m_s, for this simplifies later expressions for the scattering amplitudes.

Constitutive relations (19) signify that the medium is circularly birefringent; an electromagnetic wave $(\mathbf{E}, \mathbf{H})$ of arbitrary polarization propagates as two distinct components of specified helicity

$$\mathbf{E}\pm = \tfrac{1}{2}\left(\mathbf{E} \pm i\sqrt{\frac{\mu}{\varepsilon}}\mathbf{H}\right), \qquad \mathbf{H}\pm = \tfrac{1}{2}\left(\mathbf{H} \mp i\sqrt{\frac{\varepsilon}{\mu}}\mathbf{E}\right) \tag{21}$$

characterized by wavenumbers

$$k_\pm = n_\pm k_0 = k(1 \pm f) \tag{22}$$

in which $k = n\omega/c = nk_0$ is the mean wavenumber (or the wavenumber in the absence of chirality). The linear combinations in Eq. (21) are deducible from the relations

$$\mathbf{F} = \mathbf{F}_+ + \mathbf{F}_- \tag{23}$$

for each total field $\mathbf{F} = \mathbf{E}, \mathbf{H}$ and from Eq. (11.34)

$$\mathbf{H}_\pm = \mp i\sqrt{\frac{\varepsilon}{\mu}}\mathbf{E}_\pm \tag{24}$$

which relates the helicity components of $\mathbf{E}$ and $\mathbf{H}$.

Although the components (21) are uncoupled within a homogeneous chiral medium, they become coupled upon encountering the scatterering particle. The fields within the particle are calculated from the two Maxwell curl equations cast into the following form (by adding and substracting from each equation an appropriate vectorial quantity):

$$\nabla \times \mathbf{E} = i\mu k_0\mathbf{H} + kf\mathbf{E} - \mathbf{J}_m, \tag{25a}$$

$$\nabla \times \mathbf{H} = -i\varepsilon k_0\mathbf{E} + kf\mathbf{H} - \mathbf{J}_e. \tag{25b}$$

Note that Eqs. (25a,b) for the fields *inside* the particle manifest the electromagnetic parameters of the chiral environment—and not the parameters of the particle—except for the current densities

$$\mathbf{J}_e = [-ik_0\varepsilon(e_s - 1)\mathbf{E} - kf\mathbf{H}]P_V(\mathbf{r}), \tag{26a}$$

$$\mathbf{J}_m = [-ik_0\mu(m_s - 1)\mathbf{H} - kf\mathbf{E}]P_V(\mathbf{r}) \tag{26b}$$

in which the pulse function $P_V(\mathbf{r})$ is equal to 1 when $\mathbf{r}$ falls within the volume V of the scatterer, and is 0 otherwise. In effect, the scatterer has been replaced by equivalent electric and magnetic current sources constructed from the internal electromagnetic field.

The solution for the scattered field $\mathbf{E}^s(\mathbf{r}) = \mathbf{E}_+^s(\mathbf{r}) + \mathbf{E}_-^s(\mathbf{r})$ (and similar expression for $\mathbf{H}^s(\mathbf{r})$) takes the familiar form of a Green's function integrated over a source

$$\begin{pmatrix} \mathbf{E}_+^s(\mathbf{r}) \\ \mathbf{E}_-^s(\mathbf{r}) \end{pmatrix} = i\sqrt{\frac{\mu}{\varepsilon}} \int_V \begin{pmatrix} k_+\overline{G}_+(\mathbf{r} - \mathbf{r}') & 0 \\ 0 & k_-\overline{G}_-(\mathbf{r} - \mathbf{r}') \end{pmatrix} \begin{pmatrix} \mathbf{J}_{e+}(\mathbf{r}') \\ \mathbf{J}_{e-}(\mathbf{r}') \end{pmatrix} dV' \tag{27}$$

except that in the present case, where the fields and sources are vectors, the Green's functions $\overline{G}_\pm$ are not scalars but second-order tensors or dyads. The helicity components $\mathbf{J}_{e\pm}$ of the electric current density are obtained from $\mathbf{J}_e$ and $\mathbf{J}_m$ by precisely the same transformation, Eq. (21), leading to the components $\mathbf{E}_\pm$ and $\mathbf{H}_\pm$. The magnetic current densities $\mathbf{J}_{m\pm}$ do not appear explicitly in the solution (27) since they have been taken into account through the relation $\mathbf{J}_{m\pm} = \mp i\sqrt{\frac{\mu}{\varepsilon}}\mathbf{J}_{e\pm}$. Likewise, as a result of Eq. (24), it is not necessary to solve a separate equation for the components $\mathbf{H}_\pm^s(\mathbf{r})$; the magnetic components follow directly from $\mathbf{E}_\pm^s(\mathbf{r})$.

I leave the derivation of the Green dyads to the original paper (see note 16). For the purposes of this section, one need say only that they satisfy a type of wave equation involving dyadic operations on the scalar function $G_\pm(r) = e^{ik_\pm r}/4\pi r$.[18] It is worth noting, however, the advantage of the dyadic formalism, as opposed to the use of matrices, in performing the derivation. Dyads, like vectors, provide a coordinate-free vehicle that streamlines much of the symbolic analysis until such time as theoretical or numerical results are required for a specific experimental configuration.[19] We will see the notational advantage of dyads in the final expression for the scattered fields.

Although Eq. (27) is formally the solution to the chiral sky problem, in its present form it is not yet usefully implementable. First, one must express the current densities $\mathbf{J}_{e\pm}$ in terms of the helicity components of the fields ($\mathbf{E}_\pm$, $\mathbf{H}_\pm$). Next, to obtain true scattering solutions one makes a far-field approximation to the Green's functions $G_\pm(r)$ analogous to the derivation of Fraunhofer diffraction from the Fresnel integral. The outcome, giving a wave scattered in the direction $\hat{\mathbf{r}}$ to a detector located at $\mathbf{r} = r\hat{\mathbf{r}}$, is a moderately simple expression (at least compared with Mie theory)

$$\mathbf{E}_\pm^s(\mathbf{r}) = \frac{kk_\pm}{2} G_\pm(r)\overline{T}_\pm(\hat{\mathbf{r}})\cdot[e_s+m_s-2(1\pm f)\mathbf{P}_\pm(\hat{\mathbf{r}})+(e_s-m_s)\mathbf{Q}_\pm(\hat{\mathbf{r}})] \quad (28)$$

whose parts are interpretable as discussed next.

$G_\pm(r)$ are outgoing spherical waveforms bearing the characteristic $1/r$ spatial dependence and propagating with respective phase velocities $c/n_\pm$. The helicity dyads

$$\overline{T}_\pm(\hat{\mathbf{r}}) = \overline{1} \pm i\hat{\mathbf{r}} \times \overline{1} - \hat{\mathbf{r}}\hat{\mathbf{r}}, \quad (29)$$

where $\overline{1}$ is the unit dyad, are very useful functions. Acting on any vector $\mathbf{a}$, $\overline{T}_\pm(\hat{\mathbf{r}})$ creates circularly polarized (CP) vectors $\mathbf{a}_\pm$ in the plane perpendicular to $\hat{\mathbf{r}}$. How this works may be seen from the following chain of steps whose geometrical significance is shown in figure 13.3:

$$\begin{aligned}
\overline{T}_\pm(\hat{\mathbf{r}}) \cdot \mathbf{a} &= \overline{1}\cdot\mathbf{a} \pm i\hat{\mathbf{r}} \times \overline{1}\cdot\mathbf{a} - \hat{\mathbf{r}}(\hat{\mathbf{r}}\cdot\mathbf{a}) \\
&= (\mathbf{a} - \mathbf{a}_\parallel) \pm i\hat{\mathbf{r}} \times \mathbf{a} \\
&= a_\perp\mathbf{e}_1 \pm ia_\perp\mathbf{e}_2 \\
&= \sqrt{2}a_\perp\mathbf{e}_\pm \equiv \mathbf{a}_\pm. \quad (30)
\end{aligned}$$

The unit vectors in the figure are oriented in a right-handed sense: $\mathbf{e}_1 \times \mathbf{e}_2 = \hat{\mathbf{r}}$; the CP unit vectors $\mathbf{e}_\pm$ are defined in Eq. (9.29).

Last, the vector functions

$$\mathbf{P}_\pm(\hat{\mathbf{r}}) = \int_V e^{-ik_\pm\hat{\mathbf{r}}\cdot\mathbf{r}'}\mathbf{E}_\pm(\mathbf{r}')\,dV' \quad (31a)$$

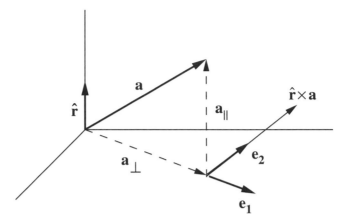

Figure 13.3. Application of the dyad $\overline{T}_\pm(\hat{\mathbf{r}})$ on the vector $\mathbf{a}$ to create the helicity vectors $\mathbf{a}_\pm$ in the plane perpendicular to $\hat{\mathbf{r}}$.

and

$$\mathbf{Q}_\pm(\hat{\mathbf{r}}) = \int_V e^{-ik_\pm \hat{\mathbf{r}} \cdot \mathbf{r}'} \mathbf{E}_\mp(\mathbf{r}') \, dV' \tag{31b}$$

take account of the differences in optical path length of waves originating at different points within the scatterer. They resemble Fourier transforms of the field components $\mathbf{E}_\pm$, but the integrations are performed only over the volume of the scatterer (whose shape, at this stage, is still unspecified).

Within the far-field approximation, Eq. (28) provides an exact solution to the problem of scattering in a chiral sky. It is a more general solution than that of Mie theory, since the ambient medium is more inclusive and the shape of the scatterer is arbitrary. Unfortunately, the integrals (31a, b) cannot be performed until one has explicit expressions for the fields $\mathbf{E}_\pm$ inside the scatterer. These fields, at least in principle, can be calculated iteratively by a series of successive approximations beginning with the incident fields $\mathbf{E}_\pm^i$, $\mathbf{H}_\pm^i$.

If the contrast in electromagnetic parameters between the material of the scattering particle and that of the host medium is not too great, then the scattered wave is relatively weak compared with the incident wave, and the fields inside the particle can be represented to good approximation by the incident fields. Such a substitution constitutes the Born approximation in quantum mechanics, introduced by Max Born for the purpose of treating electron scattering. The approximation is particularly suitable for the scattering of light by a wide variety of materials in aqueous suspension, since the refractive index of the particles (generally in the vicinity of 1.5) is close to that of water ($\sim$1.33).

Adopting the Born approximation and now specifying the particle shape to be a sphere of radius a, one finds that $\mathbf{P}_\pm(\mathbf{r})$ and $\mathbf{Q}_\pm(\mathbf{r})$ reduce to the simple analytical forms

$$\mathbf{P}_\pm(\hat{\mathbf{r}}) = G(p_\pm)V\mathbf{E}_\pm^i, \qquad p_\pm = 2k_\pm a \sin\left(\frac{\theta}{2}\right) \tag{32a}$$

$$\mathbf{Q}_\pm(\hat{\mathbf{r}}) = G(q)V\mathbf{E}_\mp^i, \qquad q = 2ka\sqrt{\sin^2\left(\frac{\theta}{2}\right) + f^2\cos^2\left(\frac{\theta}{2}\right)} \tag{32b}$$

with the geometric function G defined by Eq. (17). In effect, we have deduced the chiral generalization of the Rayleigh-Gans (RG) theory for scattering from arbitrary-sized achiral spheres within the restriction of weak phase shifts $2k_0a(n_\pm - 1) < 1$.

The outcome of the remainder of the analysis, whose algebraic details are not important here, is a linear relation between the scattered and incident fields

$$\mathbf{E}^s = S\mathbf{E}^i \tag{33a}$$

or

$$\begin{pmatrix} E_+^s \\ E_-^s \end{pmatrix} = \begin{pmatrix} S_{++} & S_{+-} \\ S_{-+} & S_{--} \end{pmatrix} \begin{pmatrix} E_+^i \\ E_-^i \end{pmatrix} \tag{33b}$$

in terms of helicity components. The scattering amplitudes $S_{\alpha\beta}$, of which there are now four, not two, are defined by

$$S_{\alpha\beta} = k^2 V \frac{e^{ik_\alpha r}}{4\pi r} s_{\alpha\beta} \quad (\alpha, \beta \text{ assume values } \pm) \tag{34}$$

in which the elements $s_{\alpha\beta}$, the principal dynamical content of chiral sky theory, are

$$s_{++} = [e_s + m_s - 2(1+f)](1+f)G(p_+)\cos^2\left(\frac{\theta}{2}\right),$$

$$s_{+-} = (e_s - m_s)(1+f)G(q)\sin^2\left(\frac{\theta}{2}\right),$$

$$s_{-+} = (e_s - m_s)(1-f)G(q)\sin^2\left(\frac{\theta}{2}\right),$$

$$s_{--} = [e_s + m_s - 2(1-f)](1-f)G(p_-)\cos^2\left(\frac{\theta}{2}\right) \tag{35}$$

for the scattering configuration illustrated in figure 13.4.

The consistency of the set of equations (35) with RG theory within their common domain can be established by setting the chiral parameter f to zero, and applying the unitary transformation matrices (9.28a, b) to obtain the scattering amplitudes in a linearly polarized (LP) basis. It is straightforward to show that

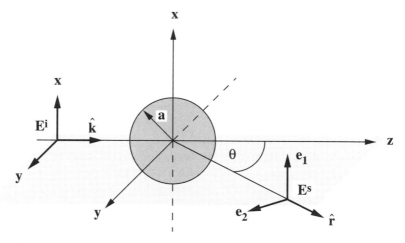

Figure 13.4. Geometric configuration of a wave E^s scattered in the y-z plane into a direction $\hat{r}$ at angle θ to the direction $\hat{k}$ of the incident wave E^i.

Eq. (33) then reduces (with p given by Eq. [18]) to the correct expression

$$\begin{pmatrix} E_x^s \\ E_y^s \end{pmatrix} = k^2 V \frac{e^{ikr}}{4\pi r}(e_s - 1)G(p)\begin{pmatrix} 1 & 0 \\ 0 & \cos\theta \end{pmatrix}\begin{pmatrix} E_x^i \\ E_y^i \end{pmatrix}, \tag{36}$$

yielding Eq. (16a,b) for the scattered intensity of incident unpolarized light.

In view of the impression often given in the literature that an achiral sphere has two independent scattering elements, it is worth emphasizing that the number of independent scattering amplitudes remains four, irrespective of whether the scattering matrix is expressed in a CP or LP basis, if the ambient medium is chiral. The scattering matrix characterizes an interaction and not merely an object, and therefore is influenced by the environment within which the scattering process occurs.

Let us first examine the effect of chirality on the angular distribution of the scattered light intensity. To make comparison with the Rayleigh-Gans formula, Eq. (16), we shall assume that the incident light is unpolarized, that is, that it comprises an equal mixture of CP intensities: $\frac{1}{2}(I_i^{(+)} + I_i^{(-)})$. Calculating from Eq. (33) the intensities of scattered light—first with $E_+^i = 1$, $E_-^i = 0$ and then with $E_+^i = 0$, $E_-^i = 1$—and averaging the two expressions yields the result

$$I_s^{(u)} \propto \frac{1}{2}(s_{++}^2 + s_{-+}^2 + s_{+-}^2 + s_{--}^2) \tag{37}$$

whose physical content is graphically depicted in figure 13.5.

The top part of the figure, illustrating the case of a nonchiral environment ($f = 0$), reproduces the salient features of RG scattering. For particles smaller than a wavelength ($k_0 a = 0.1$ for curve a in the figure), the angular distribution

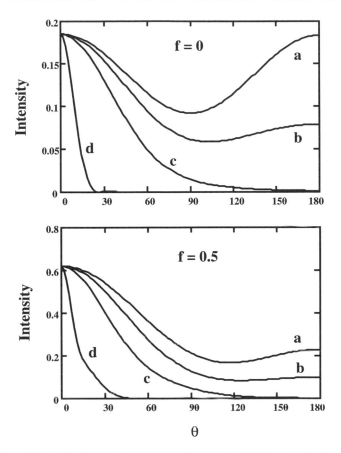

Figure 13.5. Effect of a chiral environment on the angular distribution of light scattered by nonchiral spherical particles with $e_s = 1.43$, $m_s = 1$, and size parameter $\sigma \equiv 2\pi a/\lambda = $ (a) 0.1, (b) 1, (c) 2, (d) 10. Incident light is unpolarized.

is symmetrical about $\theta = 90°$. At this angle the intensity is minimum, since the π-polarized component is suppressed, and the scattered light is 100% linearly polarized. This is the case of pure Rayleigh scattering. As the size of the spheres increases, the light becomes increasingly forward scattered. For $k_0 a = 10$ (curve d), for example, the scattered light intensity is almost insignificant beyond $\theta \sim 30°$.

The intensity of scattered light is not a sensitive function of the chirality of the environment; that is, the angular distribution would be imperceptibly altered for a small chiral parameter ($f \ll 1$). A strongly chiral environment, however, can significantly modify this distribution, as shown in the bottom part of the figure for chiral parameter $f = 0.5$. Physically, this represents an extraordinarily large, but theoretically allowable, chiral strength. In such a case, scattering toward the front is markedly enhanced, irrespective of the size of the spheres.

In comparing the curves of figure 13.4 for spheres of different sizes, one must remember that the constant omitted from Eq. (37), which includes the k^4 or $1/\lambda^4$ wavelength dependence, also includes a factor V^2. Thus, the absolute intensities increase as the sixth power of the radius of the sphere.

Of particular importance to both fundamental experiments and practical applications are the intrinsically chiral scattering functions—that is, the dynamical invariants that vanish identically in the absence of chirality. The methodological significance of these functions derives from the usually high sensitivity achievable in a "null" experiment, that is, the measurement of small departures from a zero-signal condition. One such example is the degree of circular polarization $\tau_C^{(u)}$ for incident unpolarized light. In an achiral medium, there can be no preferential handedness, and therefore $\tau_C^{(u)}$ must be zero. This is readily seen from the expressions, deducible from Eq. (9.34), for incident CP light

$$\tau_C^{(+)} = \frac{s_{++}^2 - s_{-+}^2}{s_{++}^2 + s_{-+}^2}, \qquad \tau_C^{(-)} = \frac{s_{+-}^2 - s_{--}^2}{s_{+-}^2 + s_{--}^2} \tag{38}$$

In the absence of chirality, $\tau_C^{(-)} = -\tau_C^{(+)}$, and hence the average

$$\tau_C^{(u)} = \tfrac{1}{2}(\tau_C^{(+)} + \tau_C^{(-)}) \tag{39}$$

vanishes.

Figure 13.6 shows the angular variation of $\tau_C^{(u)}$ for different sized spheres in a chiral environment. In the top part of the figure the chiral parameter is small ($f = 10^{-5}$), comparable to that of many naturally occurring organic and inorganic materials. Characteristic of the set of low-f curves is the preservation of the sign of $\tau_C^{(u)}$—negative for a positive chiral parameter, indicative of preferential -1 helicity (see Eq. [9.31])—over the entire range of scattering angles, except for forward- ($\theta = 0$) and back-scattered ($\theta = \pi$) light where $\tau_C^{(u)}$ is null. The larger the spherical particle, the greater is the degree of circular polarization, peak values occurring close to $\theta = 90°$, the angle of maximum linear polarization in a nonchiral sky. Looking up at the zenith of a chiral sky with the Sun at the horizon, one would find the light circularly polarized. The bottom part of the figure illustrates the case of very strong chirality ($f = 0.5$). Depending on sphere size, $\tau_C^{(u)}$ can be positive or negative, and large values of a can lead to multiple peaks, as in curve d ($k_0 a = 10$).[20]

Consideration of other intrinsically chiral scattering functions must again be left to the original literature. The matter of optical rotation, however, is worth discussing, for it is a significant part of the experiments described in the next section.

If a medium is chiral, the linear polarization of a transmitted beam may be thought to undergo rotation continuously at a rate dependent on the circular

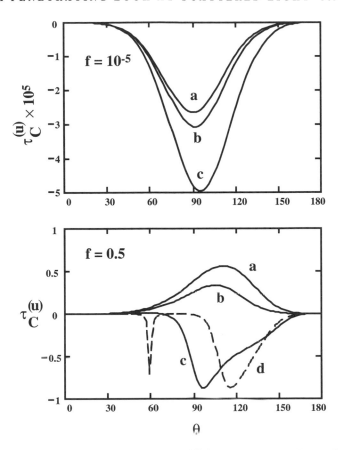

Figure 13.6. Degree of circular polarization of light scattered in a chiral environment by nonchiral spherical particles with $e_s = 1.43$, $m_s = 1$, and $\sigma = $ (a) 0.1, (b) 1, (c) 2, (d) 3. Incident light is unpolarized.

birefringence ($n_+ - n_-$). Actually, there is no linearly polarized light within the medium; the light exists in the form of CP components $\mathbf{E}_\pm$ which, over the course of their propagation, are retarded differently. It is only upon leaving the medium that these components superpose coherently (since they derive from the same incident wave) to yield again a linearly polarized wave whose plane of polarization has been rotated. While all this is quite straightforward in the case of light transmission (no scattering) as explained in chapter 9, interesting questions arise when the medium is diffusive.

Apart from forward- or back-scattered light for which the propagation axis is unchanged, how is one to define optical rotation in all other cases where the wavefront of the scattered (plane) wave is no longer parallel to the wavefront of the incident wave? In particular, with respect to what reference is optical rotation to be measured? Finally, do the spheres themselves contribute to the rotatory power of the medium, or is the optical rotation attributable exclusively to propagation through the chiral surroundings? These questions may be addressed by examining the density matrix of the scattered light $\rho^{(s)}$.

From the relations of chapter 9—specifically Eqs. (36b,d) and (47b,d)—the orientation angle ϕ of the semimajor axis of the ellipse representing a general state of elliptically polarized light is calculable from the expression

$$\tan 2\phi = \frac{\rho_1^s}{\rho_3^s} = -\frac{\text{Im}\{\rho_{+-}^s\}}{\text{Re}\{\rho_{+-}^s\}}. \tag{40}$$

Use of Eqs. (9.1) (which defines the density matrix) and (33a, b) (which defines the scattering matrix) then provides the connection between $\rho^{(s)}$ and the density matrix $\rho^{(i)}$ of the incident light

$$\rho^{(s)} = S\rho^{(i)}S^\dagger. \tag{41a}$$

For our present purposes, it is not necessary to evaluate explicitly the elements of the right-hand side of Eq. (41a); rather, it adequate to state that, for a matrix $\rho^{(i)}$ with real-valued elements, $\rho^{(s)}$ takes the form

$$\rho^{(s)} = \begin{pmatrix} \sigma_{++} & \sigma_{+-}e^{i(k_+-k_-)r} \\ \sigma_{-+}e^{-i(k_+-k_-)r} & \sigma_{--} \end{pmatrix} \tag{41b}$$

in which all elements $\sigma_{\alpha\beta}$ are real-valued. This follows straightforwardly from the fact that the elements $s_{\alpha\beta}$ in Eq. (35) (for a nonabsorbing chiral medium) are real-valued. Thus, with $\rho_{+-}^s = \sigma_{+-}e^{i(k_+-k_-)r}$, Eq. (40) leads directly to the result

$$\phi = -\tfrac{1}{2}(k_+ - k_-)r = -\frac{\pi r(n_+ - n_-)}{\lambda} \tag{42}$$

which is precisely the optical rotation expected of an unscattered wave transmitted a distance r through an optically active medium.

The negative sign of Eq. (42) can be understood by a simple heuristic argument of a kind first employed by Fresnel to account for the chromatic phenomena observed by Arago and Biot. If the chiral parameter f is positive, then $n_+ > n_-$, and the phase velocity $v_+ = c/n_+$ of the component $\mathbf{E}_+$ is less than the velocity $v_- = c/n_-$ of the component $\mathbf{E}_-$. Thus, although the two components entered the chiral medium simultaneously, the $\mathbf{E}_-$ wave "arrives first" at a given point on the exit surface. By the time the component $\mathbf{E}_+$ arrives, the $\mathbf{E}_-$ field at this point has rotated a further extent clockwise to an observer facing the light source. The superposition of $\mathbf{E}_+$ and $\mathbf{E}_-$ therefore results in a clockwise-rotated linearly polarized field. For the right-handed coordinate system used throughout this book, with the direction of light propagation along the $+z$-axis, a positive angular displacement is from the x-axis toward the y-axis, or counterclockwise. The negative sign in Eq. (42) is therefore correct.

As I noted in chapter 9, the orientation angle ϕ of the polarization ellipse is equivalent to the geometrically invariant optical rotation when determined with respect to a well-defined reference associated with the incident wave. The question of reference, therefore, is partly an empirical one, for it relates to the means by which a measurement of optical activity is calibrated. The revealing experiment of figure 13.2 suggests as a practical reference the linear polarization of the wave scattered from a nonchiral suspension of the same scatterer density as the optically active medium under investigation.

In concluding this section, it is of value to summarize explicitly the principal lessons that light scattering in a chiral sky have brought us. Although it is the host medium, and not the embedded particle, that is chiral, the scattered light carries information not only about the size of the particle, but also about the chirality of the surroundings. This information is revealed most clearly by intrinsically chiral scattering functions such as the degree of circular polarization $\tau_C^{(u)}$ or the circular intensity difference δ_C.

There is one significant exception, however. Each scattering event is predicted to contribute nothing to the optical rotation ϕ. Moreover, as this conclusion derives only from the real-valuedness, and not from the specific form, of the scattering matrix elements, it is not subject to the approximations upon which the low-contrast theory is based. Deriving the exact chiral counterpart to Mie theory would not alter the outcome.

Thus, when the first experiments to measure optical rotation of diffusively scattered light revealed that the spheres exerted a profound effect indeed, I was initially rather surprised. I was to learn how interesting turbid media can be.

13.3 SEEING THROUGH TURBIDITY

Apart from those with a proclivity toward anaerobism, the rest of us live and observe in a turbid atmosphere. The preceding statement is not a sly reference to air pollution, for even if the atmosphere were free of every particle of dust, there is still an approximate limit beyond which distant objects can not be discerned clearly This is insured by the very process that gives us the blue sky—and Rayleigh, as far as I know, was the first to estimate this limit.

Think of the atmosphere as a medium (the vacuum) in which are randomly distributed molecular-sized spheres of radius a and volume $V = 4\pi a^3/3$. The effectiveness with which each sphere scatters light in all directions is given by the total (Rayleigh) scattering cross section σ_R, obtained by integrating the intensity ratio I_R/I_0 of Eq. (16b) over a spherical surface of radius r (the distance from the scattering particle to the observer). One obtains

$$\sigma_R = \frac{8\pi k_0^4}{3} \left(\frac{n-1}{2\pi} \right)^2 V^2 \tag{43}$$

where n is the refractive index of the molecules and, as before, $k_0 = 2\pi/\lambda$. If the atmosphere contains N such scatterers per unit of volume, then the exinction coefficient

$$\gamma = N\sigma_R \frac{8\pi k_0^4}{3} \left(\frac{n-1}{2\pi}\right)^2 NV^2 \tag{44}$$

characterizes the diminution in light intensity $dI(z)$ over an infinitesimal pathlength dz in accordance with the well-known law

$$dI(z) = -\gamma I(z)dz \tag{45a}$$

leading to exponential decay

$$I(z) = I_0 e^{-\gamma z}. \tag{45b}$$

Of the preceding three equations, several essential points must be made.

First, familiar though Eq. (45b) may be, it is important to recognize explicitly the special circumstance to which it pertains—that is, single scattering only. The very structure of Eq. (45a) assumes that each scatterer in the medium is illuminated only by the incident light and directs a portion of that light out of the original beam. If, however, a molecule receives and scatters light directed toward it by another molecule, then the problem entails multiple scattering, and Eqs. (45a, b) are no longer valid. Indeed, the intensity distribution law will be markedly altered, as I shall show shortly.

Second, although we are considering light scattering by the air, the refractive index n in Eq. (44) should not be mistakenly equated with the refractive index of air ($n_{air} - 1 = 2.9 \times 10^{-4}$). Here n is the index (relative to the ambient medium) of a single particle, and not the index—which I shall designate $\bar{n}$—of the ambient medium (vacuum) as modified by the presence of particles (air molecules). Since the ambient medium without the particles is nothing but "empty space," one might think that the distinction in this case is a hollow one. However, I shall soon discuss the effect of molecular density on skylight, and the conceptual difference between n and $\bar{n}$ is critical, for it determines whether N—an enormous number, approximately 2.7×10^{19} cm^{-3} for air at STP (standard temperature [273 K] and pressure [1 atm])—ends up in the numerator or denominator of the expression for γ. Clearly, it is advisable not to make an error.

Since the fraction of a volume of medium occupied by the particles in it is NV, the phase retardation suffered by a wave traversing a distance dz through this volume is $NV(n-1)dz$. Alternatively, if one disregards the molecular composition and considers the entire medium to be a continuum of effective index $\bar{n}$, then the corresponding retardation is $(\bar{n} - 1)dz$, from which follows the (approximate) relation[21]

$$\bar{n} - 1 = NV(n-1). \tag{46}$$

It is $\bar{n}$ that represents the refractive index of the gaseous medium "air," whereas n is the index of the pure condensed matter comprising air. If, for the sake of discussion, we take air to be nitrogen (N_2), then the refractive index of liquid nitrogen is $n \sim 1.21$ (at the wavelength of the sodium D lines). Note that, apart from the use we shall make of it in light scattering, Eq. (46) provides a means of ascertaining molecular size. Substituting the previously specified values of $\bar{n} - 1$, n, and N leads to a $\sim 2.3 \times 10^{-8}$ cm for the radius of a "spherical" nitrogen molecule.

Eq. (46) is interesting from another standpoint as well. Deduced on the basis of a purely optical argument (concerning phase retardation), it is equivalent to the Lorentz-Lorenz (L-L) equation (Eq. [3.24]) ordinarily derived electrodynamically in textbooks. To see the equivalence, let ρ be the mass density of a single particle (of mass m) and $\bar{\rho}$ the mass density of the medium with embedded particles. Then the L-L equation, under the assumption that the refractive indices n and $\bar{n}$ are close to unity, reduces to

$$\frac{n-1}{\rho} = \frac{\bar{n}-1}{\bar{\rho}} \tag{47}$$

which, with $\bar{\rho} = NM$ and $\rho = m/V$, is just a restatement of Eq. (46).

Use of Eq. (46) to replace the index n by $\bar{n}$ in Eq. (44) leads to an extinction coefficient

$$\gamma = \frac{8\pi k_0^4}{3N} \left(\frac{\bar{n}-1}{2\pi}\right)^2 = \frac{32\pi^3(\bar{n}-1)^2}{3N\lambda^4} \tag{48}$$

expressed now in terms of the properties of the medium, and not of the individual particles. For air at STP the above relation predicts a mean free path $\ell = 1/\gamma$ of about 125 km (75 miles) for 600 nm light, a value not quite twice that deduced by Rayleigh whose knowledge of N (taken from Maxwell) was less accurate. According to Rayleigh, "Mount Everest appears fairly bright at 100 miles distance as seen from the neighbourhood of Darjeeling." At this distance, the fraction of light reaching Rayleigh from the mountain would have been $\exp(-100/75)$ or 26%. I have never been to Darjeeling, but I have lived in the hills bordering Hachioji on the Japanese island of Honshu and often delighted in the clear morning view of Mt. Fuji, over 100 km away. The distance is shorter, but then Fuji (3776 m) is less than half the height of Everest (8848 m).

As for the clarity with which one can view the stars overhead through clean air on a clear night, recall (from chapter 3) that the characteristic depth of the atmosphere is only about 10 km. It may seem reasonable to believe that, if molecular scattering diminishes the light reaching us through the atmosphere, then a *denser* atmosphere should result in even *less* light transmission. Surprisingly, this is not the case once multiple scattering prevails, for light scattered out of the line of view can be subsequently scattered into it again.

Let us consider multiple Rayleigh scattering under the simplified conditions schematically shown in figure 13.7. It is autumn, and the observer is standing on dark (ideally black) light-absorbing ground looking up at the sky. The source of light at the top of the atmosphere radiates an intensity I_0. At the upper

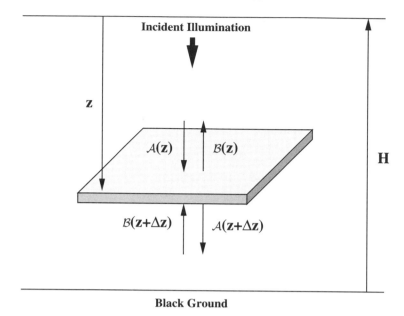

Figure 13.7. Multiple scattering in an atmosphere of thickness H. At each horizontal layer of air, half the incident flux is forward scattered and half is backscattered.

surface of each horizontal layer of air of thickness dz light incident from above is scattered so that half the flux is directed down (flux $\mathcal{A}$) and half is directed up (flux $\mathcal{B}$). (The angular distribution of light scattered by particles much smaller than a wavelength is symmetric about $\theta = 90°$ as illustrated in figure 13.5a.) Although there is no external source of light at ground level, the lower surface of each layer is bathed in an upward flux of light scattered from the layer below; again, half is forward scattered and half is backscattered. To keep matters simple, it is assumed that no light is lost by atmospheric absorption.

Under these conditions, the rate at which the upward and downward fluxes vary with penetration z is governed by the equations

$$\frac{d\mathcal{A}}{dz} = \tfrac{1}{2}\gamma\mathcal{B} - \tfrac{1}{2}\gamma\mathcal{A}, \tag{49a}$$

$$\frac{d\mathcal{B}}{dz} = \tfrac{1}{2}\gamma\mathcal{B} - \tfrac{1}{2}\gamma\mathcal{A}. \tag{49b}$$

Note that the terms on the right side of Eq. (49b) do not have the opposite sign of the corresponding terms in Eq. (49a), since flux $\mathcal{A}$ progresses in the $+z$ direction whereas flux $\mathcal{B}$ progresses in the $-z$ direction. The two equations can be solved by respectively adding and subtracting them to obtain uncoupled equations in the variables $(\mathcal{A} + \mathcal{B})$ and $(\mathcal{A} - \mathcal{B})$, which are readily integrated to yield the downward flux at ground level

$$\mathcal{A}(H) = \frac{I_0}{1 + \frac{1}{2}\gamma H}. \tag{50}$$

In arriving at Eq. (50), boundary conditions $\mathcal{A}(0) = I_0$, $\mathcal{B}(H) = 0$ were implemented.

The most immediate feature of Eq. (50), first proposed by British meteorologist Arthur Schuster in 1905,[22] is how much slower multiply scattered light diminishes in intensity compared with the exponential law for single scattering. Had multiple scattering pertained, 60% of the light, not 26%, would have reached Rayleigh in Darjeeling from Mt. Everest.

There is, however, a second implication, less obvious than the first but more striking in its visual impact. Subtracting from Eq. (50) the directly transmitted (i.e., forward-scattered) light flux (Eq. [45b]) leaves the diffusely scattered radiation $\mathcal{S}(H)$

$$\mathcal{S}(H) = I_0 \left[\frac{1}{1 + \frac{1}{2}\gamma H} - e^{-\gamma H} \right] \tag{51}$$

that constitutes the skylight seen at ground level. Assuming the depth of the atmosphere (H) to remain constant, we can ask what spectral effects would an increase in particle number density—in effect, atmospheric pressure P—engender. For this purpose Eq. (44) must be employed to estimate γ, since n (in contrast to $\bar{n}$) does not vary with N. Adopting the index $n = 1.21$ for the material particles and a particle radius of 2×10^{-8} cm yields the following relation for the extinction coefficient of a room-temperature (293 K) atmosphere as a function of pressure:

$$\gamma\,(\mathrm{km}^{-1}) = \frac{5.2 \times 10^8\,P\,(\mathrm{atm})}{[\lambda\,(\mathrm{nm})]^4}. \tag{52}$$

Based on Eqs. (51) and (52), figure 13.8 illustrates the spectral content of skylight seen through atmospheres of different pressure. For the atmosphere familiar to us (curve a), the visible light to which our eyes respond is dominated by blue, as expected. With increasing pressures, however, the maximum intensity shifts to longer wavelengths. If the surrounding air pressure were increased fiftyfold (curve d), then—physiological effects aside—we would routinely enjoy a red sky! And yet, this is all still Rayleigh scattering. The "blue-sky" law is not just a matter of λ^4; there must be single scattering as well.

Whereas the turbidity of a clean atmosphere with predominantly single scattering is hardly noticeable except over an optical path of many kilometers, the visual impact of a medium with substantial multiple light scattering is apparent immediately. Hold a glass of milk up to the light and look through it. There is not much to see besides an opaque whiteness.[23] No wonder the anonymous seven-year old cited at the beginning of this chapter thought to hide his broccoli

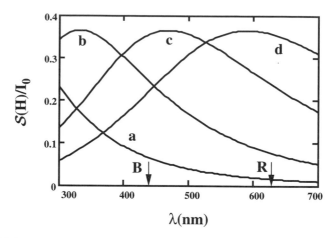

Figure 13.8. Variation of skylight intensity with wavelength for an atmosphere of thickness 10 km and pressure (in atm) of (a) 1, (b) 5, (c) 20, and (d) 50. *B* and *R* signify blue and red regions of the spectrum.

in it. Though the fact that he did not succeed probably owes less to optics than to his mother's premonitions, there is nevertheless a surprising and practically important truth to the lad's remark, as I shall demonstrate.

Incident polarized light scattered repeatedly by the particles of a turbid medium becomes spatially dispersed and depolarized to an extent depending upon the number of scattering interactions. For a sufficiently large number of collisions, photons are expected to lose all memory of their initial directions and polarization states. In gauging the conditions under which this randomization occurs, it is useful to characterize the medium by its optical density (or thickness)

$$\delta = \gamma L = (N Q \sigma_g) L \equiv \frac{L}{\ell} \tag{53a}$$

in which γ is the extinction coefficient (not necessarily given by Eq. [48]) and L is the sample thickness. In general, γ can be expressed as the product of the particle number density N, the geometric cross section of each particle $\sigma_g = \pi a^2$ (here assumed to be of spherical shape), and an efficiency factor Q calculated from Mie theory. Since the inverse of γ represents an effective mean free path ℓ, the optical thickness δ is a measure of the mean number of collisions encountered by a photon in its passage through the medium.

Outside the domain of Rayleigh scattering, individual scattering events do not distribute the light symmetrically about the scattering angle $\theta = 90°$. Rather, as illustrated in figure 13.4, with increasing particle size there is preferential scattering within the forward direction ($\theta < 90°$). This angular asymmetry is characterized by an asymmetry parameter $g \equiv \langle \cos \theta \rangle$, the mean cosine of the scattering angle, and employed to define what is termed the transport optical density

$$\delta^* \equiv \delta(1 - g). \tag{53b}$$

We shall see that δ^* plays a seminal role in the diffusive scattering and depolarization of light.

Multiple scattering—whether of light or electrons—is a notoriously difficult problem to treat theoretically from first principles. In the study of stellar atmospheres, for example, this is the problem of radiative transfer. One has but to glance through S. Chandrasekhar's classic monograph on the subject to get a feel for the mathematical complexity of even the simplest special case:[24] infinite plane waves upon a volume of isotropic scatterers. And yet, this is the only exact solution I know of. In such circumstances a few simple "tabletop" experiments can go far to elucidate the significant features of this complex phenomenon.

To understand what transpires within a turbid medium, project into it a light beam of well-defined polarization and examine the intensity and polarization of the light that emerges at various scattering angles and for different concentrations of scatterers. For this purpose the photoelastic modulator (PEM), discussed in chapter 9, is an ideal instrument, and figure 13.9 shows one of the experimental configurations I have used. Let us recall the basic features of the technique. A beam of polarized light ($\lambda = 544$ nm) from a helium-neon laser is mechanically "chopped" at a low frequency $F \sim 100$ Hz and then passed through the PEM where it is phase-modulated at a much higher frequency $f = 50$ kHZ. After being scattered by the turbid medium—an aqueous suspension of latex microspheres makes an excellent sample—the light is detected by a photomultiplier tube whose output current, or voltage across an output resistance, is Fourier analyzed electronically by lock-in amplifiers. These amplifiers furnish the signals $I(0)$, $I(f)$, $I(2f)$.

The "dc" component of the photocurrent, $I(0)$ (more precisely, the component at F), is a measure of the overall light intensity as would be seen with the naked eye. (Strictly speaking, $I(0)$ renders what the unaided eye would see only if the illumination in the experiment and the ambient lighting for viewing are equivalent. This may not be the case—a point I will illustrate later.) $I(0)$ consists of a contribution $I^{(p)}(0)$ deriving from light that has remained polarized and a contribution $I^{(u)}(0)$ from unpolarized light produced by depolarizing collisions within the turbid medium. Not every collision depolarizes light; indeed, it was one of the objectives of this study to ascertain how many collisions polarized light undergoes before it becomes effectively depolarized. Experimentally, the residual degree of polarization of the detected light can be expressed by the function[25]

$$\beta = \frac{I^{(p)}(0)}{I^{(p)}(0) + I^{(u)}(0)}. \tag{54}$$

The key advantage to using a PEM is that the harmonics $I(f)$ and $I(2f)$ derive

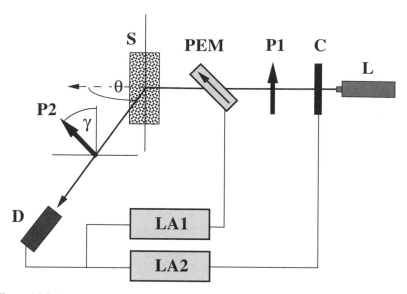

Figure 13.9. Experimental configuration for measuring the intensity, degree of polarization, and optical rotation of diffusively scattered light. The components are helium-neon laser (L), chopper (C), polarizers (P1, P2), photoelastic modulator (PEM), turbid sample (S), detector (D), and lock-in amplifiers (LA1, LA2).

exclusively from polarized light, for only light of well-defined phase can be phase-modulated.

For the configuration depicted in the figure—with initial polarizer P1 vertical (i.e., perpendicular to the scattering plane) and analyzer P2 at angle γ to the vertical—the theoretical ratio $I(2f)/I(0)$ in the absence of diffusive scattering can be shown (by the reasoning of chapter 9) to be

$$R_0 \equiv \frac{I^{(p)}(2f)}{I^{(p)}(0)} = 2J_2(\varphi_m^o) \cos 2(\phi + \gamma) \tag{55a}$$

when the modulation amplitude φ_m is set to the specific value $\varphi_m^o = 2.405$ radians. Recall that this choice, corresponding to the first zero of the Bessel function J_0, provides a good compromise between optimizing the signal and simplifying the data analysis. The angle ϕ is the optical rotation of the scattered light for an ambient medium that is optically active—an important case to be taken up shortly.

When light is depolarized by scattering, the preceding photocurrent ratio becomes

$$R \equiv \frac{I^{(p)}(2f)}{I(0)} = \beta R_0, \tag{55b}$$

a relation from which β can be obtained experimentally. One way, for example, is to determine first, as a calibration standard, the current ratio $R(N = 0)$ for the ambient medium devoid of all suspended particles. The light that reaches the detector (at any angle other than that of the incident beam) has undergone Rayleigh scattering, and the current ratio therefore takes the form of Eq. (55b) with the degree of polarization β_{Ray} (close to unity). Thus, the ratio $R(N)$ for a medium with N scatterers/cm^3 to $R(N = 0)$ yields the degree of polarization relative to that of the ambient medium alone:

$$\frac{\beta}{\beta_{Ray}} = \frac{R(N)}{R(N = 0)}. \qquad (56)$$

In a series of experiments undertaken with my colleague Wayne Strange, light scattered at various angles from aqueous suspensions of latex microspheres, each sample containing particles of a uniform size, was examined. The diameter of the particles ranged from 0.125 μm to approximately 1.0 μm, or, expressed in terms of the dimensionless size parameter $\sigma = 2\pi a/\lambda$, from $\sigma = 0.72$ (Rayleigh-Gans scattering) to $\sigma = 5.8$ (general Mie scattering).

Consider first the total intensity of scattered light (see note 15), illustrated in figure 13.10 for the case of lateral scattering ($\theta = 90°$); to avoid crowding, results are shown only for the smallest and largest particles. Irrespective of sphere size, the intensity curves reveal a similar form with increasing number density: linear growth when N is sufficiently low (as one would expect in the single-scattering domain), followed by a sharp rise to peak value (with the onset of multiple scattering), and subsequent diminution (as the medium becomes increasingly opaque). The linearity of the increase is not so easily discerned in the top figure with N plotted on a logarithmic scale but is quite obvious in the lower figures in which the abscissa is δ^* which is proportional to N. Note that whether N is "low" or "high" depends on the size of the scattering particles. Although the range of N in the figure spans six orders of magnitude, each set of data, when normalized by the maximum intensity I_{max} and plotted against the transport optical thickness δ^*, can be represented closely by a universal curve of the form

$$\frac{I(\delta^*)}{I_{max}} = \frac{1}{1 + \bar{p}\delta^*} - e^{-\delta^*} \qquad (57)$$

where the single parameter $\bar{p}$ depends on particle radius. The lower part of the figure shows the scattering curves as a function of δ^* fitted by Eq. (57) with respective values $\bar{p} = 0.05$ and 0.035 for particle sizes $2a = 0.125$ and 1.0 μm. The parameters from Mie theory needed to calculate δ^* are, respectively, $Q = 3.09$ and 0.028, and $g = 0.93$ and 0.16—indicative of the strong forward diffractive scattering of large particles and nearly isotropic weak Rayleigh scattering of small ones.

It is significant that the empirical function (57) resembles closely (although is not identical to) the theoretical expression (51) deduced earlier for one-

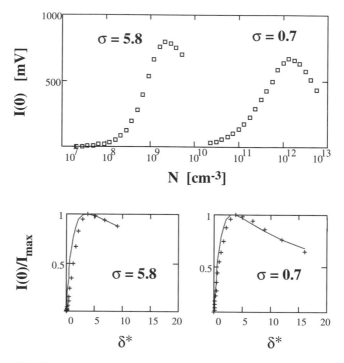

Figure 13.10. (Upper) Laterally scattered ($\theta = 90°$) intensity as a function of particle number density for particles with size parameter $\sigma = 2\pi a/\lambda$. (Lower) Relative intensity as a function of transport optical thickness with theoretical fits (solid line) to Eq. (57).

dimensional symmetric light scattering in a foggy atmosphere. Although the model does not rigorously correspond to the present experimental conditions (with light detected at 90° to the incident beam), it nevertheless suggests an interpretation of the form of the function (57). The parameter $\bar{p}$ (equal to 0.5 in the two-flux model) is a statistical measure of the probability that light incident upon a differential layer within the sample will be scattered in the opposite direction. This direction, of course, need no longer be either 0° or 180°. The exponential term represents the portion of coherent light remaining after extinction by single scattering. Substracting this term as before leaves the intensity of diffuse radiation.

With regard to the matter of residual light polarization, figure 13.11 shows the variation in β for the same turbid samples and scattering configuration. The data reveal several striking features. First, when N is sufficiently low and single scattering dominates, the scattered light remains 100% polarized. This result is not surprising—but it is always gratifying to see confirmed what one thinks he understands. Second, it takes many more small particles to depolarize light

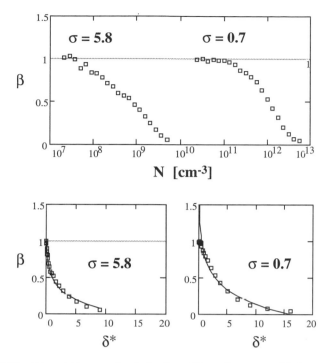

Figure 13.11. (Upper) Degree of polarization of laterally scattered light as a function of number density. (Lower) Same data as a function of transport optical thickness with theorctical fits (over multiple scattering domain) to Eq. (58).

than large particles. For example, to depolarize light to $\beta - 50\%$ takes 1000 times as many 0.125 μm diameter particles per cm^3 as 1.0 μm particles. Third, with the onset of multiple scattering the decrease in β appears markedly linear with the logarithm of N. Indeed, as in the case of light intensity, the degree of polarization can be fitted over the multiple scattering regime to a general function of δ^*

$$\beta = -K_1 \log(\delta^*) + K_2. \tag{58}$$

The excellence of the fit is illustrated in the lower part of the figure.

Displayed in terms of the transport optical density, the data indicate that $\delta^* \sim 10$ is an approximate upper limit at which laterally scattered light retains a degree of polarization measurable with the PEM. At this level of opacity, the mean number of collisions is $\delta \sim 150$ for the sample with 1 μm diameter spheres, but only $\delta \sim 12$ for the sample with 0.125 μm diameter spheres. In the first case, however, each collision preferentially scatters light in the direction of the incident beam, whereas in the second case the scattering is nearly isotropic. The universal dependence of β on δ^*, rather than on δ, suggests that the mean

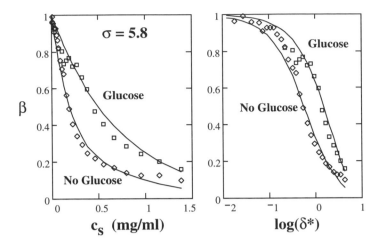

Figure 13.12. Degree of polarization of backscattered light from a suspension of 1 μm particles in either pure water or glucose solution (0.5 g/ml). Solid lines are fits to Eq. (59) as described in the text.

number of depolarizing collisions incurred by the light detected at 90° is about the same for both samples.

The capacity to observe in detail the effect of scattering on light polarization permits one to explore a conceptually interesting question whose answer lies at the core of exciting new possibilities for imaging and optical information extraction. Are different states of light randomized at different rates? In this regard, it is instructive to compare light scattering by a nonchiral aqueous suspension of particles with scattering from particles of identical size and concentration in water containing chiral molecules, such as d-glucose. The left half of figure 13.12 shows the results of β as a function of scatterer concentration c_s for backscattering from relatively large spheres ($\sigma = 5.8$). The depolarization of light is considerably less in the optically active solution than in pure water. Why?

This curious feature may be attributed to the chirality of the medium and to the difference in rates of randomization of linear and circular polarizations. The light issuing from the PEM and incident on the samples in these experiments is in general elliptically polarized, that is, it contains a superposition of LP and CP states. Both polarizations propagate with the same phase velocity within an isotropic nonchiral medium. However, the optical eigenstates of a chiral medium are states of definite helicity; LP waves entering an isotropic optically active medium decompose into independently propagating RCP and LCP waves, as first proposed by Fresnel.

In the absence of a comprehensive analytical theory of multiple light scattering in chiral media, one can attempt to interpret the backscattering data by assuming, in analogy to a diffusion process, that the degrees of linear (β_L) and circular (β_C) polarization decrease exponentially with scatterer concentration or, more pertinently, with the transport optical density: $\beta_X = b_X \exp(-q_X \delta^*)$. In this model the depolarization rates q_X depend specifically on polarization (X = L or C), and the composition coefficients b_X are taken to be independent of δ^*. Then the total degree of polarization (see Eq. [9.15c]) is given by $\beta = \sqrt{\beta_L^2 + \beta_C^2}$, or equivalently by

$$\beta(\delta^*) = \sqrt{(1 - \alpha) \exp(-2q_L \delta^*) + \alpha \exp(-2q_C \delta^*)} \tag{59}$$

where the constants $b_C = \alpha$, $b_L = 1 - \alpha$ satisfy the initial condition $\beta(0) = 1$ of the experiments. (No scattering $\Rightarrow$ no depolarization.)

For the chiral sample, it follows that $\alpha = 1$, and the depolarization is characterized by q_C alone. Under these circumstances a linear fit to $\ln \beta(\delta^*)$ yielded $q_C = 0.98$. With q_C known, application of Eq. (59) to the data taken in the absence of glucose led to the ratio $q_L/q_C = 3.22$. The resulting theoretical curves, shown as solid lines in the figure, match the data sets reasonably well. Moreover, when expressed as a function of $\log(\delta^*)$, as illustrated in the right half of figure 13.12, the model reproduced both the apparent linear variation of β over the domain of multiple scattering and the gradual approach to unity in the region of single scattering.

That linearly polarized light is depolarized after fewer scattering events than circularly polarized light is perhaps not surprising in this example. All collisions that alter the azimuthal orientation of the electromagnetic field in the plane normal to the propagation vector contribute to depolarizing LP light, whereas only those collisions that reverse the sense of field rotation within that plane depolarize CP light.

The effective preservation of light polarization in a chiral medium, coupled with the capacity of the PEM to select those photons which remain polarized, encourages the prospect of observing chiral asymmetry by diffusive light scattering. As discussed previously, the use of light transmission and specular reflection to observe different manifestations of optical activity has been limited largely to homogeneous materials. Heretofore, the thought of observing the optical activity of turbid chiral media must have appeared unworkable in view of the anticipated randomization of light polarization, for I know of no previous experimental efforts in this direction, successful or otherwise.

That it can be done, however, is illustrated in figure 13.13, which summarizes the variation in optical rotation with glucose measured by forward, lateral, and backscattering. For each configuration the optical rotatory power was deduced at two concentrations of 1 μm diameter particles ($c_{s1} = 2.5 \times 10^{-4}$ g/cm^3 and $c_{s2} = 1.38 \times 10^{-3}$ g/cm^3) by a "null detection" procedure based on Eq. (55a). With an achiral aqueous suspension for a standard sample ($\phi = 0$), the orientation of polarizer P2 in figure 13.9 was set to a value γ_0 (close to $-45°$) at which the photocurrent component $I(2f)$ vanished. The standard was then

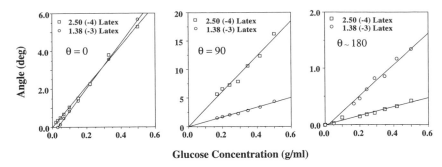

Figure 13.13. Variation of optical rotation with glucose concentration for forward, lateral, and backscattered light from samples containing different concentrations of 1 μm diameter latex spheres.

replaced by a chiral sample of the same particle concentration, and P2 adjusted to a value γ at which a null signal at the first harmonic was again attained. The optical rotation, which can be measured to a precision of 0.02° by an angular micrometer, follows from the relation

$$\phi = \gamma_0 - \gamma. \tag{60}$$

This method has the advantage of being largely unaffected by the mean light flux or degree of polarization.

As in the familiar case of light transmission through homogeneous transparent media, the optical rotation of diffusively scattered light was found to vary linearly with concentration of the chiral component—but at a rate that depended upon the concentration of the achiral particles. Although individual scattering events do not (according to the theory of section 13.2) contribute directly to the optical rotation, the particles have the effect of lengthening the path of the scattered light through the turbid medium. Moreover, the nature of this modification depends sensitively on the scattering asymmetry and angle of observation.

In the case of forward scattered light (detected within a 10 mrad cone of acceptance centered on the incident beam), the variation in optical rotation with glucose ($\sim$12.71°/[g/ml]) was weakly dependent on the particle density and larger by a factor of only $\sim$1.4 than the corresponding rate ($\sim$8.93°/[g/ml]) in the absence of all particle scattering. Such a modest increase is consistent with Mie theory, which predicts in this case predominantly forward scattering. Thus, along the incident direction, the optical pathlengths of multiply scattered and unscattered (transmitted) light did not differ greatly.

With laterally scattered light however, where no unscattered photons are received, the optical rotation was considerably greater ($\sim$30.8°/[g/ml]) for the less turbid sample than for the sample with higher particle density ($\sim$8.5°/[g/ml]). In the first case, the detector received light that for the most part had been multiply scattered from deep within the cell interior, thereby incurring an effective pathlength considerably longer than the actual cell width. By contrast, in the second case less light penetrated the interior and the optical pathlength was shorter. This interpretation is supported by the visual appearance of the cell which was brightly illuminated throughout the sample c_{s1}, but exhibited radiance only near the point of entry of sample c_{s2}.

The case of backscattering is of especial interest, since this is the geometry of choice for materials analysis and biomedical diagnostics. Thinking in terms of light reflection (as Tyndall had done), one might have anticipated that no optical rotation would be observed, for light specularly reflected at normal incidence carries no imprint of the chirality of the medium, a point discussed in chapter 11. This is not so, however, for diffusive scattering. The bottom part of figure 13.13 shows the optical rotation of light detected within a 20 mrad cone of acceptance centered on a scattering angle within 6° of exact backscattering. Rotatory powers are smaller than for forward and lateral scattering and in the opposite sense (since the helicity of light is reversed).

In contrast to lateral scattering, it is now the suspension of greater turbidity c_{s2} that manifested a larger optical rotation (2.78°/[g/ml]) than the suspension of lower turbidity (0.79°/[g/ml]). As a result of primarily forward scattering, most photons escaped undetected into the cell interior. However, because of the greater number of scattering events in sample c_{s2} compared with sample c_{s1}, those photons which ultimately returned (nearly) antiparallel to the incident beam incurred on average a longer optical pathlength in the optically thicker medium. In effect, the larger rotatory power signified a greater entrapment of light.

The important lesson of these chiral studies is that diffusively scattered light carries information about both the chiral structure of the ambient medium and the size and density of the scattering particles. In particular, the recording of a significant optical rotation in backscattering suggests that detected photons may have undergone several scatterings into the medium, but only one reverse scattering into the detector (in keeping with the large asymmetry parameter for large particles). Despite the fact that the suspensions were highly turbid, the use of the PEM to select photons that had preferentially preserved their states of polarization made it possible, in a manner of speaking, to see inside the material and deduce signficant facts about its chemical structure.

It is natural to inquire next whether one can literally "see" inside a turbid medium—that is, to discern the presence of embedded objects hidden from view by multiple scattering. Turbid media are everywhere—fog, clouds, seawater, skin, blood—and the capacity to penetrate such an optical barrier visually has far-reaching consequences for science, technology, and medicine. Herein lies one of the most remarkable and practically useful outcomes of the study of light scattering by phase modulation.

Earlier studies, in which black plastic absorbing objects were immersed in a turbid medium,[26] have shown that the presence of the objects could be revealed by the small fraction of transmitted or forward scattered light that followed the shortest paths through the medium—that is, paths close to the axis of the incident light beam. These "short-path" photons, which suffered the fewest collisions and therefore the least depolarization, could be distinguished from "long-path" photons by means of phase modulation and synchronous detection.

Practically speaking, however, forward scattering is not the most useful configuration. More likely, an interested observer will need to illuminate a turbid medium and view it by means of the light that returns. In contrast to largely undeviated short-path photons, *this* light undergoes large-angle scattering and contains in principle no undeviated portion of the incident illumination. Can imaging be done this way?

Recent experiments with a variety of objects immersed in the opaque milkiness of a dense latex suspension have shown that it can.[27] The guiding principle in these experiments is to detect nearly instantaneously the inequivalent reflection from the same object of two orthogonal states of polarization. If the linear or circular intensity difference (LID, CID) of light returned by the embedded object differs from that diffusively scattered by the ambient medium, the intensity profile obtained by scanning the light source (a helium-neon laser beam in the experiments described here) should vary in correspondence with structural features of the target.[28]

To convert the PEM to an imaging device, one simply removes polarizer P2 from the experimental configuration of figure 13.9, as redrawn in figure 13.14. The dc photocurrent component $I(0)$ corresponds to what a polarization-insensitive detector sees, whereas the first harmonic $I(2f)$ is proportional to the LID and the fundamental component $I(f)$ is proportional to the CID. (Theoretical justification of these correlations is demonstrated in the appendix, a calculation instructive in its own right as an example of the use of the two-dimensional density matrix formalism in precisely the capacity for which four-dimensional Mueller matrices are traditionally deemed unavoidable, namely, diffusive scattering.)

As examples of the use of phase-modulated light to delineate objects hidden within a turbid medium, let us examine four distinct types of targets illustrated in figure 13.15: (1) an absorbing slab with slotted apertures on a reflective metal base ("slotted" target), (2) a metal slab with two contiguous sections of orthogonally oriented rulings ("grooved" target), (3) two linearly polarizing plastic strips mounted side by side, with transmission axes at right angles, on a reflective metal base ("laminar" target), and (4) a small rectangular metal plate with a centrally placed round hole ("aperture" target). The distinguishing feature of the slotted target is its edges, the grooved target its surface texture, the laminar target its polarization-selective absorption, and the aperture target its transparency. All targets were positioned midway between the front and back surfaces of a sample cell 1 cm deep and scanned horizontally across the illuminated face.

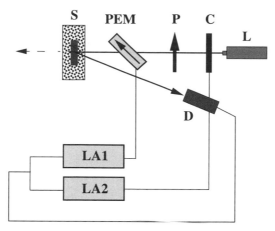

Figure 13.14. Experimental configuration for observing the LID and CID from an object embedded in a turbid medium. Labeling of components is the same as for figure 13.9.

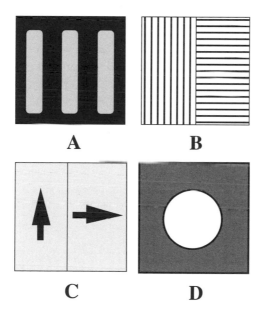

Figure 13.15. Target types: (A) slotted, (B) grooved, (C) laminar (where the arrows signify transmission axes of the polarizing films), and (D) aperture.

Consider first the slotted target comprising three vertical slots of width ~2 mm and separation ~3 mm. Scans of $I(0)$, $I(f)$, and $I(2f)$ taken in the absence of scattering particles (i.e., the target surrounded only by air) are shown in figures 13.16a–b. The asymmetries in peak height are attributable

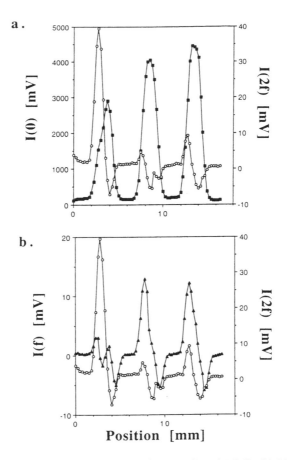

Figure 13.16. Scans of the slotted target in air: (a) $I(0)$ and $I(2f)$; (b) $I(f)$ and $I(2f)$. [$I(2f)$ plotting symbol = open circle.]

to the asymmetric location of the detector (at approximately 7° from exact backscattering). Of particular interest here, however, are the lateral shift and the sign of the signals $I(f)$ and $I(2f)$ relative to the signal $I(0)$. Each $I(0)$ peak corresponds to the center of a slot where reflectance from the metal base is greatest. By contrast, the $I(f)$ and $I(2f)$ scans—which correlate closely with one another—attain maximum values (both positive and negative) near the steepest slopes of the $I(0)$ scan, that is, at the edges defining the slots. Thus, the LID and CID of a target with sharp boundaries are particularly sensitive to the location of these boundaries.

The signs of the LID and CID signals—the fact that the left edge (with positive LID) reflects more σ- than π-polarized light, whereas the right edge (with negative LID) does the reverse—derives in part from shape of the slot

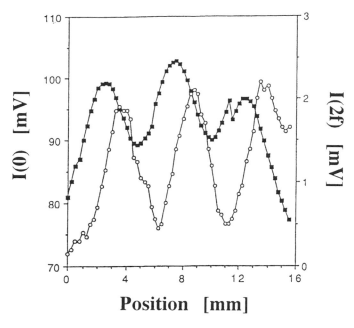

Figure 13.17. Scans of $I(0)$ and $I(2f)$ across the slotted target in a 2.8×10^{-3} g/cm³ suspension of 1 μm particles. [$I(2f)$ plotting symbol = open circle.]

profile. Slot walls are not exactly perpendicular to the base, but slightly inclined to the vertical at an angle of approximately $+5°$ at the right edge and $-5°$ at the left edge. As I will relate in a moment, shape sensitivity can lead to some unexpected and puzzling results.

The outcome of immersing the target in a dense (2.8×10^{-3} g/cm³) suspension of 1 μm diameter latex particles ($\sigma = 5.8$) is recorded in figure 13.17. Without the target the optical densities of the medium are $\delta \sim 125$ and $\delta^* \sim 9$. Examined by the unaided eye under conditions of ambient illumination, the entire cell (with target positioned) reveals only a featureless milky white opacity; the presence of the target can not be discerned. The LID, however, obtained from the $I(2f)$ scan marks the edges of the slots with a contrast of $\sim 63\%$. It is to be noted in this regard that the dc component of the photocurrent, obtained from pointwise laser illumination of the sample, also renders the variation in surface topography with a greater degree of visibility (contrast $\sim 14\%$) than the naked eye. Thus, contrary to what one might have assumed, the signal $I(0)$ is not necessarily equivalent to direct viewing.

Curiously, although a strong CID signal was detected from the slotted target in air, immersion in the latex suspension resulted in no photocurrent component (apart from noise) at frequency f. This observation was at first surprising, for it seemed to indicate that diffusive scattering by large particles ($2\pi a/\lambda > 1$)

had depolarized circularly polarized light to a much greater extent than linearly polarized light—in marked contrast to the results of earlier experiments. Actually, though this may sound paradoxical, it is precisely because the circularly polarized light remained polarized that the $I(f)$ signal vanished.

In air, where there is no multiple scattering, the narrow pencil of laser light illuminates each edge individually as the sample cell is translated across the fixed beam. As a consequence of the high point resolution, the detector receives light reflected from only one edge. In the turbid medium, however, suspended particles diffuse the incident radiation, thereby illuminating not only the edge in direct line with the beam, but also the other edge. Light from both edges reaches the detector.

Multiply scattered by large particles, LP photons are more strongly depolarized than CP photons—and so the edge illuminated by scattered light returns radiation of a much weaker degree of linear polarization than the light reflected from the directly illuminated edge. The detector receives both these fluxes, but only the linearly polarized component contributes to the signal $I(2f)$. In effect, the detector again sees only one edge at a time in linearly polarized light. On the other hand, both edges return radiation of a comparable degree of circular polarization, except for opposite sign. The superposed fluxes therefore lead to a smaller net signal $I(f)$, which, at sufficiently high particle concentrations, vanishes entirely.

To test this explanation, a "step" target with only one edge (the right edge with positive slope) was constructed and imaged under conditions of increasing particle concentration. The resulting $I(f)$ signal broadened in width and diminished in intensity as the medium became more and more cloudy—but it did not vanish even at the optical density of figure 13.17. When the step target was then replaced by the original slotted target of figure 13.15, $I(f)$ again vanished completely. Such replacement, however, had little effect on the $I(2f)$ signal.

The "anomalous" disappearance of the signal $I(f)$ graphically illustrates a dual aspect of light polarization in PEM-based imaging. Without polarized light one will of course see nothing (except noise) at the fundamental and first harmonic frequencies. Yet, if the light scattered from all points of an object remains polarized to the same degree, then one may again see nothing. Particle size, as well as concentration, has a marked influence on the extent to which light is depolarized and whether LID or CID (or both) is suitable for seeing through a turbid medium.

Experiments with the grooved target illustrate the sensitivity of PEM-based imaging to surface features. The target was made by ruling parallel slits of width ≤ 1 μm with a spacing of ~ 50 μm; vertical striations (normal to the plane of scattering) were ruled on the left half and horizontal striations on the right. Unlike the wide slots of the previous target, the rulings are much narrower than the cross section of the probe beam, but not greatly in excess of the wavelength ($\lambda = 0.544$ μm). One might expect this target to behave

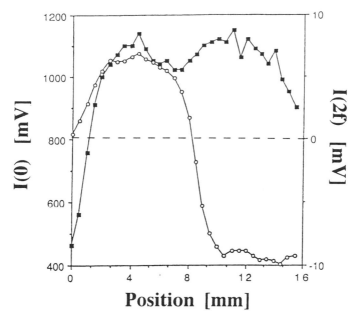

Position [mm]

Figure 13.18. Scans of $I(0)$ and $I(2f)$ across a scratched aluminum target in a 1.4×10^{-3} g/cm^3 suspension of 1 μm diameter particles. Lines are oriented vertically on the left half and horizontally on the right half of the figure. [$I(2f)$ plotting symbol = open circle.]

somewhat like a conducting grating and impart a preferential linear polarization to the back-diffracted light parallel to the orientation of the lines.[29]

Figure 13.18 shows the results of scanning the backscattered light from the grooved target immersed in a 1.4×10^{-3} g/cm^3 suspension of 1 μm particles ($\delta \sim 63$). The $I(0)$ scan, which is insensitive to polarization, exhibits virtually no structure except for a slight dip at the junction between the two grooved segments. In sharp contrast, however, the $I(2f)$ scan takes the approximate form of a step function with positive plateau on the left (signifying a predominance of σ-polarized light) and negative plateau on the right (signifying a predominance of π-polarized light)—just as predicted. No CID was anticipated, and correspondingly no signal $I(f)$ was observed either in air or in turbid liquid.

Consider next the smooth laminar target immersed in the same turbid suspension as the grooved target. Like the rulings of the latter, the transmission axes of the polarizing films were oriented vertically on the left half of the figure and horizontally on the right half. The results (see figure 13.19) are very similar to those of figure 13.18: The dc signal is flat across the target (dropping off at the edges) and reveals no information, whereas the variation in $I(2f)$ corresponds to preferential σ-polarization on the left and π-polarization on the

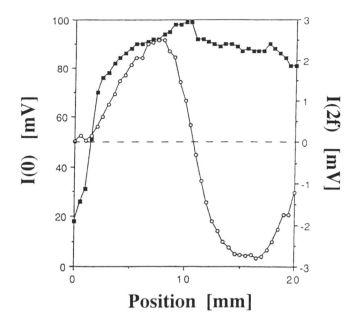

Figure 13.19. Scans of $I(0)$ and $I(2f)$ across a smooth polaroid target immersed in the same turbid medium as the grooved target. Transmission axes are oriented vertically on the left half and horizontally on the right half of the figure. [$I(2f)$ plotting symbol = open circle.]

right. The interaction of the target with light, however, is completely different here than in the preceding example where surface structure created unequal fluxes of backscattered polarized light. Now the surface is entirely smooth, and a LID is created by selective absorption of light transmitted twice (forward and reverse) through the laminae.

To confirm that the spatial asymmetry revealed by the $I(2f)$ scans of the grooved and laminar targets indeed arose from structural properties of the target and not from asymmetric placement of the detector, the experiments were repeated with (1) the detector repositioned at its mirror-image location on the opposite side of the beam axis, the target orientation being unchanged, and (2) the target rotated 180° about the incident beam axis, the detector location being unchanged. As anticipated, variation (1) had no effect on the previous outcomes, whereas variation (2) reversed the sign of the $I(2f)$ scans, indicating a reversal in LID.

Last, consider the aperture target, which was in fact the first one to be examined in establishing the feasibility of seeing through cloudy media with phase-modulated polarized light (see note 15). The target consists of a small hole 4.1 mm in diameter drilled in a metal plate approximately 25 mm high and 6.2 mm wide. In contrast to the slotted target, there is no reflective backing, and

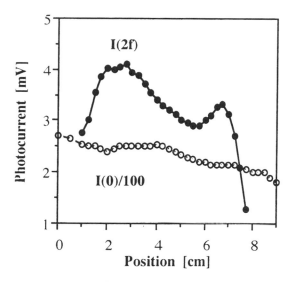

Figure 13.20. Scan of aperture target in an aqueous suspension (3.3×10^{-3} g/cm³) of 0.125 μm diameter particles.

it is this feature that provides an indication of the extent to which LID or CID measurements subtract out diffusive scattering from the surrounding medium.

Figure 13.20 shows scans of $I(0)$ and $I(2f)$—there is again no signal $I(f)$—for the target immersed in an optically dense ($\delta \sim 11$, $\delta^* \sim 9$) aqueous suspension of 0.125 μm diameter particles; that is, particles are small in size compared to the wavelength. Recall that small particles scatter light nearly isotropically, in which case the backscattering from the medium now constitutes a much larger proportion of the detected light. Scanning a light beam across the front surface of the sample cuvette at the level of the hole diameter should produce in the absence of diffusive scattering two peaks of approximate width 9.4 mm separated by a 4.1 mm wide valley. The degree of turbidity is such, however, that neither direct viewing in ambient light nor the dc component of backscattered laser light reveals any presence of a target. Nevertheless, the $I(2f)$ signal still shows the profile of the object with sufficient contrast for recognition.

There is an interesting sidelight to this research that has strongly impressed upon me—and *not* for the first time—the "ingenuity" of Nature. While these experiments were in progress, I came across a news article concerning the retina of the sunfish,[30] a sea creature that must wend its way through murky ocean water. How does this fish see where it is going?

Examined under the microscope, a retinal cell displays an ellipsoidal shape, the long axis of which is oriented in one of two mutually perpendicular positions within the cellular network. According to the report, the orientation of each cell determines which one of two orthogonal states of linear polarization a cell responds to, and the fish's brain is thought to register an image by subtracting the signals of the one group of cells from the other. In other words, the fish navigates through its turbid environment by monitoring in its own way the linear intensity difference of backscattered light, that is, by precisely the principle of the preceding experiments. How marvelous it is that through natural selection this creature has solved a complex imaging problem millions of years before the same principles were rediscovered by humans.

The human visual system, as I remarked in chapter 9, is virtually indifferent to polarized light. But this insensitivity is not total. There is, in fact, an intriguing optical phenomenon known as "Haidinger's brush," the occurrence of an elongated yellowish image (with bluish clouds on either side) produced in the unaided eye by linearly polarized light.[31] The brush, which is attributable to dichroism of the fovea and requires a certain amount of practice to discern, is oriented perpendicular to the direction of light polarization. For example, upon viewing unpolarized light Rayleigh-scattered at 90°, the brush would appear in the scattering plane. A trained human eye *can* see the polarization of skylight.

Since learning about the sunfish, I have wondered whether a person, after sufficiently long use of cross-polarized spectacles, could eventually manage to see through turbid media in an analogous way. I suspect not; our retinal cells are not shaped like those of the sunfish. In any event, for those who dislike broccoli, know that henceforth you cannot hide it in a glass of milk.

APPENDIX: DENSITY MATRIX ANALYSIS OF LIGHT SCATTERING

To employ phase modulation and synchronous detection in light scattering, one can place a PEM either before ("pre" configuration) or after ("post" configuration) the scattering cell. The outcomes of the two configurations are not, in general, the same, although I have seen the same term ("degree of polarization") applied in both cases to the quantities of interest. The term is not expressly incorrect, but rather not sufficiently precise to convey the subtle and highly useful distinctions involved. It is the purpose of this appendix to show exactly what pre and post configurations yield in the specific case where the only optical elements, besides the sample, are the PEM and one polarizer.

Consider the configuration of figure 13.9 with polarizer P2 removed. The density matrix of the light source, in effect the radiation from polarizer P1, is simply

$$\rho^{(1)} = \begin{pmatrix} 1 & 0 \\ 0 & 0 \end{pmatrix} \tag{A1}$$

indicative of the pure "vertical" state $\mathbf{x}$. The light subsequently passes through the PEM (oriented at 45° to the vertical) whose system matrix (Eq. [9.83]) is

$$M_{\text{PEM}} = \frac{1}{2} \begin{pmatrix} e^{i\varphi} + 1 & e^{i\varphi} - 1 \\ e^{i\varphi} - 1 & e^{i\varphi} + 1 \end{pmatrix} \tag{A2}$$

where $\varphi(t) = \varphi_m \sin(2\pi f t)$ is the time-dependent retardation. After scattering by the turbid sample, which can be described by a matrix $\mathbf{S}$ with elements S_{ij} (i and $j = 1, 2$), the light is detected and the photocurrent Fourier analyzed electronically.

From the principles outlined in chapter 9, the density matrix of the detected light is calculable from the relation

$$\rho = \overline{(SM_{\text{PEM}})\rho^{(1)}(M_{\text{PEM}}^{\dagger}S^{\dagger})} \tag{A3}$$

in which the overbar denotes an average over the random variables that characterize the diffusive scattering. The observed physical quantity is the light intensity I, which is proportional to the trace of ρ and expressible in the following way:

$$I = \tfrac{1}{2}\text{Tr}\{\rho\} = \tfrac{1}{2}\text{Tr}\left\{(M_{\text{PEM}}\rho^{(1)}M_{\text{PEM}}^{\dagger})\overline{S^{\dagger}S}\right\} \equiv \tfrac{1}{2}\text{Tr}\{A\rho^{(s)}\}. \tag{A4}$$

From the form of Eq. (A4) one may interpret the intensity as the expectation value of an output operator[32]

$$A \equiv M_{\text{PEM}}\rho^{(1)}M_{\text{PEM}}^{\dagger} = \frac{1}{2}\begin{pmatrix} 1 + \cos\varphi & -i\sin\varphi \\ i\sin\varphi & 1 - \cos\varphi \end{pmatrix} \tag{A5}$$

in a state of scattered light represented by an effective density matrix

$$\rho^{(s)} = \overline{S^{\dagger}S} = \begin{pmatrix} \rho_{11} & \rho_{12} \\ \rho_{21} & \rho_{22} \end{pmatrix}. \tag{A6}$$

Note carefully: although the flux through P1 is the original light source, we can regard the light issuing from the scattering cell as the effective source of illumination with density matrix $\rho^{(s)} = \overline{S^{\dagger}S}$. In the post configuration, already treated in chapter 9, whereby the light from the sample passes through the PEM and then P1—that is, in reverse order to the pre configuration—the initial density matrix takes the canonical form $\rho^{(s)} = \overline{SS^{\dagger}}$.

Evaluation of the trace in expression (A4) yields (up to an unimportant numerical factor) a detected light intensity of the form

$$I \sim (\rho_{11} + \rho_{22}) + (\rho_{11} - \rho_{22})\cos\varphi - 2\text{Im}\{\rho_{12}\}\sin\varphi. \tag{A7}$$

From the Fourier-Bessel expansions of chapter 9 it should be evident that the coefficients of $\sin\varphi$ and $\cos\varphi$ constitute the signals at the fundamental frequency f and first harmonic $2f$, respectively; with the modulation amplitude set to φ_m^o (first zero of J_0), the first term in expression (A7) is the only contribution to the dc signal.

To interpret these coefficients note that the effective density matrix has the explicit form

$$\rho^{(s)} = \overline{S^{\dagger}S} = \begin{pmatrix} \overline{|\mathbf{s}_1|^2} & \overline{\mathbf{s}_1^* \cdot \mathbf{s}_2} \\ \overline{\mathbf{s}_1 \cdot \mathbf{s}_2^*} & \overline{|\mathbf{s}_2|^2} \end{pmatrix} \tag{A8}$$

where

$$\mathbf{s}_1 = S\mathbf{e}_1 = \begin{pmatrix} S_{11} & S_{12} \\ S_{21} & S_{22} \end{pmatrix} \begin{pmatrix} 1 \\ 0 \end{pmatrix} \tag{A9a}$$

and

$$\mathbf{s}_2 = S\mathbf{e}_2 = \begin{pmatrix} S_{11} & S_{12} \\ S_{21} & S_{22} \end{pmatrix} \begin{pmatrix} 0 \\ 1 \end{pmatrix} \tag{A9b}$$

are the hypothetical fields emerging from the scattering cell for respective incident waves of σ- and π-polarization. Relations (A9a,b) resemble the reflection vectors $\mathbf{r}_1$ and $\mathbf{r}_2$ of Eq. (11.42). However, in the present case of multiple scattering with attendant depolarization, it is only "ensemble-averaged" bilinear products of the fields that relate to physical observables. Comparing expressions (A6) and (A8), one sees that the first term in Eq. (A7) is proportional to the total flux of polarized and unpolarized light

$$I(0) \sim \rho_{11} + \rho_{22} \tag{A10a}$$

and the second term corresponds to the difference in reflection of σ- and π-polarizations

$$I(2f) \sim \rho_{11} - \rho_{22}. \tag{A10b}$$

The off-diagonal elements of expression (A8) may be understood by considering hypothetical fields emerging from the scattering cell for incident circular polarizations

$$\mathbf{s}_\pm = S\mathbf{e}_\pm = \begin{pmatrix} S_{11} & S_{12} \\ S_{21} & S_{22} \end{pmatrix} \begin{pmatrix} 1 \\ \pm i \end{pmatrix}. \tag{A9c}$$

It is then not difficult to show that the photocurrent component

$$I(f) \sim -2\mathrm{Im}\{\rho_{12}\} = \overline{\mathrm{Im}\{\mathbf{s}_1 \cdot \mathbf{s}_2^*\}} = \overline{|s_+|^2} - \overline{|s_-|^2} \tag{A10c}$$

corresponds to the difference in intensities of $+1$ and -1 helicity states.

Although the individual photocurrent components were the physical quantities of interest in the experiments employing the pre configuration to image objects embedded in turbid media, one could also have deduced from the ratios of these components the linear and circular intensity differences

$$\delta_{\mathrm{L}} = \frac{\rho_{11} - \rho_{22}}{\rho_{11} + \rho_{22}}, \tag{A11a}$$

$$\delta_{\mathrm{C}} = \frac{-2\mathrm{Im}\{\rho_{12}\}}{\rho_{11} + \rho_{22}} \tag{A11b}$$

defined in chapter 9.

Analysis of the post configuration with the polarizer preceded by the modulator leads to an expression, Eq. (9.87), similar to that of expression (A7) above. In that case, however, the density matrix of the source of light is $\rho^{(s)} = \overline{SS^\dagger}$, and the coefficients of the photocurrent components at f and $2f$ are the degrees

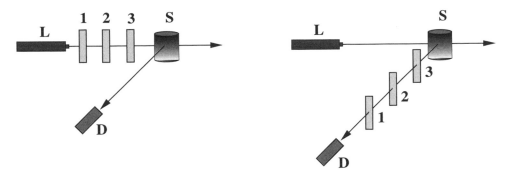

Pre Configuration   **Post Configuration**

Figure 13.21. Pre and post configurations of optical elements, the latter in reverse order of the former.

of polarization τ_C and $\tau_L^{(0)}$, which involve different bilinear combinations of the scattering matrix elements.

The foregoing discussion is readily generalized to pre and post configurations of arbitrary numbers of optical elements as illustrated in figure 13.21. For the case where the elements of the latter are in the reverse order of the former, there is a simple relationship between the expressions for observed light flux:

$$I_{\text{Pre}} \sim \text{Tr}\left\{(M_n \ldots M_1)(M_1^\dagger \ldots M_n^\dagger)\overline{S^\dagger S}\right\} = \text{Tr}\{(MM^\dagger)\overline{S^\dagger S}\}, \qquad \text{(A12a)}$$

$$I_{\text{Post}} \sim \text{Tr}\left\{(M_n^\dagger \ldots M_1^\dagger)(M_1 \ldots M_n)\overline{SS^\dagger}\right\} = \text{Tr}\{M_{\text{rev}}^\dagger M_{\text{rev}})\overline{SS^\dagger}\} \qquad \text{(A12b)}$$

in which

$$M = M_n \ldots M_1, \qquad M_{\text{rev}} = M_1 \ldots M_n. \qquad \text{(A13)}$$

To convert from pre to post configuration one merely replaces S and each component matrix M_i by its adjoint.

NOTES

1. Taken from H. Jackson Brown Jr., *Live and Learn and Pass It on* (Nashville, Tenn.: Rutledge Hill Press, 1992), p. 33.

2. G. Mie, "Beiträge zur Optik trüber Medien, speziell kolloidaler Metallösungen," *Annalen der Physik* 25 (1908): 377–445.

3. Both Tyndall's and Rayleigh's remarks are taken from Lord Rayleigh, "On the Light from the Sky, Its Polarization and Colour," *Philos. Mag.* 41 (1871): 107–20. Rayleigh quoted from Tyndall's paper, *Philos. Mag.* 37 (1869): 388.

4. Like the case of specular reflection, the designations π and σ refer, respectively, to waves oscillating parallel and perpendicular to the scattering plane defined by the incident and scattered wave vectors.

5. Lord Rayleigh, "On the Transmission of Light through an Atmosphere containing Small Particles in Suspension, and on the Origin of the Blue of the Sky," *Philos. Mag.* 47 (1899): 375–84.

6. The result was first obtained by P. P. Ewald (in his 1912 dissertation) for crystalline media and independently by C. W. Oseen (1915) for isotropic media. See M. Born and E. Wolf, *Principles of Optics*, 4th ed. (New York: Pergamon, 1970), pp. 98–108.

7. W. T. Doyle, "Scattering Approach to Fresnel's Equations and Brewster's Law," *Am. J. Phys.* 53 (1985): 463–68.

8. G. B. Parrent Jr. and B. J. Thompson, *Physical Optics Notebook* (Redondo Beach, Calif.: SPIE, 1969), pp. 7–10.

9. O. Heaviside, *Electromagnetic Theory* (London: Electrician, 1912), vol. 3, ch. 9; see W. T. Doyle, "Graphical Approach to Fresnel's Equations for Reflection and Refraction of Light," *Am. J. Phys.* 48 (1980): 643–47.

10. H. C. van de Hulst, *Light Scattering by Small Particles* (New York: Dover, 1981).

11. Excerpted from the *Treasury of World Science*, ed. D. D. Runes (Patterson, N.J.: Littlefield, Adams, 1962), p. 138.

12. M. P. Silverman, J. Badoz, and B. Briat, "Chiral Reflection from a Naturally Optically Active Medium," *Opt. Lett.* 17 (1992): 886–89.

13. M. P. Silverman, W. Strange, J. Badoz, and I. A. Vitkin, "Enhanced Optical Rotation and Diminished Depolarization in Diffusive Scattering from a Chiral Liquid," *Opt. Comm.* 132 (1996): 410–16.

14. R. Cowen, "Meteorite Hints at Early Life on Mars," *Science News* 150, no. 6 (1996): 84.

15. M. P. Silverman and W. Strange, "Light Scattering from Turbid Optically Active and Inactive Media," in *Optics and Imaging in the Information Age* (Springfield, Va.: Society for Imaging Science and Technology, 1997), pp. 173–80.

16. I. V. Lindell and M. P. Silverman, "Plane-Wave Scattering from a Nonchiral Object in a Chiral Environment," *J. Opt. Soc. Am. A* 14 (1997): 79–90.

17. C. F. Bohren, "Light Scattering by an Optically Active Sphere," *Chem. Phys. Lett.* 29 (1974): 458–62; G. W. Ford and S. A. Werner, "Scattering and Absorption of Electromagnetic Waves by a Gyrotropic Sphere," *Phys. Rev. B* 18 (1978): 6752–69.

18. In the formulation of the chiral sky scattering problem through integral equations there are initially electric and magnetic Green's dyads. These are ultimately superposed to form $\overline{G}_\pm$ which can be shown to satisfy the dyadic equation

$$\overline{G}_\pm(\mathbf{r}) = \left(\bar{1} \pm \frac{1}{k_\pm} \nabla \times \bar{1} + \frac{1}{k_\pm^2} \nabla\nabla \right) G_\pm(r).$$

19. A detailed discussion of the use of dyads in chiral scattering theory may be found in two monographs: A. Lakhtakia, V. K. Varadan, and V. V. Varadan, *Time-Harmonic Electromagnetic Fields in Chiral Media* (Berlin: Springer, 1989); I. V. Lindell, A. H. Sihvola, S. A. Tretyakov, and A. J. Vitanen, *Electromagnetic Waves in Chiral and Bi-Isotropic Media* (Boston: Artech House, 1994).

20. For sufficiently large value of the size parameter k_0a, the weak-phase condition will be violated (as is the case with curve d), and the applicability of the theory becomes

questionable. The results may yet be valid, but there is at present neither a more comprehensive theory nor an experiment with which to compare them.

21. Eq. (46) is approximate because the right-hand side effectively assumes that the electromagnetic field within the particles is that of the incident wave, a reasonable assumption when $n - 1$ is not too large.

22. A. Schuster, "Radiation through a Foggy Atmosphere," *Astrophys. J.* 21 (1905): 1–22.

23. Actually, if a few drops of milk are added to a transparent container of water so that the resulting liquid is turbid but not opaque, the visual appearance is rather interesting. Viewed from opposite the source of illumination (forward scattering $\theta = 0$), the liquid has a reddish-orange character since long-wavelength light is scattered least; viewed from the side (lateral scattering $\theta = 90°$), the liquid has a greater blue content and appears whiter.

24. S. Chandrasekhar, *Radiative Transfer* (New York: Dover, 1960).

25. The symbol β has been adopted, rather than τ, since Eq. (54) is a measure of the degree of polarization not necessarily identical to one of the specific functions defined in chapter 9. Moreover, in discussions of multiple scattering τ is often used to represent optical thickness. Although δ (optical density) is used here for that purpose, it is perhaps best to avoid any confusion.

26. J. M. Schmitt, A. H. Gandjbakhche, and R. F. Bonner, "Use of Polarized Light to Discriminate Short-Path Photons in a Multiply Scattering Medium," *Applied Optics* 31 (1992): 6535–46.

27. M. P. Silverman and Wayne Strange, "Object Delineation within Turbid Media by Backscattering of Phase-Modulated Light," *Optics Communications* 144 (1997): 7–11. See also note 15.

28. This basic idea was discovered independently by two groups of researchers employing very different experimental approaches. In one—that of M. P. Silverman and W. Strange, reported at the 1996 Meeting of the Optical Society of America (see note 15)—the PEM with laser illumination and synchronous detection yielded the LID and CID directly on a lock-in amplifier. In the other—that of M. P. Rowe, E. N. Pugh Jr., J. S. Tyo, and N. Engheta, *Opt. Lett.* 20 (1995): 608–10—σ- and π-polarized intensities of incoherent light were separately measured by a CCD camera, and the digitized signals subtracted electronically.

29. A classroom demonstration that usually reveals how unintuitive some aspects of light polarization may be is the transmission of polarized microwaves through a grid of parallel wires. Students who picture the electric vector **E** as something actually "protruding" perpendicular to the rays will inevitably predict that a beam of microwaves polarized parallel to the wires will be transmitted to a greater extent than a beam polarized perpendicular to the wires, for in the former case the waves can "slip through," whereas in the latter they are "blocked" or "impeded." In reality, it is the reverse that occurs, as the parallel electric field induces currents along the wires, and energy is thereby lost in part to Joule heating. The waves, however, are not for the most part "absorbed," as students frequently think. If a detector were placed on the same side of the screen as the microwave source, it would be readily apparent that the waves are backscattered with the electric field parallel to the wires (which radiate like antennas).

30. "Sunfish Shows the Way through the Fog," *Picked Up for You* (a publication of CERN), no. 13 (1996), p. 3.

31. M. Minnaert, *The Nature of Light and Color in the Open Air* (New York: Dover,

1954), pp. 254–57.

32. The interpretation is borrowed directly from quantum mechanics. The expectation value $\langle A \rangle$ of an operator A in a quantum system represented by the density matrix ρ is $\langle A \rangle = \text{Tr}\{A\rho\}$.

Part Five

PLAYING WITH WAVES

CHAPTER 14

Voice of the Dragon

It occurred to one of the authors (EMP) that if an
electron passes close to the surface of a metal
diffraction grating, moving at right angles to the
rulings, the periodic motion of the charge induced on
the surface of the grating should give rise to radiation.
(S. J. Smith and E. M. Purcell, 1953)[1]

A SIMPLE, yet unusual, child's toy, in effect a rotating corrugated resonator,
generates a wide range of tones by spatial modulation of air flow and illustrates
some of the most basic features of the physics of resonance, waves, and fluids.
Although production of sound and electromagnetic waves is quite different in
detail, the toy calls to mind an equally unusual light source.

14.1 "HEARING" THE LIGHT

One of the most remarkable light sources I ever learned of was reported in
one of the shortest scientific papers I ever read: less than a single page in *The
Physical Review* (a sizable fraction of which was a photograph).

The opening line of the paper (cited above) reveals in its entirety Purcell's
clever idea: charged particles moving at a *constant* velocity can give rise to
electromagnetic radiation. Passing at speed v close to the corrugated surface
of a metal diffraction grating with ruling interval d, individual electrons of
an electron beam produce image charges that effectively oscillate normal to
the grating surface, as illustrated in figure 14.1. In conformity with classical
electrodynamics, the oscillating image charges should radiate—and a simple
Huygens construction predicts the fundamental wavelength to be

$$\lambda = d \left(\frac{c}{v} - \cos\theta \right) \tag{1}$$

where θ is the angle of observation relative to the forward direction of the beam.
For a grating with spacing $d = 1.7\mu\text{m}$ and electrons of energy 300 keV, the
predicted radiation falls within the visible spectrum.

Using a small Van de Graaf generator, Smith and Purcell found that, like
"real" charges, accelerated image charges do indeed produce light. Viewing
their grating from a forward position $10°$ to $20°$ from the beam, they saw

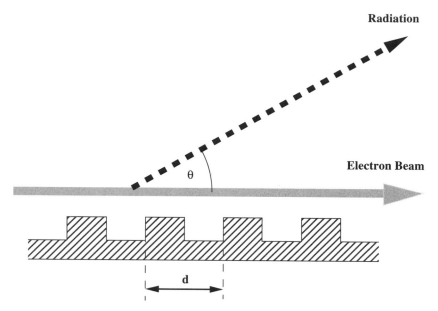

Figure 14.1. Configuration of the Smith-Purcell light source. An electron beam passing over a grating with spacing d generates an oscillating image charge that radiates.

a "sharp, luminous, colored line on the surface of the grating" whose color varied, as expected, with the observation angle and electron energy.

The authors, as far as I know, never resumed this work—or at least never reported on it again. And when, having discovered this paper some fifteen years or so after its publication, I endeavored to know more, a dip into the physics literature then available to me produced but a paltry harvest of references. Even today, when on some rare occasion I mention Smith-Purcell (SP) radiation to a colleague, the latter is more likely than not to look at me quizzically. Within the physics community at large, the SP light source still seems to be a fairly well-kept secret.

I had intended at one time to develop a quantum theory of SP radiation, but, as one project led to another, I continually put off the undertaking, eventually dropping the subject altogether. I never returned to it, but many years later I came across a fascinating device that brought this early interest to mind.

One day, while living in Japan as a guest scientist of the Hitachi Advanced Research Laboratory in Tokyo, I emerged with my family from a train station to encounter the eerie strains of a most unearthly symphony. A dozen or so Japanese children, soon joined by my own, were feverishly grabbing long flexible plastic tubes from the stand of a streetside vendor and twirling them above their heads like lariats. The burst of tones that emerged from each musical pipe soared and dropped with rotational speed over what seemed like a good portion of the range of a flute.

Designated "The Voice of the Dragon" by a sign in English, each musical tube was basically a right circular cylinder roughly 3 cm in outside diameter and 70 cm in length. The tubes, however, were not smooth; rather, corrugations with a depth of about $\frac{1}{4}$ cm and a periodicity of 2 per cm spanned the full length. The overall effect of this loud, wavering, rich-toned chorus of dragon voices left an unforgettable impression.

Despite the outward simplicity of the toy, the details of how it worked were by no means clear. Watching the dragon in action, I puzzled over such questions as the following. Are the characteristic modes of the rotating tube the same as those of a comparable stationary open-ended pipe? Are the modes excited by air rushing past the moving end transversely or through the tube longitudinally? What relation is there between the pitch of a tone and the rate of rotation? Are the corrugations necessary? Can the dynamics of the tube be quantitatively understood, even in part, as a simple application of Bernoulli's principle?

A cursory examination of the tube clarified immediately at least a few points. First, there was a strong air current passing though the rotating tube from the stationary end toward the moving end. By grasping the tube about 30 cm from one end and rotating the remaining length, I heard the air flow and felt the suction as I progressively covered the stationary end with the palm of the other hand; small bits of tissue paper placed in my palm were sucked up into the tube and discharged from the rotating end. Holding the tube near the middle with both hands and causing the two ends to rotate simultaneously led to no tones, as there was apparently no net air flow through the tube.

Second, the corrugations were definitely critical to the production of sound. Inserting into the tube a rolled sheet of paper that formed a smooth inner cylindrical wall eliminated the musical sounds without obstructing the flux of air; one then heard only the whoosh of white noise. Moreover, it did not matter where the paper was inserted: Whether at the moving end, stationary end, or middle, the dragon voice remained silent if corrugations were covered. I found, in fact, that merely covering a small part of the last few corrugations of the stationary end with my thumb was sufficient to reduce the sound intensity greatly at any angular frequency of rotation, although it barely affected the volume of flow. The resonant production of sound apparently depended sensitively on the entire structure and was severely attentuated by even small impediments.

Intrigued by the unexpected melodic qualities and behavior of the dragon voice, and mindful of the unusual physical effects of rotational air flow in other devices I have studied (such as the vortex tube),[2] I decided to investigate this unusual toy more thoroughly with an undergraduate colleague upon my return to the United States. The experiments, which are simple and instructive, employ such basic procedures as Fourier analysis of wave forms and measurement of air speed and pressure, and may easily serve as thought-provoking and

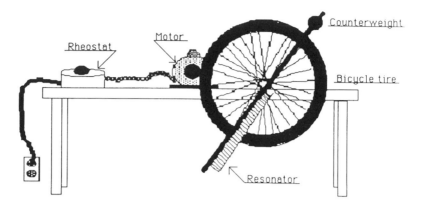

Figure 14.2. Apparatus for producing distortion-free rotation of the resonator. The tube is fixed to a wooden slab, balanced by a counterweight, and mounted to a wheel whose rim is driven by friction contact with a rheostat-controlled motor.

interest-generating laboratory projects in courses on general physics, waves, or acoustics.[3] Indeed, these experiments are ideally suited to such purpose, for the musical toy seems to stimulate the curiosity of all who hear its lyrical tones.

This is a book about light—and, in truth, the present chapter has little really to do with light except, as Maxwell said, "to see the instruments." Nevertheless, it does have to do with waves and with learning.

14.2 DRAGON NOTES

To anwer the questions posed in the preceding section, my student and I constructed an apparatus by means of which the musical tube could be rotated in a controlled, reproducible way free from centrifugal distortions. As shown in figure 14.2, the corrugated resonator was mounted on a thin wooden slab which was attached to a spoked wheel free to rotate in a vertical plane; a counterweight was fixed at the opposite end of the slab so that the center of mass of the system lay on the axis of the wheel. A motor, whose speed could be varied by a rheostat, was placed in direct contact with the outer rim of the wheel.

The first experiments were designed to determine the frequencies of the acoustic resonances of the rotating tube and to compare them with the modes of an identical stationary tube mounted horizontally on a table and activated by the air blast of a blower, such as is commonly used in air-track experiments. The air flux into the stationary tube was controlled by varying the distance between the mouth of the tube and the blower.

Assuming the dragon to be a sort of open-ended organ pipe of length L, whose vibrational modes have displacement antinodes (maxima) at the two

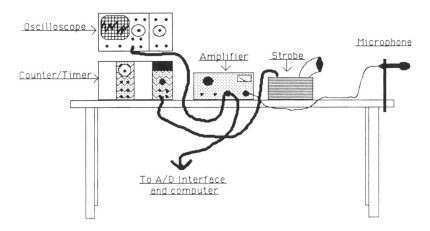

Figure 14.3. Apparatus for recording sound amplitude against time plots of the acoustic resonances. The sound traces are picked up by a microphone, amplified, displayed on an oscilloscope, and sent to a computer (not shown) for Fourier analysis into composite frequencies. The rotation rate is determined by means of a stroboscope and counter/timer.

ends, one would predict the spectrum of mode frequencies to be

$$f_n = \frac{nv_s}{2L} \qquad (n = 1, 2, 3 \ldots) \qquad (2)$$

in which v_s is the speed of sound in air. Actually, since the air is moving in and out of the open ends, the effective length L' of the resonating air column is a little longer than L and depends on the radius R of the pipe; for a pipe with two open ends (and $R < \frac{1}{25}$-th of a wavelength) this approximate correction is[4]

$$L' = L + 1.22R. \qquad (3)$$

Figure 14.3 illustrates schematically how the musical tones were obtained and processed. A microphone, aligned along the axis of rotation, recorded the tones at the different angular frequencies at which the wheel was spun. These spin rates were measured by means of a stroboscope and counter/timer. The sounds from the dragon were amplified and then Fourier analyzed into their composite frequencies by an Apple IIe computer. In preliminary studies of the frequency spectrum, the computer sampled 2048 points at 50 ms per point; later a higher resolution (and therefore slower) program with a 200 ms sampling time per point was employed for more precise work.

An example of the temporal recording of sound intensity for the acoustic mode $n = 6$ is shown in figure 14.4. Theoretically—that is, according to Eqs. (2) and (3)—the frequency of this mode should be 1350 Hz. The weak

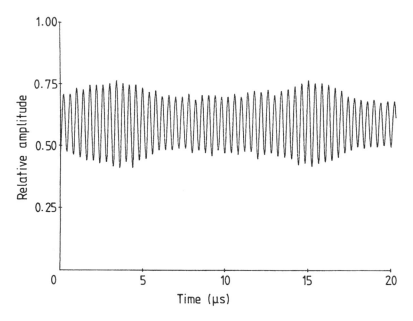

Figure 14.4. Sound amplitude against time for the mode $n = 6$ as recorded by a microphone slightly offset from the axis of rotation.

amplitude modulation which one sees in the figure could be produced by a nonuniform rotation of the motor shaft, or a small distortion of the wheel, or offset of the microphone from precise alignment along the axis of rotation. Such modulation can be useful by providing an independent confirmation of the rotation rate. The recording in the figure was made with a slightly offset microphone. However, the data which were actually used to interpret the mechanism of sound production were obtained with a uniformly rotating motor, undamaged wheel, and centrally placed microphone.

Plots of amplitude (in the musical sense, i.e., sound intensity) against time for each mode were recorded at a spin rate just slightly beyond that for threshold production of the tone. Increasing the rate shifted the resonance frequency upward in accordance with expected intramode frequency shifts characteristic of organ pipes;[5] a given mode would be extinguished by about 10 Hz beyond threshold. Further increase in rotation rate then produced low-level white noise until the sudden onset of the next mode. For the stationary tube excited by a blower, over whose air stream there was less precise control, the intermode transitions occurred smoothly and continuously, rather than discretely, with increase in air flux.

A composite spectrum of resonance frequencies, obtained from Fourier analysis of sound variations like that of figure 14.4, is illustrated in figure 14.5 which

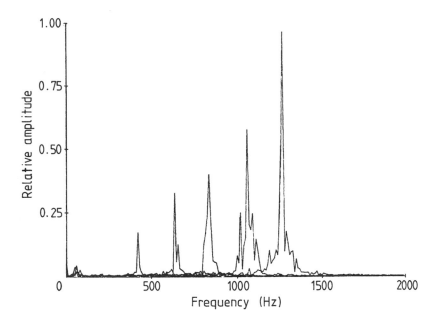

Figure 14.5. Frequency spectrum of the rotating tube for modes $n = 2$–6 as obtained by Fourier analysis of temporal plots like that of figure 14.4.

shows the modes $n = 2$–6 for the rotating tube. The fundamental ($n = 1$) mode is absent, for it was found to be difficult to produce and sustain this mode whether the tube was rotated by hand or by motor.

The resonance data ("dragon notes") for a tube of length $L = 74$ cm, outside radius $R = 1.5$ cm, and corrugation spacing $d = 0.6$ cm are summarized in table 14.1. At the ambient temperature of approximately $16°C$, the speed of sound in air was $v_s = 342$ m/s. The first column of the table records the theoretical resonance frequencies of an ideal open-ended organ pipe (with end correction applied); for the stated geometry the fundamental is 225 Hz. In the second and third columns are the corresponding experimental frequencies of the stationary and rotating tubes, respectively, measured with an estimated precision of ± 3 Hz. The modal frequencies of the stationary tube were about 3–4% below the theoretical values, and the frequencies of the rotating tube are approximately 3% below those of the stationary tube. That the resonance frequencies of the dragon voice, whether stationary or rotating, are lower than the comparable theoretical values of a smooth, open-ended pipe results, I suspect, from the corrugations which give the tube an effective length longer than L'. The reduction in pitch of the modes of the rotating tube below comparable values of the stationary tube seems to be a consequence of the lower-threshold air speeds through the rotating tube, as indicated in the fifth and sixth columns.

TABLE 14.1
Threshold Mode Frequencies and Air Speeds

	Threshold Mode Frequencies[a]			*Air*	*Speeds*[b]
Mode n	Theory[c] (Hz)	Stationary Tube (Hz)	Rotating Tube (Hz)	Stationary Tube (m/s)	Rotating Tube (m/s)
1	225	—	—	—	—
2	450	432	419	6.3	5.4
3	675	646	632	9.2	8.2
4	900	861	839	11.8	10.7
5	1125	1083	1051	14.9	13.3
6	1350	1310	1263	18.9	17.1

[a] Uncertainty of ± 3 Hz
[b] Uncertainty of ± 0.3 m/s
[c] Eqs. (2) and (3) with $v_s = 340$ m/s, $L = 0.74$ m, $R = 0.015$ m

I will discuss the determination of air speed and pressure difference (across the tube) shortly.

The initial experiments established that, for purposes of modeling, it was not unreasonable to regard the dragon as a kind of rotating organ pipe (although one whose length is a bit slippery to pin down precisely). But what makes the dragon sing?

14.3 A BERNOULLI PUMP

To probe the dynamics of the dragon voice further, it is necessary to examine different aspects of the air flow through the tube. Despite the outward simplicity of construction, the actual pattern of air flow is complex, and it is useful to consider separately the hierarchy of scales at which the flow takes place. The largest-scale flow, as perceived in the rest frame of the tube, is vortical, centered on the pivoted end and moving perpendicular to the axis of the tube with a tangential air speed ideally given by

$$V(s) = 2\pi us \qquad (4)$$

where u is the rotation rate (in rev/s) and s is the distance (ranging from 0 to L) along the tube. Next in scale is an axial flow through the tube produced by the rotationally induced pressure difference between the ends. Finally, at the smallest scale is the internal flow over the corrugations which becomes increasingly turbulent as the rotational rate and mode number increase.

The objective here is not to provide a rigorous hydrodynamical analysis of the dragon voice, but rather to understand some of the prime features of its behavior at as basic a level as possible, so that students being introduced to general physics can appreciate it.

Let us consider first the question of how the resonant pressure difference across the length of the tube and the rotation rate are related. For a smooth, incompressible mesoscale flow of fluid with mass density ρ, Bernoulli's principle (with neglect of gravity) leads to the relation

$$p(s) + \tfrac{1}{2}\rho[v(s)^2 + 4\pi^2 u^2 s^2] = \text{constant} \qquad (5)$$

in which $p(s)$ is the pressure and $v(s)$ is the axial velocity at location s along a streamline. Air, of course, is a compressible fluid, but it is assumed that compression can be neglected for the range of gentle pressure differences created by tube rotation. Under the same conditions, therefore, the conservation of mass flow through a tube of constant cross section yields a uniform axial velocity within the tube, that is, $v(s) = \text{constant}$ for $0 \leq s \leq L$. It then follows that the pressure difference p between the stationary and rotating ends is given by

$$p = p(0) - p(L) = 2\pi^2 L^2 u^2. \qquad (6)$$

Equation (6) is consistent with one of my earliest observations, namely, that the pressure at the rotating end is lower than that at the stationary end (which accounts for why the bits of paper were sucked out of my palm). The dragon is a kind of "Bernoulli pump."

Consider next the question of how the resonant tones and the rate of rotation are related. Here the microscale flow plays a critical role. Moving past the corrugations, the air is perturbed at a frequency proportional to the axial air speed and inversely proportional to the corrugation spacing d. Let us assume that when this frequency coincides with a resonant frequency the resulting sound is amplified by the tube. Thus the resonant frequency f_n of a given mode n may be related to the axial air speed v_n of the same mode by an expression of the form

$$f_n = a^{-1}\frac{v_n}{d} \qquad (7)$$

where a is a proportionality constant. In the ideal case of frictionless flow, application of Eq. (5) at the stationary end of the tube and at a point just outside the rotating end (where the streamline is assumed to be perpendicular to the tube) leads to a uniform axial speed

$$v_n = 2\pi u_n L = V_n(L) \qquad (8a)$$

which is equal to the tangential speed at the rotating end. The flow, however, is not perfectly frictionless, or there would be no sound. It seems reasonable to assume, however, that v_n is proportional to $V_n(L)$, in which case one can write

$$v_n = b^{-1} 2\pi u_n L \qquad (8b)$$

where b is another proportionality constant.

Combining Eqs. (7) and (8b) leads to the expression

$$f_n = (ab)^{-1} \left(\frac{2\pi L}{d} \right) u_n \qquad (9)$$

which predicts a linear proportionality between the threshold acoustic and rotational frequencies.

To test the preceding model of the dragon, in particular the relationships

1. $V(L) = 2\pi u L$
2. (axial air speed)/(acoustic frequency) = constant (Eq. [7])
3. (rotation rate)/(axial air speed) = constant' (Eq. [8b])
4. (acoustic frequency)/(rotation rate) = constant'' (Eq. [9])

my student and I measured the resonant acoustic frequency, rotation rate, air pressure difference across the tube, tangential air speed at the rotating end, and axial air speed at the fixed end for the modes $n = 2$–6.

The determination of air pressure and tangential air speed at the rotating end—which, by $n = 6$, was whipping around at over 20 m/s—posed at first a difficult task, but it was eventually accomplished in the following way. A narrow-bore rubber sampling tube was mounted along the outside of the corrugated resonator to sample the air stream at the rotating end. Near the stationary end, the sampling tube was connected to an air-tight swivel joint—a spring-loaded device with high-density carbon seal designed primarily for high-speed cooling rolls on printing presses, paper converters, plastic sheeting machinery, and metal foil–processing equipment. But it served admirably on the dragon. The static end of the swivel joint was connected to another segment of rubber tubing that attached directly to an inclined tube manometer capable of measuring pressures up to 300 Pa with a resolution of 2 Pa. The frame of the manometer was calibrated for both pressure (in Pa) and air speed (in m/s), the two scales being related as follows: air speed = [2 × (pressure)/(mass density of air)]$^{1/2}$.

To measure the axial air speed for each mode, a short sampling tube attached to the swivel joint was inserted a few centimeters inside the fixed end of the resonator where the macroscale tangential flow should ideally be null. Since the inserted probe largely extinguished the resonances, and one could not hear the dragon voice while making measurements of internal air speed, the rotation rate for each mode was set close to the threshold value previously established in the determination of modal frequencies.

The outcome of all these measurements of "dragon breath" are recorded in the following three tables. Table 14.2 summarizes the pressure measurements

TABLE 14.2
Threshold Pressure Differences and Bernoulli's Law

			Pressure Difference[a]		
Mode n	Resonance Frequency (Hz)	Rotation Rate (Rev/s)	Experiment (Pa)	Theory[b] (Pa)	Fractional Deviation[c] (%)
2	419	1.47	28	28.5	0.2
3	632	2.33	73	71.6	−2.0
4	839	3.14	134	130.0	−3.1
5	1051	3.75	198	185.5	−6.7
6	1263	4.40	266	255.5	−4.1

[a] Uncertainty of ±1 Pa
[b] Eq. (6) with mass density of air $\rho = 1.22$ kg/m^3
[c] 100× (Theory − Experiment)/Theory

as a function of rate of rotation. The agreement of these experimental results with the predictions of Eq. (6) is very good. For modes $n = 2$ and 3, with rotation rates under 3 rev/s, the theoretical pressure differences are within the estimated experimental uncertainty of pressure measurement. As one might expect, however, higher rotation rates led to greater turbulence and therefore to departures from streamline flow. Nevertheless, for the five modes accessible to study, the fractional deviation between theory and experiment was under 7%.

Table 14.3 compares the measurements of tangential air speed at the rotating end (i.e., effectively the end pressure relative to atmospheric pressure) with the theoretically expected speed based on tube length and rotation rate. Agreement is again excellent at low speeds with a fractional deviation of less than 1%; for modes $n = 5$ and 6 the fractional deviation increases to 2–3% as a result of increased turbulence. The measurement of tangential air speed, or pressure, provides a third method (in addition to stroboscope and Doppler effect) of determining the rotation rate.

Last, Table 14.4 recapitulates the results of pitch, rotation rate, and axial air-speed measurements. The fifth column shows the effective constancy of the ratio of mode frequency to axial air speed; the axial air speed is found to be about 80% the tangential air speed for all modes sampled. Similarly, the sixth column confirms the effective constancy of the ratio of axial air speed to rotation rate, and the seventh column establishes the effective constancy of the ratio of mode frequency to rotational frequency. (Any two of these three columns are independent.)

The results, it would seem, are gratifyingly consistent with the hypotheses underlying the proposed mechanism of the dragon voice.

TABLE 14.3

Tangential Air Speeds from Pressure and Rotation

Mode n	Pressure Difference (Pa)	Rotation Rate (Rev/s)	$V(L)$[a] Experiment (m/s)	$V(L)$[b] Theory (m/s)	Fractional Deviation[c] (%)
2	28	1.47	6.8	6.83	0.4
3	73	2.33	10.9	10.83	−0.7
4	134	3.14	14.8	14.60	−1.4
5	198	3.75	18.0	17.44	−3.2
6	266	4.40	20.9	20.46	−2.2

[a] $V(L) = \sqrt{2p/\rho} = 1.28\sqrt{p}$
[b] $V(L) = 2\pi u_n L = 4.65 u_n$
[c] $100 \times$ (Theory − Experiment)/Theory

TABLE 14.4

Proportionality Relations

Mode n	Resonance Frequency f (Hz)	Rotation Rate u (Rev/s)	Axial Air Speed (m/s)	a[a]	b[b]	ab[c]
2	419	1.47	5.4	2.1	1.3	2.7
3	632	2.33	8.2	2.2	1.3	2.9
4	839	3.14	10.7	2.1	1.4	2.9
5	1051	3.75	13.3	2.1	1.3	2.8
6	1263	4.40	17.1	2.3	1.2	2.8

[a] Eq. (7) with $a = v_n/f_n d$
[b] Eq. (8) with $b = 2\pi u_n L/v_n$
[c] Eq. (9) with $ab = (2\pi L/d)(u_n/f_n)$

14.4 THE DRAGON CAUSES FRICTION

The investigations outlined in the previous section were long completed—and indeed I had already returned from a second invited stay in Japan—when a letter arrived one day from a physicist in California. Unknown to my student and me, he had himself experimented on the rotating resonator some sixteen years earlier[6] and enjoyed our paper about the dragon voice—although he would have enjoyed it more had he been referenced, which of course was understandable. I replied politely, expressing our regret and pointing out that sixteen years was, after all, a long time and that most scientific papers, once published, ordinarily fade from communal awareness in a much briefer period than that.

The author had arrived at much the same conclusions as my student and I had, although his experimental methods were simpler and devilishly clever. To ascertain the longitudinal air speed at which the dragon (then marketed in the United States as a "Hummer") produced different tones, he attached it to his car and took it for a scientific joy ride, exciting the eleventh harmonic at approximately 80 miles per hour. Whether the California Highway Patrol intervened or not, I cannot say, but he eventually devised an alternative procedure. One end of the corrugated tube was inserted into the bottom of a cylindrical wastebasket, and the inverted basket was pressed into a large tub of water. With the water as a piston to force air (assumed incompressible) through the resonator, he could readily determine the air flow through the latter from the rate of linear displacement of the basket in water. The author had, in fact, made several musical instruments in this manner and enclosed with his letter of priority a photograph from a local newspaper with the title "Professor Blows His Own Horn," which somehow seemed quite fitting under the circumstances.

With other projects at hand and a lengthy stay in New Zealand impending,[7] thoughts of the dragon again receded from my mind as quickly as before. From time to time since the dragon paper was first published, letters from interested readers continued to arrive from around the globe, although most merely requested reprints or posed simple questions that did not require much thought on my part. This was fortunate, for the apparatus had long ago been dismantled. Then, one day, there arrived in the mail a request for help from a *Gymnasium* student in Germany that began

> Dear Mr. Silverman,
> I am a German student and I have to write my research paper about the rotating corrugated resonator. ... Now I have two problems with it and it would be very pleasant of you, if you could send me an answer.

The first problem was not difficult, but the second was puzzling indeed:

> You found out that $f_n = a^{-1} v_n/d$ and that the value of the proportionality constant a is more than 2 [actually, it was just about 2], but another article ... shows that $a = 1$. ... Is there anything wrong about the consideration or is it perhaps your measurement?

The article to which the student referred was none other than that of the physicist who had written me some six years earlier. Having seen from a rapid perusal of the text that our conclusions were similar, and intrigued at the time more with his methods and musical instruments, I had not noted that there was, in fact, this curious discrepancy. Was the resonant axial air speed identically equal to, or merely proportional to, the product of acoustic frequency and corrugation spacing?

Although I was in the midst of other projects which I was reluctant to interrupt, I could not put the matter aside. It was not that the distinction between the two possibilities was of earthshaking importance, but rather many years of doing physics had taught me that it is often in pursuit of small departures from

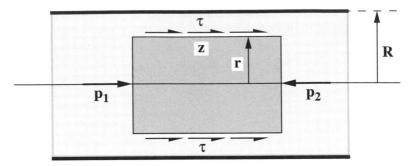

Figure 14.6. Viscous flow through a cylindrical tube. When the flow rate is constant, the difference in pressure forces is balanced by the frictional shear force.

expectation that one learns interesting things. And so it was in this case as well.

Stimulated by the student's letter, I considered the model of the dragon voice once more, and thereby rediscovered what Isaac Newton had already known several centuries ago: the consequences of friction. By means of simple flow experiments, Newton reached two conclusions fundamental to nearly everything that pertains to the mechanics of real fluids. First, a fluid does not slide along a solid, but adheres to it; thus, both the tangential and normal components of flow velocity vanish at a bounding surface. Second, the frictional force (shear stress) at any internal point of the flow is proportional to the velocity gradient (flow shear) normal to the flow. The proportionality constant μ, the coefficient of shear viscosity, is independent of the flow geometry.

That the flow through the corrugated resonator entailed friction was clear to me from the outset of the investigations, and I in fact commented on this point in the dragon paper. I had not, however, taken explicit account of the following glaring implication: If the air flow is maximum at the center of the tube and vanishes at the cylindrical boundary, then clearly there must be a radial velocity gradient; to have assumed a uniform flow was not strictly correct.

Figure 14.6 shows the forces acting on a designated cylindrical volume of air of radius r and length z moving at constant speed v through a cylindrical tube of radius R. Since the volume of air is not accelerating, the force from unequal pressure (p) on the circular ends is balanced by the force of the shear stress (τ) along the sides, and the condition of dynamic equilibrium becomes

$$\pi r^2 \frac{dp}{dz} + 2\pi r \tau = 0 \tag{10}$$

in which Newton's friction law, applied to an axially symmetric system, is simply

$$\tau = \mu \frac{dv}{dr}. \tag{11}$$

Combining Eqs. (10) and (11) results in a first-order differential equation for the longitudinal speed which can be readily integrated to yield a solution of the form

$$v(r) = v_m \left(1 - \frac{r^2}{R^2} \right) \tag{12}$$

with v_m the maximum speed along the tube axis ($r = 0$).

Now, in the experiments with my student, the narrow-bore sampling tube measured in effect the axial air speed v_m. On the other hand, the speed inferred from riding with the dragon attached to a car, or pumping it with a water piston, corresponded to the average speed

$$\bar{v} = \frac{1}{\pi R^2} \int_0^R v(r) 2\pi r \, dr. \tag{13}$$

Performing the integration of Eq. (13) with the velocity law (12) leads to the interesting result $\bar{v} = \frac{1}{2} v_m$. Thus, upon replacing v_n in Eq. (7) by the equivalent expression $2\bar{v}_n$ and substituting the empirically determined value $a = 2$, one obtains the result $f_n = \bar{v}_n/d$. The two factors of 2 have canceled—and the discrepancy has vanished.

Although the preceding analysis neglects the corrugations of the resonator and rigorously applies only to a smooth tube, it incorporates the primary effect of friction, which is to generate a nonuniform flow. Presumably the velocity law deriving from a more comprehensive model would resemble that of Eq. (12) except in the immediate vicinity of the corrugated walls. For what it is worth, I note that if the power of the term (r/R) were some general value k, then Eq. (13) would lead to a mean speed $\bar{v} = v_m(1 - \frac{2}{k+2})$, which yields $\bar{v} = \frac{1}{3} v_m$ for $k = 1$ and $\bar{v} = \frac{3}{5} v_m$ for $k = 3$. As k increases, the mean speed approaches the maximum axial speed. Thus, for relatively low values of k, the mean and maximum axial air speeds are sensitive to the effects of friction, and their measurement can be useful in refining further the theoretical model.

In any event, I believe the dragon is now reasonably tamed. At the most basic level, it can be understood as an open-ended organ pipe blown by an air stream pumped by a rotationally induced pressure difference following Bernoulli's principle. Resonant amplification of sound is initiated at a rotation rate for which the axial air flow is perturbed by corrugations at a frequency corresponding to one of the characteristic frequencies of the tube. It is worth noting that the relationship $f_n = \bar{v}_n/d$, with mean air speed replaced by electron speed, is in fact the same expression that results from the nonrelativistic limit ($v/c \to 0$) of Eq. (1) for the SP light source.

As for the German student, I heard from him again a year later and am happy to say he did very well:

Dear Mr. Silverman,
Now I am in the gap between *Gymnasium* and university and finally have time to

thank you for your help and send you my research paper. Indeed your letter was a great help. . . . Unfortunately I made a mistake . . . and used Bernoulli's law in a wrong way. Although my teacher didn't understand this law either, he recognized the mistake, which lead to the result that I "only" got 14 out of 15 points, which was still one of the best results. . . . Anyway, I thank you very much for answering my letter and giving me some tips for my work. Enclosed there is a shortened version of my work (the longer version for my teacher contains some babble just to make it longer) I used in a special contest called "*Jugend forscht*" in which I won 100 DM.

In its charming simplicity, the letter reminded me in various ways how much of science education transcends national boundaries!

Those with a questing spirit could undoubtedly find in the Voice of the Dragon further intricacies to explore that will enhance their understanding and enjoyment of physics. Someday, perhaps prompted by other letters, I may return to the subject myself. But for now I have had enough of dragons.

NOTES

1. S. J. Smith and E. M. Purcell, "Visible Light from Localized Surface Charges Moving across a Grating," *Phys. Rev.* 92 (1953): 1069.

2. M. P. Silverman, "The Vortex Tube: A Violation of the Second Law?" *Eur. J. Phys.* 3 (1982): 88–92; I have written more comprehensively about the vortex tube and the principles of thermodynamics in my book *And Yet It Moves: Strange Systems and Subtle Questions in Physics* (Cambridge: Cambridge University Press, 1993).

3. M. P. Silverman and G. M. Cushman, "Voice of the Dragon: The Rotating Corrugated Resonator," *Eur. J. Phys.* 10 (1989): 298–304.

4. L. L. Beranek, *Acoustics* (New York: McGraw-Hill, 1954), p. 133.

5. W. Barnes, *Contemporary American Organ* (New York: Fischer, 1937), p. 113.

6. F. S. Crawford, "Singing Corrugated Pipes," *Am. J. Phys.* 42 (1973): 278–88.

7. I describe my activities in New Zealand and the fascinating work of the University of Canterbury Ring Laser Laboratory in *And Yet It Moves*, ch. 5, "Two Worlds, Large and Small: Earth and Atom."

Part Six

SCIENCE AND LEARNING

CHAPTER 15

A Heretical Experiment in Teaching Physics

> Hardly anyone can understand the importance of an
> idea, it is so remarkable. Except that, possibly, some
> children catch on. And when a child catches on to an
> idea like that, we have a scientist. . . . It's too late for
> them to get the spirit when they are in our
> universities, so we must attempt to explain these
> ideas to children.
> (*Richard Feynman, 1958*)[1]

ALTHOUGH I do not believe that it is actually futile to attempt to communicate the importance of great ideas to students in our universities, I think there is nevertheless much truth in Feynman's emphasis on the need to reach children. I know this firsthand, because I have taught children—my own—from elementary school through high school. Together with my wife, we covered the full range of required academic subjects—and then some. This was not a "hit or miss" activity, but a serious responsibility which I carried out with daily regularity simultaneously with college/university teaching and pursuing a career as a research scientist. To say the experience was challenging and required unwavering dedication is to state a truth as blatant as the blueness of a bright summer sky. Certainly no other research scientist I know personally has ever cared to follow my example. Public schools are in no danger of emptying because of people like me!

It was from teaching my own children, however, that I learned what learning was all about and what a good teacher could do to foster it. I learned that healthy young children (brought up without a television in the house) are never at a loss for activity; they are always learning—and, because they value their own time highly, they resent and resist being bored. I learned that the most important lesson to teach is that learning is a lifelong activity; it does not end with the end of a term or a textbook. I learned that it was far more important to show my children how to find information and use it, than to require them to memorize it; and I saw that the information they actually needed was more often than not committed to memory anyway. My children taught me that evidence of learning cannot be found in makeshift paper exercises, but in the actions and attitudes of their daily lives, in their character and creativity. They showed me that thinking—and therefore learning—more readily occurred in stretches of unhurried leisure than in desperate adherence to schedule. And they impressed upon me the critical distinction between schooling and education.

I have taught my children for many years—and they have taught me. And I wondered: Why can't students in our universities learn in the same way?

15.1 SELF-DIRECTED LEARNING

Science is difficult enough without the added burden of being perceived as boring or irrelevant. It becomes relevant—and consequently interesting—when it addresses questions arising out of the learner's own curiosity. How does a teacher tap that curiosity and elicit each student's natural inclination to learn? I would like to discuss here an educational framework that I believe accomplishes this; I call it "Self-directed Learning."

What follows is an account of an experiment in the teaching of physics at the college and university level. The underlying principles and methods of instruction are equally applicable, however, to the teaching of any science at any level. Nevertheless, as the most fundamental of the natural sciences, physics is widely perceived as a subject of exceptional difficulty. Whether warranted or not, this perception and the heightened anxiety it generates give urgency to the need of more humane and effective ways to teach physics, if not all sciences. I have applied the principles of self-directed learning for many years in the instruction of my own children, and I know that they work. What qualifies as a "heretical experiment" in the present context is the implementation at institutions of higher education where traditional modes of teaching are ordinarily strongly antipathetic to those recounted in this chapter.

The essential idea underlying the self-directed learning of science is in principle very simple, as I have written before:[2] Science as it is taught should more closely resemble science as it is done by professional scientists. This requires recognizing what good science is and how it is practiced—two points over which even professional scientists may hold divergent opinions. But one must start somewhere, and I draw my own inspiration and hypothetical basis from some thirty years of personal experiences as a research physicist and chemist.[3]

Since the above key assertion colors strongly the attitudes and activities herein described, it is worth asking at the outset why the teaching of science should embrace activities of a practicing scientist. Would the statement make sound educational sense if, as one reviewer asked when these ideas first entered the physics literature,[4] "science" and "scientists" were replaced by other disciplines and other practitioners? Is there no art of appreciation of science distinct from the practice? "I can imagine," wrote the reviewer, "taking a good art appreciation course in which one looks at paintings and never paints, oneself."

Although the rationale for my assertion is addressed in one way or another throughout the entire chapter, in the main it follows from the belief that science is meaningful and its learning of lasting value only to those motivated by their own curiosity to study it. This observation is undoubtedly true (at least to me) for *any* subject; one *could* replace "science" by "art" in the preceding sentence. However, when it comes to motivation, introductory courses like art appreciation are often overflowing with eager students; physics by contrast is taken, often grudgingly, by large numbers of students who need it to satisfy the requirements of other programs that interest them far more. Were it not for these requirements, most students enrolled in physics, I suspect, would not be there.

Yet physics is intensely interesting, as any practicing physicist knows. What makes it so? There is no unique answer, but a reasonable reply would certainly recognize that the study of the natural world raises intriguing questions that arouse a researcher's curiosity, and that there is an intrinsic personal satisfaction in the quest to answer these questions.[5] I doubt whether students who never study physics or who study it unwillingly are aware of this. Does it not seem plausible that students could be motivated by engaging in activities similar to those that have made scientists devoted to their work?

Furthermore, I am concerned with *learning* physics—and not just learning *about* physics. Certainly one can appreciate art without ever painting, oneself, and similarly one could study the history or philosophy of physics without ever solving an equation or making a measurement. But one does not learn physics this way. As is characteristic of any real science, physics is concerned with the ways in which information is sought and tested, knowledge extracted, conclusions drawn and verified. To learn science one must learn at some level, however rudimentary, to analyze, synthesize, and experiment—that is, to perform activities routinely undertaken by practicing scientists.

A seminal element of the approach to teaching science advocated here is that the major activity of scientists is inquiry. However, it has been my overwhelming experience in examining programs of science instruction that inquiry does not play a significant part. The aspect of the "hard" sciences (physics, chemistry, astronomy, geology, biology, etc.) most emphasized is the network of supposedly solidly known and accepted facts and principles. It is this image of science as a "storehouse of knowledge"—rather than as a dynamic activity by means of which knowledge is acquired and tested—that characterizes nearly all science instruction I have observed in the United States and other technologically developed nations. Moreover, even in curricula which incorporate to some extent a component of student inquiry, the major objective of the investigations—for example, the required exercises to be performed in an instructional laboratory—is to reproduce predetermined items from this storehouse of knowledge. Authentic inquiry in science, however, is driven largely by the investigator's own curiosity.

The perception of science as a rigid and infallible source of information percolates the institutions within society and can have serious repercussions.

Thus, for example, in his essay "Uneasy Bedfellows: Science and the Law," forensic scientist H. J. Walls eloquently describes the deleterious effects of misunderstanding the nature of science in the courtroom: "One of the difficulties of communication between science and the law is that the law, or at least its practitioners, seem to have a wrong idea of what science is all about and how it advances. Lawyers appear ... to think that science was a corpus of exact knowledge firmly established for all time and that its main business was the collection of facts. If they did think so, they could scarcely be more wrong on both counts." One important societal consequence, as Walls points out next, of regarding science as a corpus of exact knowledge is that

> lawyers seem to imagine that all scientific measurements are made with absolute precision, whereas all scientists know that this is impossible. The scientist is quite happy to accept this, as long as he knows how large the error is liable to be, or what the mathematical probability is that the true value differs by less than some known amount from the measured one. The conscientious expert witness, however, who tries to put this fundamental idea across from the witness box is likely to provoke a pained rejoinder: 'Dr. So-and-So, you are here to give us the facts, not probabilities.'[6]

Walls's comments were written some twenty-five years ago, but there is every reason to believe from more recent accounts of the presentation of scientific evidence in the courtroom that little has changed.[7] It is not the purpose of this chapter to comment in depth on the misunderstanding and misuse of science in the courtroom. The principal point here, as illustrated by this one example, is simply that the public perception of science diffusing out of the academic communities is no mere academic matter, but has practical consequences that reach to the very foundations of a democratic society, such as its judicial system. Lawyers and judges presumably constitute one of the most educated sectors of a democratic society. If their understanding of the content and purpose of science is faulty, one can only imagine how much less satisfactory is the understanding of the general public.

Self-directed learning is a general approach to education that attempts to remove (or at least minimize) the element of coercion in instruction and correspondingly to give students wide latitude over what they choose to learn. It is based upon a framework in which (1) curiosity-driven inquiry is recognized as an essential ingredient of both science and science teaching; (2) the principal role of the instructor is to provide students the incentive to learn by helping raise questions that they find personally meaningful; (3) mistakes are considered a natural part of learning and are not to be penalized; (4) the laboratory is seen as a place for minimally restrictive free exploration rather than rigid adherence to "cookbook" recipes; (5) research skills are developed through out-of-class projects that involve literature search, experiment, and analytical modeling of real-world phenomena; (6) articulate communication, both written and oral, is regarded as seminal to science (as it is in the humanities) and encouraged; and (7) performance is evaluated on the basis of a portfolio of accomplished work rather than on the outcome of formal testing.

15.2 THE "STANDARD MODEL"

Although a number of innovative approaches to the teaching of physical science have been proposed over the years, most to my knowledge concerned how best to master the factual content of a particular science, rather than how that content came to be known and how the learner could use it. There is nothing necessarily improper to mastery of detail as part of an overall program of instruction, for clearly students, like scientists, must have a base of knowledge with which to operate. The difficulty arises when detail in and of itself is the aspect of science instruction most accentuated. This is the situation which occurs in almost all classroom instruction I have witnessed or learned about vicariously. Under these circumstances, despite variations among instructional methods and philosophies, the interactions between students and teacher largely share the following common characteristics:

1. The content of the classroom lectures largely repeats the information in an adopted textbook.

2. Assigned problems, which involve primarily formula substitutions, usually test one basic idea or principle at a time. Indeed, textbook problems at the end of each chapter are frequently arranged according to chapter sections.

3. Examinations of various kinds (quizzes, hour exams, laboratory tests, final exam) are frequent, and the final course grade is effectively a measure of cumulative performance on these tests.

So entrenched and universal is this mode of instruction that I refer to it as the "standard model," an appellation familiar to all physicists. Within the standard model of science instruction, education consists principally of "teaching"—an activity that the skilled (the teacher) does "to" the presumably unskilled (the students)—and testing.

When I think of all the time spent by teachers in constructing and grading tests, and by students in preparing for and taking them, I cannot help but believe that such time could be used more productively by all participants to understand better what science is all about. I have often posed the question "Why is it necessary to test students?" to various scientists and educators. The replies were usually obvious: Testing enables instructors to determine what a student has learned. But why should it be necessary for an instructor to know what a student has learned—if indeed that is even possible? This question elicits astonishment, yet the responses basically distill down to the following belief. Unless instructors ascertain what students have learned, they cannot evaluate students fairly—and without such evaluations it is not possible to determine which students should be permitted or denied access to various courses, programs, schools, and so on. In brief, testing and evaluating serve to lubricate the administration of education.

15.3 BASIC PREMISES

The motivational basis of self-directed science learning differs radically from the standard model of instruction. For one thing, the purpose of education is to foster "learning," which—in contrast to "teaching"—is something a person does for him- or her*self*. Second, the purpose of testing—which does serve an important educational function—is for the exclusive benefit of the individual student and not of the teacher or of educational administration in general. It is a means by which students can assess, for their own knowledge and subsequent personal action, the extent of their mastery of course material. However, within the framework of self-directed learning, tests are not graded by the instructor, and, most importantly, the outcome of testing—either directly by examinations, quizzes, and the like, or indirectly through out-of-class (i.e., homework) assignments—is not a basis upon which student performance is evaluated.

To avoid at this point any semantic confusion, I would like to stress that the designation "self-directed learning" does not imply that students are expected to learn science entirely on their own with no assistance from an instructor, or that the role of the instructor is in some sense a minor and possibly dispensable one. Quite the contrary, a skilled science instructor—and preferably one who has personally engaged in authentic scientific activities involving research and possibly publication or public presentation—is of vital importance.

In self-directed learning the instructor's principal role is to motivate the study of science in such a way that students do not find the process of learning discouraging and threatening. The justification of this role derives from the experience of many educators, particularly those involved in programs of non–school-based instruction,[8] that *students who understand the significance of what is being taught and who see the personal need to learn it will master on their own, sooner or later, what they believe is important to know.* For the remainder of this section I would like to elaborate on the italicized portion of this paragraph, which constitutes, in essence, a sort of credo of self-directed learning.

The preceding assertion regarding a student's capacity to learn independently is not simply a paraphrase of Gibbons, that "The power of instruction is seldom of much efficacy except in those happy dispositions where it is almost superfluous."[9] Quite the contrary, the underlying idea is that the power of instruction can be very effective, when it is utilized in an appropriate way. It is worth emphasizing that those happy dispositions of Gibbons are rare, and that most students are not able on their own to see the significance of what they are studying or inclined to undertake its mastery voluntarily. This is particularly true in the case of physics, and quite possibly in other sciences as well, where fewer and fewer students enroll for introductory or general survey courses to learn science, but rather to fulfill the requirements of other educational programs.

Under circumstances such as occur all too commonly in schools, in which an instructor faces a captive audience who are unlikely to follow through with additional science courses of the same kind and who believe they will rarely, if ever, use the facts and principles of the present course in their future occupations, emphasis on memorization, drill, and problem solving for its own sake is educationally unproductive and difficult to justify. It is precisely then (as well as under the more congenial setting of having interested participants at the outset) that students need to be helped to understand the universal significance, personal relevance, and intellectual excitement of the scientific studies upon which they are embarking voluntarily or otherwise. Such motivation is not easily effected and, without an adept instructor with both knowledge and enthusiasm, is unlikely to be achieved at all. Clearly, then, self-directed science learning does not minimize the contribution of teachers, but instead redefines their primary role.

There are two additional related points essential to the foundation of self-directed learning that are neither trivial, nor widely recognized, nor ordinarily implemented in educational practice. The first of these concerns the holistic nature of normal (i.e., out-of-school) learning, in contrast to the linear nature of teaching within the standard model. The second point, to be discussed at the end of this section, concerns broadly the issue of performance evaluation.

A heuristically apt metaphor for holistic learning is that provided by neural networks in which various network connections are simultaneously modified (the network "learns") by repeated exposure to diverse stimuli and subsequent comparison of response with desired results. This is, in fact, not unlike the way a young (preschool) child learns. No one ordinarily instructs preschool children in the rudiments of their native language; a healthy child, motivated by the need and desire to communicate, simply "picks up" language by exposure to others who speak it (one important reason, therefore, for parents to exercise good judgment in monitoring the influences to which their children are exposed). Unfortunately, once formal schooling begins, the holistic way of learning is discouraged, if not suppressed, and is supplanted by a linear mode of instruction.

Consider, for example, the skill of reading. One might imagine that normal children who can learn to speak can also, with suitable guidance and exposure to books, teach themselves to read. And yet at many primary schools within the United States reading is regarded as an ability which students acquire only through mastery of a progression of preliminary skills that only trained and certified instructors are capable of teaching. This partitioning of a holistically learnable activity, reading, into a formal sequence of artificially devised steps can be extreme. According to one 1977 newspaper report cited by educator John Holt

> It has been ten years in the making, but Chicago school officials now believe they have in place a complete sweeping program to teach children to read—a program that may be the pacesetter for the nation. . . . For some years, a Board of Education reading expert . . . has been putting together a package of the reading skills children need to learn in elementary school. At one point, [the] list topped 500 elements. It has since been reduced to 273 over grades 1 through 8."[10]

To such a linear and fragmented approach to education, I share Holt's reaction:

> This might be rather comic if it were not so horrifying. Five hundred skills! What in the world could they be? When I taught myself to read, I didn't learn 500 skills, or even 273; I looked at printed words, on signs, in books, wherever I might see them, and puzzled them out, because I wanted to know what they said. Each one I learned made it easier for me to figure out the next. I could read before I went to school. . . . Most people who read, above all those who read well, were never taught 273 separate skills.

Indeed, that has been my own experience as well. When, some years ago, I was invited to be Chief Researcher at the Hitachi Advanced Research Laboratories in Tokyo, I realized that at the very least I would have to teach myself to read the Japanese language. There are two finite sets of syllabaries (Hiragana and Katakana) and a virtually open-ended set of Chinese characters (Kanji). If it were truly the case that 500 skills are required to read English, then the number of skills needed to read Japanese ought to be astronomical. This is, of course, not the case. Motivated by the strong desire not to remain illiterate, I, like Holt, puzzled out primers, journals, newspapers, traffic signs, shop windows—any source of printed words—until I could understand them. Motivation, rather than external instruction, was all that was needed.

The holism that characterizes the way in which both professional scientists and young children learn is the most natural way for college and university students to learn as well, if they have the opportunity. Holistic learning involves the integration of diverse threads of knowledge (synthesis) and the recognition of simplifying elements (modeling) in the development of tractable solutions to real-world problems. Although it is true that the conceptual structure of a science like physics is to a large degree hierarchical—one could not, for example, understand Newton's laws of motion without first understanding the concept of acceleration, nor Maxwell's equations without first knowing what electric and magnetic fields are—this does not necessarily require that the teaching of science be partitioned into a linear sequence of nearly mutually exclusive units of information.

Consider university-level physics again. A course of study in general physics, which may last two to four semesters, ordinarily begins with mechanics and then embraces in more or less the following order the topics of gravity, fluids, heat, electromagnetism, optics, and various aspects of twentieth-century physics including relativity and quantum physics. These different topics are treated separately in different parts of the textbook and in widely separated classroom presentations. For example, mechanics and electromagnetism often fall into different courses taught in different semesters.

There is, of course, a certain sense to this organization; no sane instructor would begin a course with the most advanced material. However, most of the above topics are arranged not by complexity, but by the specific forces that apply. In the "real" world of science, however—as opposed to the academic environment of most classrooms—it is rare that a physicist would encounter an interesting physics problem so simple that it could be adequately described by a set of principles falling into only one of the units taught as an independent entity in introductory physics. In my own case, for example, a "typical" problem to be studied might involve the internal structure of an atom (quantum mechanics) subject to external static fields (electricity and magnetism) interacting with a laser beam (waves and optics) and exchanging energy with a heat reservoir (thermodynamics).

A successful scientist learns how to build appropriate models of real physical systems in order that the mathematical analyses be tractable and experimental tests lead to comprehensible results. It is generally the case that systems *worth* investigating involve many threads from the far-reaching fabric of physics. *That* is what in part makes an interesting science problem interesting. Conceivably, it is also what would make a course of instruction in science interesting. And yet, in the traditional presentation of physics, a student might well pass an entire semester listening to lectures and working homework and test problems involving electric and magnetic fields with hardly any connection to the mechanics course of the previous term or to the modern physics course to follow. One consequence of this strict partitioning of material is that students do not learn to make synthetic associations. Having analyzed, for example, the behavior of a small-amplitude pendulum in the mechanics part of Physics 101, the same students a semester later in Physics 102 might look upon the flow of electric charge in a circuit with inductance and capacitance as a totally foreign phenomenon. And yet the two different systems are both examples of harmonic oscillation.

By contrast, one of the goals of a physics course following the principles of self-directed learning is to help students recognize the interconnectedness of the physical world. The task faced by the instructor is at least twofold: First, to create—since textbooks are rarely of help in this regard—classroom problems, not of the substitution variety, that induce students to draw upon as much as possible of the previous physics they have encountered. Second, to formulate—

or, better still, to help students formulate—independent out-of-class projects that engage each student in one of the principal activities of research scientists: realistic modeling of physical systems. Neither of these two tasks is simple. Indeed, to develop educationally meaningful problems and projects appropriate to an introductory course of study requires skill and commitment on the part of the instructor.

Although an important part of self-directed learning centers on building motivation through realistic and purposeful problem solving, students do not necessarily end up with an idiosyncratic or "spotty" introduction to physics where the beauty of the whole is obscured by the details of individual investigations. Quite the contrary. What is it about the whole of physics that physicists find aesthetically satisfying? Is it not in part the universal validity, internal consistency, and wide application of a few general principles—in effect, the recognition that there is reason in nature amenable to discovery and explanation or prediction and verification? Is the beauty of this whole more likely to be perceived through comprehensive, if not overwhelming, coverage of separate, sequentially presented topics, or through selected investigations, each of which reveals the simultaneous and coherent interplay of different physical laws? I would choose the latter, having learned from years of teaching that coverage by itself and unrelated to students' own questions has a deadening effect which, far from exposing the beauty of physics, obscures it under a veneer of disconnected fact and mathematical formulas. Moreover, in implementing self-directed learning, an instructor is certainly at liberty to present unifying class lectures to complement the personal out-of-class investigations undertaken by students.

Last, a holistic approach to education does not imply, as some might fear, the daunting prospect that one must learn everything in order to learn anything. Clearly this is not possible. In self-directed learning, as with traditional teaching methods, classroom discussion focuses on a particular topic at hand. Nevertheless, by recurrent and varied encounters, facilitated by the instructor, with principles and phenomena beyond their present ability to understand fully, students can become familiar with the broader content of physics and see much of that content as logically interconnected instead of fragmented. The details of the logic will be sharpened progressively if continued interest leads to further study. I will give examples in the following section of how holistic learning can be fostered in the classroom. It is worth illustrating by a personal example, however, that this is a natural way in which scientists, themselves, learn.

As a physicist whose graduate studies centered on atomic and optical physics, I had occasion early in my career to be faced with a problem involving application of general relativity, an esoteric subject in which I was not explicitly "trained." With this problem as my incentive—and no longer in a position to take courses—I turned directly to the science literature to learn what I could about the system that interested me. What I first came to realize, however, was the existence of whole areas of mathematics and physics of which I was ignorant, but now needed to know. In pursuit of the answer to a problem in which I had a personal stake, I had to inform myself about the elements of non-Euclidian geometry, the manipulation of tensors, the theory of continuous mathematical groups, the approximate solution of nonlinear differential equations, and the principles of astrophysics—none of which I had studied in school.

I emerged from this problem—certainly no expert in any of these areas—but more knowledgeable than before. And, as the need again arose from time to time to investigate systems where general relativity was applicable, I immersed myself again in the arcana of the above topics, each time emerging with a broader and deeper knowledge and more confidence in my problem-solving abilities. I am still no expert, but I can understand the significance of general relativity, the conceptual connections among different domains of physics to which it allows access, and basic problems at the frontiers of astrophysics and cosmology to which real experts devote their attention. The research I have done in this self-taught subject has even received several distinctions from the Gravity Research Foundation.

The above (and other similar) experiences have impressed upon me three significant lessons: (1) I did not have to learn *all* of general relativity in order to learn to use some of it, (2) I would not likely have learned *any* of it had not an express purpose motivated me, and (3) what I did was not exceptional, but part of the normal way scientists teach themselves. No accomplished scientist I know would have sat through a series of courses to be taught step by step all the skills necessary to study some physical phenomenon of interest. (Long before such a course of study was completed, the original problem would undoubtedly have been solved by someone else.) These lessons apply not only to the way physicists "do" physics, but also to the way students can learn physics.

I conclude this section with the thorny question of assessing performance, or, more specifically, issues related to error, penalty, and student self-esteem—matters that are handled more humanely, in my opinion, in self-directed learning than by traditional modes of teaching. As a scientist and teacher, I am often reminded of the advice given by the Princesses Rhyme and Reason to young Milo in Norton Juster's delightful adventure story for children *The Phantom Tollbooth*:

> "You must never feel badly about making mistakes," explained Reason quietly, "as long as you take the trouble to learn from them. For you often learn more by being wrong for the right reasons than you do by being right for the wrong reasons."
>
> "But there's so *much* to learn," [Milo] said, with a thoughtful frown.
>
> "Yes, that's true," admitted Rhyme; "but it's not just learning things that's important. It's learning what to do with what you learn and learning why you learn things at all that matters."[11]

Most scientists, I surmise, would hold feelings similar to my own that, in the course of their research, errors are critical to advancement. Indeed, it is precisely when one corrects a mistaken idea or measurement that learning begins again.

If science is to be taught as it is practiced, then students must be allowed to commit errors and learn from them without the fear of punishment or failure. "Essentially, it seems, our brains learn best—and grow to learn more—" said educator John Abbot, "when we exercise them in highly challenging, but low-threat environments." This is decidedly not the case in the standard model of instruction where students are tested and evaluated continually. Judgmental monitoring of this sort is one of the most pernicious ways of suppressing creativity and transforming what ought to be a pleasurable activity, namely learning, into a nightmarish experience.

Aside from the anguish that testing and the threat of failure may engender, the traditional methods of evaluation are often misguided and do not generally provide educationally useful information. Students frequently do poorly on exams who, under less stressful circumstances—such as in personal discussions with the instructor—show good understanding of course material. Conversely, many students receive high marks in a course for the wrong reasons (in the words of Princess Reason) by "plugging and chugging" their way through prepared material, yet (in the words of Princess Rhyme) having little idea what to do with it or why it was worth learning.

Effective learning is not an activity that can be adequately assessed over the short duration of an academic term, but is demonstrable only over the long run—perhaps years—by the ways in which a person meets the challenges of life including, among others, those of career, family, and society. This applies as well within the narrower domain of science. When I consider the many students who have worked on research projects with me over the past two decades, I do not find course grades, determined in the conventional manner, to be a reliable indicator of a student's potential abilities to use scientific facts, principles, and modes of reasoning in constructive and creative ways. And is this not, in fact, what science teachers are ultimately trying to achieve—to prepare students for meeting the challenges of the future?

Self-directed learning recognizes at the outset that errors are part of learning, that mistakes made while learning should not be penalized, that it is fruitless to try to measure—as one would a volume of liquid—the volume of material absorbed by a student in the course of an academic term, and that the judgment of students is best based on what they actually do (i.e., on their effort), rather than on what they supposedly learn.

15.4 IMPLEMENTATION

Self-directed learning is more an attitude than a uniquely formulated program or procedure. Once the underlying framework and goals discussed in the previous sections are accepted, an instructor can undoubtedly find a variety of ways to achieve them. For illustrative purposes I will describe the principal features of an approach I followed in teaching the second half of a four-term sequence of introductory physics. The students had already completed the first two terms comprising basic mechanics and thermodynamics; these courses were taught by other instructors in strict conformity with the standard model. The final two terms, during which the ideas of self-directed learning were implemented, were nominally devoted to electricity and magnetism (Term 1) and optics and "modern" physics (Term 2). The course format, fixed long ago by the physics department in accordance with basic institutional requirements, consisted of three one-hour lecture sessions and one three-hour laboratory session per week. As at many other schools, the individual instructor had no control over this general format.

Although the following account uses examples drawn from the teaching of physics, any natural science can be taught within the same philosophical framework. The central issue here is the *reshaping* of instructors' attitudes toward teaching and students' attitudes toward learning.

It is worth stressing that what transpires during initial contact with students the first day of class is critical to the smooth and successful progression of the course throughout the rest of the semester. It is essential that students have an understanding of the new mode of instruction to be followed, the rationale for it (as just outlined), and what precisely is expected of them (to be elucidated later in this section). With no exception that I have ever encountered in many years of teaching, the students entering my courses have all been educated traditionally—that is, in accordance with the standard model—for a span of some twelve or thirteen years from the day they first entered primary school. Under this system the principal motivation is external, originating, at least among the more conscientious students, in securing high marks and teacher approval by doing well on tests and other graded exercises. Since only a few receive the highest grades, the system fosters competition among students, rather than an internally motivated drive to do well for the sole sake of personal intellectual advancement. The reader can well imagine, therefore, that it is not an easy mental adjustment for students to make initially when, in contrast to all they have experienced previously, their physics instructor now tells them at the start of term the following:

1. This is a course based on *self*-directed learning. *I* will talk to you (in the morning lectures) about what I think is conceptually important, practically useful, historically interesting, philosophically challenging, or (to me) personally meaningful—but *you* are free to take away from this course as much or as little as you choose.

2. There will be *no* scheduled class exams, and the quizzes that will be given periodically will not be graded; they are for your personal guidance in keeping you informed of what I think you ought to know.

3. Homework problems will be assigned throughout the term and solutions will be subsequently provided and discussed—but they are exclusively to help you learn physics; they will not be graded, and you will not be penalized for working problems incorrectly.

4. Your grade for this course will be determined on the basis of your *effort*, rather than on the outcome of testing, as demonstrated by a portfolio of work to be prepared throughout the term and submitted by the last day of class.

Grading, according to either standard or nontraditional modes of instruction, always involves considerable subjectivity, and a full discussion of the matter is not possible here. Personally, I would have preferred to teach physics without assigning letter grades—or on a simple pass-fail basis—but this was not an available option. Suffice it to say that, since the fear of doing poorly is one of the principal reasons why in many cases student performance in physics actually is poor, I decided to alleviate that fear at the outset in the following way. Every student whose physics portfolio met the minimum criteria of acceptability was guaranteed a grade (B−) representative of "good" work. Since a course grade now depended more upon the individual conscious decision to work or not to work rather than on the vicissitudes of testing, students had more control than before over their destiny in this course.

The physics portfolio, which was to be prepared in accordance with specified guidelines that I will discuss shortly, represents concretely what students have actually done in the course and serves as a more meaningful basis for evaluation than the inference from tests of how much or how well they may have learned. Each physics portfolio was to contain the following items:

1. A written account (original or recopied) of class activities, which includes, for example, lecture notes, solutions to in-class problems, interpretive discussions of physics demonstrations, and summaries of class "workshop" sessions and student-led presentations.

2. The student's own written solutions to assigned or suggested homework problems.

3. Class quizzes, reworked correctly if necessary.

4. The laboratory notebook with originally submitted experimental reports and any subsequent revisions.

5. The account of one (or more) independently researched projects during the course of the term.

Although the portfolios are to be submitted and evaluated at the end of the course, it is essential that students understand they are to be preparing their portfolios throughout the term. Indeed, the very nature of the contents almost assures this—although instructors would be prudent to remind their classes periodically.

Within the general framework of (1) class work, (2) laboratory work, and (3) out-of-class work, I will comment on the above portfolio items to the extent they reflect the distinctive objectives and methods of self-directed learning.

15.5 CLASS WORK

In courses taught traditionally, class notes serve the principal purpose of summarizing the information students think most likely to appear on examinations. Correspondingly, the notes, which are taken at the sole discretion of each student since the instructor rarely has interest in them, are usually consulted only shortly before a scheduled test. In many cases, they are discarded upon the termination of a course, since they contain little more than reworked examples and exercises from the textbook.

In self-directed learning students are not faced with tests, and a record of classroom activity serves a different purpose. First, it is clearly not possible for students to take personal class notes unless they show up for class. In this regard, the requirement of note taking is tantamount to a requirement of course attendance. Obligatory attendance, I believe, is a reasonable part of the required "honest effort" upon which a guaranteed grade is contingent. More importantly, however, the responsibility to present a legible set of notes in the portfolio has a beneficial influence on study habits. It necessitates that students review, and possibly recopy, their notes each night (or perhaps several times per week) if the procedure of note preparation is not to become too burdensome. Students who take notes attentively and review them regularly can become aware quickly of what they do not understand—and can seek assistance in a timely way. By contrast, when students take and consult notes for exams only, they frequently do not realize how little they understand until it is too late.

The purpose of keeping records, however, is not merely—or even mainly—to write down what the teacher puts on the blackboard. If students are required to attend class and take notes, then instructors are under a particular obligation to utilize class time effectively—in particular not to repeat what students can readily acquire from their books. I do not dispute that discussions of textbook concepts and applications, including worked numerical examples, as typically comprise the lectures of traditionally taught science courses can contribute to productive use of time (presuming students know why they are doing these things). Nevertheless, there are other educational opportunities that, if not unique to self-directed learning, are certainly facilitated by it.

I have at times begun a class by posing a problem suggested to me by some recent report in a science periodical, physics conference, or even daily newspaper, and have the students break up into small groups to attack it. Or I have assigned different problems to separate groups, each group reporting on its progress in the class discussion that later ensued. At other times, in the course of my lecturing, one of the students might raise a particularly thought-provoking question. Rather than answer it myself and continue with prepared material, I would use that question as the focal point of a class discussion exploring the different aspects—physical, mathematical, philosophical—of the new issue. In self-directed learning, as in home schooling, a teacher can proceed flexibly without the frantic rush through a syllabus. The goal is not so much to cover a subject, as to "uncover" it.

In traditional modes of instruction, students frequently perceive new material as threatening—particularly if it lies outside the scope of the textbook—since they will often see it again on an examination. Thus, instead of eliciting intense curiosity and animated classroom discussion, the presentation of intellectually challenging and significant ideas evokes instead the apprehensive query "Do we really need to know this?" An advantage of self-directed learning is that instructors can enrich course content without simultaneously arousing anxiety within their classes.

I would like to give one specific example of how, by exploiting the opportunities of self-directed learning, I have been able to make students aware of an issue of great import to physics and direct relevance to the thematic content (electricity and magnetism) of their course (Term 1). This seminal topic, which is ordinarily ignored in introductory physics, is the Aharonov-Bohm (or AB) effect (introduced in chapter 10). The significance of this illustration lies not only in the physics, but also in how it promotes holistic learning by exposing students in a nonthreatening way to the essential principles, historical development, and philosophical underpinnings of a far-reaching contemporary controversy in science.[12]

Following the researches of Faraday and Maxwell, it has become an integral part of the philosophical and mathematical foundations of physics to regard all interactions between matter as being mediated by fields of influence that are directly related to forces and propagate through space from one material particle to another. Each particle interacts locally with a field in its immediate vicinity. Thus, in regard to the presentation of classical electromagnetic phenomena, virtually all elementary general survey textbooks I know take electric and magnetic fields as the fundamental theoretical constructs. An introductory course of electricity and magnetism is concerned primarily with the characteristic motions of charged particles accelerated by specified electric and magnetic fields, and, reciprocally, the configurations of the electric and magnetic fields created by charged particles in different states of motion.

One can perhaps imagine the consternation among many in the physics community when, in 1959, Yakir Aharonov and David Bohm demonstrated theoretically that the state of motion of a charged particle can be altered by the presence of a static magnetic field through which the particle never passes and which therefore could exert no local classical force on it.[13] The effect on the particle's motion, according to AB, derives instead from a local interaction with other fields (vector and scalar potentials) that had been introduced into electromagnetism by Maxwell to facilitate calculation and were considered by most twentieth-century physicists to be devoid of "reality" since they exerted no forces. Consider, for example, the following brief eyewitness account of the reaction of Niels Bohr—foremost among quantum exegetes—and his disciples upon first learning of the AB paper:

> Our tranquility was rudely disrupted when in the fall of 1959 a paper entitled "Significance of Electromagnetic Potentials in Quantum Mechanics" came to our attention. The authors of this paper had the audacity to claim that in quantum mechanics, unlike classical physics, the potentials can directly affect the motion of charged particles, although they are "merely" mathematical instruments introduced in the 19th century to simplify the calculation of electric and magnetic fields that are "truly" the quantities that one observes.
>
> Many physicists, including Bohr and his associate Léon Rosenfeld, were profoundly taken aback by this suggestion. For a while Bohr was extremely skeptical; in the lunchroom Rosenfeld declared the Aharonov-Bohm proposal ... to be not only an affront to the Copenhagen concept of a physical observable, but also "contrary to the spirit of Galileo."[14]

Subsequent theoretical analyses and difficult experiments have confirmed the existence of the AB effect, and—after some thirty years of controversy—most physicists interested in the foundations of physics have reconciled themselves to this strange but calculable behavior of charged particles. Indeed, this behavior is not by any means an unimportant side curiosity, but absolutely essential to the internal consistency of quantum mechanics.[15]

Although the AB effect reflects the quantum, rather than classical, behavior of charged particles—a subject that students had not at the time encountered—it still concerns only classical electric and magnetic fields. The quantum nature of electromagnetic fields (the concept of the photon) does not enter at all. Thus, the conceptual implications of the AB effect were of direct relevance to this introductory class. These implications are far-reaching. The AB effect forces physicists to accept that either (1) the electric and magnetic fields can have nonlocal influences on the state of motion of charged particles, or (2) the potentials, which exert no forces and cannot be uniquely specified, are more fundamental than the electromagnetic fields.

All this is fairly "heady" material, but contains the ingredients—surprise, paradox, controversy, action (experiment), resolution, deeper understanding—of an intellectual drama that can attract and sustain a student's interest in a physics course. It provided an occasion to review the basic philosophical premises of classical physics—matters relating to cause and effect, prediction, precision, and the local nature of laws expressed in the form of differential equations—as well as interesting biographical details of some of the principal contributors (e.g., Newton, Faraday, and Maxwell). Likewise, it permitted me to expose students in advance of the traditionally prescribed sequence of courses to the phenomenology and concepts of quantum physics and to the historic personages (e.g., Planck, Einstein, Schrdinger, Heisenberg) upon whose work this edifice was built. And—since the AB effect was an area which has occupied my own attention as a research physicist—I could convey to students personally the exhilaration of research and discovery.

In discussing a topic like the Aharonov-Bohm effect, one must of course take care to present it at a level that physics students can follow to a sufficient extent to reap some benefit from attentive listening. But I have seen that with self-directed learning this was both possible and advantageous. Students came to appreciate that (1) the content of science is intricately interconnected, and a discovery in one area of it can have far-reaching major repercussions (the AB effect led to a reformulation of classical electromagnetism and a reinterpretation of the roles of fields and potentials),[16] and (2) science is a *human* activity,[17] where not only personal creativity but also articulateness and persuasiveness matter. (The AB effect was actually reported about ten years earlier by British physicists Ehrenberg and Siday;[18] their paper was narrowly directed to electron microscopists, and the full conceptual implications of its content were not widely appreciated.)

The basic message of the preceding example is not limited to physics teachers. Metaphorically speaking, every field of natural science has its AB effects—that is, critical points of development beginning with universal disbelief and culminating in major reversals of contemporary thought. In the earth sciences, for example, the idea that continents "float" upon a plastically deformable planet was first ridiculed and rejected (in part because the idea originated with a meteorologist rather than with a geologist) but was ultimately sustained by a confluence of evidence from diverse fields of science, especially physics. This makes another gripping and instructive story interweaving the principles of physics and the vagaries of human nature.

In short, one of the most effective ways a teacher can utilize class time to help students learn is by the regular presentation of historically and conceptually significant material beyond the textbook that fosters a holistic and challenging learning experience within a nonthreatening environment.

15.6 LABORATORY WORK

From discussions with many science instructors, I would say that the laboratory in introductory science courses is ordinarily meant to serve three principal purposes: (1) to allow students to witness basic phenomena, (2) to introduce students to practically important apparatus and measurement techniques, and (3) to give students a flavor of scientific research. Although it is possible to realize the first two objectives in a traditionally taught laboratory, I have grave misgivings concerning the extent to which any course taught in accordance with the standard model could approach the third objective which reflects the activity of inquiry and is, I believe, the most essential part of both science and science education. As I have written elsewhere,

> as usually constituted, [an instructional laboratory] does not in the slightest way represent what a scientist experiences when he is performing experimental research; for teachers to portray it as such, as is frequently done, is very misleading. A science instructor deceives himself who believes that any laboratory programme designed from the outset to yield clean, unambiguous data in a reasonably short time on previously well-studied phenomena with low probability of failure could in any serious way reflect what experimental science is like! I attest to this as one who has engaged in experimental work for over a quarter century.[19]

Although practices vary, of course, among institutions, the divers modes of laboratory instruction that I have witnessed usually have the following attributes in common. First, to the extent that lectures and laboratories are coordinated, a particular experiment is usually meant to illustrate material that has already been discussed in some detail by the instructor. Second, the students are given information sheets detailing the experimental objectives, the apparatus to be used, the instructions to follow in executing the experiment, what physical quantities to measure or observe, what formulas to use for analysis, and ultimately what experimental or theoretical end result to deduce and compare with some known standard result. Third, students are expected to report on their exercise according to a formal prescription covering everything from the content of the introduction (title, objectives, schematic design of apparatus, etc.) to how and where to record data, calculations, results, interpretations, and conclusions. Finally, the students' notebooks are collected weekly and graded, often on the basis of the accuracy of their results.

Although the foregoing mode of laboratory instruction might turn out functional technicians or engineering assistants, I do not believe it is accomplishing what ought to be one of the principal objectives of science education—namely, to *interest* students in science. It is worth noting again that many, perhaps most, of the students in an introductory physics course are not there by choice, but by force of circumstances. Assuredly the above ritualized procedure does not foster curiosity-based inquiry, for it scarcely permits students to exercise their curiosity. Worse still, it may be destructive of whatever creative instincts lie dormant in the students. The widespread emphasis, especially in physics courses, on detailed random-error analyses may also border on the ludicrous, given that apparatus available for instructional laboratories is generally far below that of research laboratories in quality, and poor results are more often attributable to systematic problems than to randomness.

In a laboratory organized around the principles of self-directed learning, students are encouraged to observe, question, and experiment under conditions that provide adequate time for inquiry and that do not penalize them for the conclusions they reach, even if these conclusions are at variance with accepted results. In short, the purpose of a scientific experiment is to satisfy the curiosity of the experimenter; it is not another type of academic test upon which students' grades are based.

Irrespective of the exact format of laboratory organization, the essential ingredient to an educationally productive laboratory experience is that, to the extent possible, experiments not be of the cookbook variety. The laboratory exercises should be so designed that students can employ their own ingenuity in executing them and must at times go beyond the information provided in class—such as, to use the resources of the library or computer center—to analyze and interpret them. It is important to emphasize—since state-of-the-art apparatus is ordinarily out of the financial reach of many colleges and universities, not to mention secondary and primary schools—that such apparatus is not required, or even necessarily desired, for a successful learning experience. Indeed, I have often found that experiments flawlessly performed and analyzed by expensive equipment for which the students' role consists largely of pushing the start button and subsequently collecting the computer printouts have little, if any, educational significance. As any research scientist knows, the learning experience in experimental work comes from solving the problems attendant to assembling the apparatus and making and interpreting the observations. Thus, the apparatus for experiments which are meant to interest and educate students should not be "black boxes," but rather open to inspection and capable of comprehension. In this regard it is worth noting that in Japan, where for decades students have traditionally excelled in the sciences, interest in science has been declining, according to the Japanese Science and Technology Agency, in part as a result of the black-box nature of consumer electronic products.[20] "Gone are the days when youngsters acquired a taste for science from their investigations of the innards of clocks and wirelesses," since it is no longer possible to tell from the external appearance of a device how it works.

I would like to give one example of the implementation of a laboratory based on self-directed learning in which students examined the nature and effects of light polarization. Historically, the investigation of light polarization by French "opticians" such as Arago, Biot, Malus, and especially Fresnel contributed seminally to the wave theory of light and the demise of the corpuscular theory associated with Newton. Although many instructors neglect the historical evolution of scientific concepts in their struggle to cover in their lectures the technical material of the textbook, it is noteworthy that the laboratory, and not just the lecture hall, can provide an appropriate forum for discussing issues of historical significance. Also, given the times in which they were performed, historically significant experiments generally employed relatively simple, intellectually accessible apparatus (as opposed to black boxes) of the kind suitable for educational purposes today.

At different work stations over a period of two weeks students had the apparatus and materials they needed to investigate the effects of linear polarizers, to measure the intensity of light of different polarizations specularly reflected from a smooth surface, and to examine the properties of birefringent media. At the start of the first session topics like Malus's law, the Fresnel relations, and light interference had not been discussed in lecture, and so most students (who do not read ahead in the text) did not know the explanations for what they were "supposed" to find. Indeed, they were not required to find anything, but rather to experiment, observe, and at some future time (three weeks later) to account for what they saw and measured.

In the course of the experiments students encountered a number of enigmatic results. For example, if a polarizer acts to filter out light, why does inserting a third polarizer between two orthogonal polarizers lead to greater light output? If the polarization of a light source is a measure of the "degree of order" of the light waves, why does reflection of unpolarized (and therefore totally disordered) light at Brewster's angle lead to a 100% polarized reflected beam? Does this not somehow violate the Second Law of thermodynamics that natural processes always occur in the direction of increasing disorder? Why does placing certain transparent colorless materials (like cellophane) between two neutrally colored polarizers give rise to a profusion of bright colors? Does the thin cellophane layer refract the light like a prism? (Subsequent experimentation readily showed that this was not the case.)

Eventually the students came to understand by their own experimental work and out-of-class research the reasons behind the phenomena. By the time the sessions devoted to light were completed, the class had progressed in several ways. First, they learned something about the behavior of light (e.g., interference)—not by the traditional mode of sitting through lectures, but in the course of pursuing answers to questions (Where *did* the colors come from?) that personally intrigued them. Second, they became more skilled in analysis (disengaging the various processes that contribute to an observed phenomenon) and synthesis (developing a model for interpreting an observed phenomenon by combining various processes). Third, and perhaps most important, they came to appreciate that the laboratory itself can serve to introduce new ideas. This is an important point illustrative of one of the principal objectives of self-directed learning: *One can teach oneself by observing.*

Although no claim is made that a laboratory employing the principles of self-directed learning replicates all the experiences of authentic scientific research, I do believe that it approaches this ideal much more closely than laboratories taught in the traditional format as previously described. For one thing, freed from the burden of test-based evaluations, students could approach experimental work with a more positive attitude and a greater willingness to explore and just "try things out." When grades are at stake, every false step in conducting an experiment is regarded as wasted time rather than as a useful learning experience. (Thomas Edison, having failed repeatedly to make a durable lamp filament, is alleged to have replied that he had not wasted his time at all, but simply learned of ten thousand ways *not* to make an electric lamp.) For another, the students themselves raised many questions for investigation directly from their observations and, with their curiosities aroused, set about to answer them. To be sure, they were helped by the ensuing lectures and by conversations with the instructor and were therefore able to continue the experiments with greater understanding of the phenomena they were investigating. Nevertheless, in stark contrast to the sense of futility that people often experience when perfunctorily executing some mandatory procedure in an attempt to answer questions they never posed themselves, self-directed learning permits students a greater sense of personal control over their laboratory work.

15.7 OUT-OF-CLASS WORK

In traditional methods of instruction, the principal focus is on classroom lectures. So entrenched is this standard model of education that many teachers cannot believe or admit that students are capable of learning independently outside the classroom. At one institution where I have taught, for example, the faculty voted to eliminate a week of respite from classes in the apparent belief that little good can come to either students or teachers from having "free time." Yet the leisure to observe and to think is precisely what *is* necessary for learning. In this regard the reflection of educator Jerome Bruner is pointedly relevant:

> The will to learn is an intrinsic motive, one that finds both its source and its reward in its own exercise. The will to learn becomes a "problem" only under specialized circumstances like those of a school, where a curriculum is set, students confined, and a path fixed. The problem exists not so much in learning itself, but in the fact that what the school imposes often fails to enlist the natural energies that sustain spontaneous learning—curiosity, a desire for competence, [and] aspiration to emulate a model.[21]

Bruner was concerned with the formative years of schooling, but his words are no less applicable to education at the level of college and university as well. And they are especially pertinent to the teaching of science.

What makes science a critically important and fascinating subject in the first place is that it is drawn from, and relates directly to, the "real" world. In physics, particularly, the basic principles are of universal applicability, as valid within the remotest galaxy as they are on Earth. The manifold implications of an abstraction like "universality" are not easily communicated to students, or grasped by them, within the confines of a classroom (even including instructional laboratory). To appreciate the notion that "science is everywhere"—that natural phenomena can be readily observed and their underlying causes understood— requires that one actually go outside the classroom and examine the world with an open and curious mind. A simple walk through the woods or along the seashore can provide to the receptive mind more of a science education than would come from hours spent listening to lectures. This is not simply empty rhetoric. I have myself encountered nearly the entire phenomenological content of a standard course in optics—polarization, refraction, reflection, interference, diffraction, scattering (Rayleigh, Mie), and dispersion—by puttering in my kitchen,[22] observing submerged objects near the seashore, and gazing at the sky.[23] Teachers and administrators who would "improve" education by lengthening class time and diminishing opportunities for outside contacts and leisurely contemplation do their students no genuine service.

Those who would teach science in accordance with the principles of self-directed learning—although they are constrained to operate under the same specialized circumstances (to use the words of Bruner) as all other teachers—must nevertheless find educationally meaningful ways for students to learn from their own experiences and research outside the classroom. One way I have tried to do this in physics is through "self-directed projects." As explained to students the first day of class, a self-directed project is an independent investigation of questions pertaining to the subject matter of the course (which, in the spirit of a holistic approach to science, can be broadly construed). Depending on the particular topic to be investigated, the research may entail any or all of the following: analytical calculation, numerical analysis, computer simulation, library research, or simple experiment. What is of importance is not that the students arrive at some predetermined right answer—there may, in fact, be no single solution—but rather that an effort be made to obtain results and to support them in a coherently written, grammatically acceptable report.

The projects are an integral—indeed, most essential—part of the course. They represent, more closely perhaps than any other course-related activity, the elements of curiosity-driven inquiry which lie at the heart of science. Students are permitted—in fact, encouraged—to select their own topics for investigation. Although some do, others, particularly those not initially interested in physics, prefer to select from among twenty or more suggestions provided by the instructor. In compiling my proposals, I look for projects that (1) introduce students to contemporary scientific developments as reported, for example, in high-quality general science periodicals like *Nature, New Scientist, Science, Scientific American*, and *American Scientist*, (2) engage students in simple, out-of-class experiments or observations, and (3) give students practice in constructing and testing theoretical models of physical systems.

The development of predictive models constitutes a key part of scientific research and education. In terms reminiscent of Hermann Hesse's *Das Glasperlenspiel* (The Glass Bead Game), physicist David Hestenes has colorfully characterized modeling the real world as "The great game of science": "The object of the game is to construct valid models of real objects and processes. Such models comprise the core of scientific knowledge. To understand science is to know how scientific models are constructed and validated. . . . It has aptly been said that the purpose of a scientific experiment is to ask a question of Nature. This being so, the *answer is a validated model*, not merely a mound of data."[24]

Faced with diverse phenomena, multiple possible mechanisms, and the complexities of real apparatus, a practicing scientist must produce an analytically or numerically tractable explanation amenable to testing. The more detailed the model, the better may be its predictions, but also the more arduous it may be to obtain analytical or numerical results. Thus, model building requires making judicious compromises.

In the traditional linear mode of instruction students rarely have occasion to develop the skills of the "game." Homework problems provide all the data needed to solve a problem usually by substitution into readily available formulas. Laboratory exercises are designed to elucidate a narrowly specified process or principle with an analytical procedure (in effect, the model) already provided. Yet, as Nobel laureate and accomplished expositor of the nature of science Peter Medawar has written, science is the "art of the soluble."[25] Self-directed projects help students practice that art. (I find it interesting that scientists should discuss their work in terms of games and art, whereas students so often see it as anything but recreation.)

For the projects to serve their purpose, students should work on them throughout the semester, and not as a last-minute obligation to be satisfied just before the course portfolio is due. Nevertheless, as one measure of the vital role self-directed projects assume, the course syllabus specifically allows for two weeks—at points approximately one-third and two-thirds through the term—to be devoted to research and writing without scheduled classroom lectures or laboratories. In addition, the final week of term includes the oral presentation of selected projects by those students who volunteer to do so.

In the remainder of this section I will give several examples of self-directed projects that students pursued. As before, it is not the intrinsic physics that is the focus of interest in this account, but the way in which physics is (1) introduced through commonplace activities, (2) associated with contemporary discoveries, or (3) made relevant to contemporary issues of concern. Attention should also be directed to the way in which individual topics usually (4) comprise diverse phenomena and concepts (rather than fall into specifically identifiable units of a textbook), (5) call for the development of simple, tractable theoretical models, and (6) foster learning in a holistic (rather than linear) way.

Topic 1: *Lightning*   Investigate the subject of natural lightning. What is lightning? How is it generated? What accounts for the zigzag shape of lightning bolts? What is ball lightning? How does a lightning rod work? How would you design a lightning-protection system for (1) a single-family dwelling, and (2) the campus Engineering and Computer Science Building?

Comments   For all its familiarity, lightning is a subject that is still fascinating and in some ways mystifying. In the course of library research, students were challenged to understand fundamental ideas drawn not only from electricity and magnetism, but also from chemistry, thermal physics, and atomic physics. There are also significant aspects of lightning which remained controversial even after years of investigation. Even to so outwardly primitive a device as the lightning rod have been attributed various mechanisms of operation. The study of a topic like this one provides a needed counterweight to the overwhelming impression of scientific certainty conveyed by introductory textbooks.

In addition to simply reading about lightning, students were asked to make practical use of their findings by designing lightning-protection systems for two structures very different in size and content. In the case of the computer science building, for example, one must not only protect against direct lightning strikes, but consider as well the effects of strong fields on the functioning of sensitive electronic apparatus. Although the research for this project was presumed to be largely theoretical in nature (I did not want a self-styled Benjamin Franklin standing in the middle of a sports field flying a kite in a thunderstorm!), students could, if they wished, construct a tabletop model and test it with a small Van de Graaff generator as a lightning source.

Topic 2: *Mystery Top*   There is a physics toy called the Top Secret.[26] A small top spins seemingly forever on a plastic pedestal with a small convex nub at the center. Examine the device and explain how it works.

Comments   A remarkable creation of the inventor Roger Andrews (first constructed when he was only fifteen years of age), the Top Secret generated utter amazement when demonstrated to my classes. Once spun on its plastic base, the top executes a complex orbital motion about the central nub. If left undisturbed, the top can spin for hours, apparently defying the laws of thermodynamics by drawing energy from a hidden source to compensate for frictional losses. Even after the specific nature of the top and the hidden source of energy are revealed, the detailed mechanism of its operation is by no means trivial. In fact, the toy remains undeniably intriguing even when it is fully understood. Without divulging here the secret of the top, I would nevertheless say that it provided one of the strongest incentives I have ever found for students to come to grips with conceptual and practical implications of Faraday's law of induction. To explain the device is, in effect, to construct a model of its operation. The model, in this case, is not necessarily a mathematical one, but instead calls for a coherent explanation of how specific electronic and mechanical components, functioning in accordance with basic physical laws, give rise to an apparent perpetual motion.

It has been my experience in general that science toys and demonstrations can be used with great effectiveness to motivate students at any level to want to understand material that would otherwise be boring if encountered primarily in the form of classroom lectures and cookbook laboratory exercises. I collect such devices, have integrated them at times into my research program with undergraduates,[27] and use them for teaching purposes whenever possible.[28] Toys bring the abstract principles of science into the domain of a student's personal experience, not by shallow emphasis (as instructors are wont to give) upon their manifold implications for technology and society—but simply by raising in each observer's mind the burning question "How *does* that work?" A student personally driven to answer that question is well on the way to learning science.

Topic 3: *Conduction*   In his acclaimed textbook on electricity and magnetism Nobel laureate Edward Purcell states, "In metals Ohm's law is obeyed exceedingly accurately up to current densities far higher than any that can be long maintained. No deviation has ever been clearly demonstrated experimentally. According to one theoretical prediction, departures on the order of 1 percent might be expected at ... more than a million times the current density typical of wires in ordinary circuits."[29] Is this really the case? Look into the subject of mesoscopic rings, that is, normal metal rings of about 1 micron in diameter with wire thickness of about 0.05 micron.

Comments   Ohm's law, of course, is not really a law in the same sense as the universally applicable laws of conservation of electrical charge and conservation of mass-energy. Nevertheless, as indicated by Purcell's remark, it is a relation thought to apply with great accuracy to ordinary (i.e., not superconducting) metals. In pursuing the questions raised by this project, students learn about the remarkable electrical properties of normal metal rings that, though small (a human blood cell may be about 10 microns in diameter), nevertheless contain billions of atoms and so would not usually be regarded as a system of quantum mechanical size. A low current passes through a ring—not a current one million times greater than ordinarily encountered—and yet Ohm's law does not hold. In the textbook and classroom, the electrical conductivity of metals is commonly explained by means of the classical electron model in which electrons are treated as small, hard billiard balls of charge. The mesoscopic ring project helps make students aware that the wavelike behavior of electrons can fundamentally alter the electrical responses of a seemingly classical object.

   A project of this nature, which clashes with textbook assertions, suggests to students that they should be willing to question what they read and believe— even in an outstanding textbook like Purcell's. This is not a trivial point. If science is a self-correcting process of inquiry, then a course that leaves students with the impression that scientific knowledge is fixed and certain is communicating the wrong message.

Topic 4: *Health Implications of E&M Fields*   Can low-level nonionizing electromagnetic fields affect human health? Look into the hazards allegedly presented by electromagnetic fields in everyday living. For example, is it dangerous to live near a power line or power transformer? Is it dangerous to sit in front of a computer terminal? What evidence supports and refutes these claims? What frequencies and fields (electric or magnetic or both) are under suspicion? What mechanisms are proposed for the alleged hazards?

Comments   This topic (and a similar one concerned with the potential dangers of radon in the home) provided pointed reminders of the relevance of physics to problems of everyday living. Indeed, the personal concerns of some students in regard to health issues were strong incentives for their studying physics (and biology) to a far greater degree through self-directed projects than they would otherwise have been inclined to do through assigned reading and classroom lectures alone. For example, to the extent that it bore on the questions that interested them, these students devoured outside sources of information (such as special reports by government, industry, and science academies) addressing topics in basic physics such as the power and frequency of electromagnetic fields, sources of electromagnetic fields, factors determining the strength of these fields, methods of measurement, and observations of these fields in the state where the students resided. Much of the same material, presented in the textbook for "academic" purposes, had previously been of little interest.

One salutary educational outcome of this project is that students understand better the fact that different scientists employing the same universal principles can nevertheless arrive at different results. This is not a trivial point, for too often the pat way in which academic problems—those encountered in class, homework, and tests—are neatly and definitively resolved reinforce the caricature of scientific method that students carry with them out of school into society at large. According to this caricature, the steadfast accumulation of sufficient data accompanied by the ritualized exercise of logic inexorably leads to the correct scientific conclusions. Real-world problems are rarely so simple. Most are like those of this project: controversial issues about which many contradictory things have been written. It is not definitive answers that science provides in these cases, but rather the tools with which to make informed decisions in the face of uncertainty and risk.

The preceding are four representative samples of self-directed projects that I, the instructor, suggested. Space does not permit discussion of others, among which were studies of optical activity, optical and electron interferometry, image recognition and enhancement by spatial filtering, holography, the Einstein-Podolsky-Rosen paradox, highly excited (Rydberg) atoms, lasers, nonlinear optical phenomena, and pulsars. In succeeding years an increasing number of students proposed their own projects concerning, for example, the physics of how a nerve cell fires, application of piezoelectric devices in the design of downhill skis, the optics of the "glory" (a diffractive phenomenon), the origin of the auroras, magnetic resonance imaging, and the development of photovoltaic devices to harness solar energy.

Apart from their purely scientific benefits, the self-directed projects help develop language skills, especially among science and engineering students whose regimen of technical courses provided few, if any, occasions to speak or write. Upon completion of their research, students were expected to prepare a written report for their course portfolios. The essays must be grammatically acceptable and, ideally, display a certain measure of eloquence and personal style. During the last week of term, those who desired to do so gave an oral presentation of one of their projects.

The importance of developing both writing and speaking skills can hardly be overestimated, yet these activities are rarely promoted in traditionally taught science courses that emphasize the technical (as opposed to cultural) aspects of science and the mechanical details of problem solving. Students who concentrate in science, engineering, or medicine frequently never have to write again in school (beyond the preparation of lab reports) after they fulfill some core English requirement. Professionally active scientists and engineers, however, are engaged in writing throughout their careers, as for example, in preparation of research papers, internal progress reports, and proposals for research funding. They also frequently speak at conferences and schools, and before civic groups and government committees. Communication, together with research, lies at the heart of science.

15.8 ASSESSING THE UNASSESSABLE

There is, I believe, an important core of truth to B. F. Skinners paradoxical remark that "Education is what survives when what has been learned has been forgotten."[30] True learning is a long-term affair—ideally, lifelong. Unless one intends to become a physicist, for example, the details of calculating the moment of inertia of a rolling cylinder or the oscillation frequency of a particular circuit are likely to be of little future importance and soon forgotten. If this is the only kind of information a student takes away from a physics course, then, for all practical purposes that course has been a waste of time. By contrast, the legacy of a science course that opens students' minds to the fact that nature is both marvelous and comprehensible—and instills in students the desire and confidence to seek understanding on their own—is of lasting value. *That* is the goal of self-directed learning. However, it is not a simple matter to evaluate an experimental teaching method that measures its success not in the "here and now" but in the capacity of students to cope with challenges of the future.

Although surveys of student opinion solicited at the end of term run the risk of providing hastily formulated judgments, it was nevertheless gratifying to find that the overwhelming majority of students who participated in this educational experiment expressed their satisfaction with the way the class was taught. In their anonymous replies, many pointedly remarked on the sense of freedom they felt for the first time in more than a dozen years of schooling in being able to learn without the fear of tests. Others commented favorably on the opportunities to explore interesting subjects outside the classroom. Still others appreciated classroom discussions of contemporary issues not treated in the textbook. Most were thankful for opportunities to improve their writing and speaking under conditions that did not subject them to embarrassing criticism. There were also one or two students each term who would have preferred the traditional regimen of graded tests and graded homework as a stimulus to "work harder." A teacher cannot, of course, please everyone. The need of some students for an externally imposed motivation is all too comprehensible, for the educational legacy of a dozen years can not be entirely undone in one or two terms.

In assessing the impact of this self-directed learning experiment, it is instructive to note that not one student in either course intended to become a physicist. All were engineering majors enrolled in physics exclusively for the purpose of fulfilling the requirements of an engineering degree. The engineering department had, in fact, dropped the requirement of my Term 2 course in the misguided belief that modern engineers do not need modern physics—and, in fact, the department barely encouraged its students to take the Term 1 course in the equally perverse view that only prospective electrical engineers needed a physics course in electromagnetism. In any event, the consequences to the physics department of this curricular revision in engineering were drastic. No longer mandatory, course enrollment had plummeted to such an extent that the physics department considered eliminating modern physics altogether from the college catalog. This was the prevailing situation when I taught these courses for the first time according to the principles herein described. At the end of Term 1—in marked contrast to the enrollment pattern of the previous years— 90% of the students either registered for the sequel, or, unable to do so because of course conflicts, planned to register the following academic year.

It should be said that the sudden and unusual "popularity" of introductory physics cannot be attributable to an easy work load or generous grade distribution, for neither was the case. For most students the effort required to complete out-of-class projects and in general to prepare a satisfactory course portfolio was at least as great as that which they ordinarily expended in studying for exams. And yet, although no longer forced to do so, nearly all students concretely expressed their desire to continue the study of physics. Clearly these students must have believed they were learning something of value.

With respect to final evaluations, the majority of grades fell within the "average" and "good" categories as is typical of most university courses. What is especially significant, however, is that, of the students receiving the highest marks, all but one had previously done relatively poorly during the first two terms of introductory physics taught in the traditional way. Their low grades derived principally from lack of interest and poor test scores. On the other hand, the quality of their portfolios during the second two terms suggested that these students had the ability to use available resources to find information and comprehend it, to refine complex problems into tractable questions and solve them, and to communicate their findings cogently and articulately—in short, to do creative work analogous to scientific research. Several, in fact, were subsequently awarded summer research fellowships to work with individual faculty members in science or engineering. Had the last two semesters been taught in the standard way as were the first two, it is likely the performance of these students would have been equally low and their interest in physics further diminished. Assuredly, they would not have received fellowships or perhaps even encouragement to continue in engineering. *Seen directly in terms of lost opportunities for potentially talented and motivated students, the manner in which science courses are taught and individual ability measured is a matter of the highest consequence.*

15.9 PROGRESSIONS

In the years following this experiment in education, I continued using self-directed learning techniques in other courses with students whose career goals were not predominantly in engineering or the physical sciences, but in biology and medicine. In all cases, the outcomes were the same: Students felt a deep appreciation, a sense of gratitude, for having an opportunity to explore physics in a sympathetic environment created principally to inspire, rather than to judge, them. I was not surprised to be told by many of my students—both personally and through the anonymous end-of-term questionnaires—that, though they had always disliked physics, this physics course was the highlight of their academic studies. Such remarks are not surprising if one believes, as I do, that physics is intrinsically a fascinating subject.

Despite the encouraging results with students, I cannot say that I ever managed to convince colleagues at the same institution to adopt similar methods. The need to test, grade, and rank students was considered indispensable— although admittedly not because it helped students learn better—and the work involved in creating projects and reading portfolios was regarded as prohibitively time-consuming.

In the world beyond, however, dissemination of these ideas through journal articles and invited talks brought a large and favorable response. One 1995 article in *American Journal of Physics*[31] alone elicited more personal letters— all supportive—than anything else I had ever written up to that time. There were requests to translate the paper into a variety of European and Asian languages, and invitations to visit and lecture, both at home and abroad. In Finland, for example, where I was a guest at the Helsinki University of Technology, one of my hosts had adapted the methods of self-directed learning to accommodate more than 350 students (ten times the number in my own classes) studying introductory electromagnetic field theory. Slowly, but progressively, I suspect, attitudes can be changed.

"At college age," one teacher has remarked, "you can tell who is best at taking tests and going to school, but you can't tell who the best people are. That worries the hell out of me."[32] Indeed, that disturbs me, too, but I believe that there *are* more appropriate ways than testing to identify the best people. To me, the outcome of this educational experiment in self-directed learning is both heartening and instructive. It confirms, along with all other experiences I have had, Einstein's inspiring advice: "Teaching should be such that what is offered is perceived as a valuable gift and not as a hard duty."

NOTES

1. R. P. Feynman, "The Value of Science," reprinted in *The Shape of This Century*, ed. D. W. Rigden and S. S. Waugh (New York: Harcourt Brace Jovanovich, 1990), pp. 352–57. Quotation on p. 355.

2. M. P. Silverman, "Two Sides of Wonder: Philosophical Keys to the Motivation of Science Learning," *Synthèse* 80 (1989): 43–61.

3. M. P. Silverman, *And Yet It Moves: Strange Systems and Subtle Questions in Physics* (Cambridge: Cambridge University Press, 1993), provides an account of the author's experiences as a scientist and their bearing on his philosophy of education.

4. M. P. Silverman, "Self-directed Learning: A Heretical Experiment in Teaching Physics," *Am. J. Phys.* 63 (1995): 495–508.

5. R. H. Romer, "Reading the Equations and Confronting the Phenomena—The Delights and Dilemmas of Physics Teaching," *Am. J. Phys.* 61 (1993): 128–42.

6. H. J. Walls, *Scotland Yard Scientist* (New York: Taplinger, 1973), pp. 177–78, 180.

7. E. Gerjuoy, "Improving Courtroom Presentations of Scientific Evidence," *Physics and Society* 22 (October 1993): 6–9.

8. The interested reader might consult reports in *Growing without Schooling* (Cambridge, Mass.: Holt Associates).

9. Cited in R. P. Feynman, R. B. Leighton, and M. Sands, *The Feynman Lectures on Physics* (Reading, Mass.: Addison-Wesley, 1963), p. 5.

10. J. Holt, *Teach Your Own: A Hopeful Path for Education* (New York: Dell, 1981), p. 17.

11. Norton Juster, *The Phantom Tollbooth* (New York: Random House, 1961), p. 233.

12. See M. P. Silverman, "Raising Questions: Philosophical Significance of Controversy in Science," *Science & Education* 1 (1992): 163–79, for a discussion of the role of controversy in the advancement and teaching of science.

13. Y. Aharonov and D. Bohm, "Significance of Electromagnetic Potentials in the Quantum Theory," *Phys. Rev.* 115 (1959): 485–91.

14. E. Merzbacher, "The World through the Master's Eyes," *American Scientist* 80 (1992): 484–85. This is a review of A. Pais, *Niels Bohr's Times, in Physics, Philosophy, and Polity* (Oxford: Oxford University Press, 1991).

15. I discuss the Aharonov-Bohm effect and related quantum interference phenomena in my book M. P. Silverman, *More Than One Mystery: Explorations in Quantum Interference* (New York: Springer-Verlag, 1995), as well as in chapters 1 and 2 of Silverman, *And Yet It Moves*.

16. T. T. Wu and C. N. Yang, "Concept of Nonintegrable Phase Factors and Global Formulation of Gauge Fields," *Phys. Rev. D* 12 (1975): 3845–57.

17. M. P. Silverman, "Science as a Human Endeavor," *Am. J. Phys.* 53 (1985): 715–19.

18. W. Ehrenberg and R. E. Siday, "The Refractive Index in Electron Optics and the Principles of Dynamics," *Proc. Phys. Soc. Lond. B* 62 (1949): 8–21.

19. M. P. Silverman, "Two Sides of Wonder," page 57 of Note 2.

20. "Ready-made Entertainment Is Killing Japanese Curiosity," *New Scientist* 140, no. 1903 (1993): 7.

21. J. S. Bruner, *Toward a Theory of Instruction* (Cambridge, Mass.: Harvard University Press, 1966), p. 115.

22. M. P. Silverman, "Interference Colors with 'Hidden' Polarizers," *Am. J. Phys.* 49 (1981): 881–82.

23. M. P. Silverman, "How Deep Is the Ocean/How High Is the Sky? Some Thoughts on Imaging by Parallel Plates and Gravitationally Stratified Media," *Eur. J. Phys.* 11 (1990): 366–71.

24. D. Hestenes, "Modeling Games in the Newtonian World," *Am. J. Phys.* 60 (1992): 732–48. Quotation from pages 732 and 738.

25. P. B. Medawar, *The Art of the Soluble* (London: Methuen, 1967). Quotation on p. 7.

26. The Top Secret is manufactured by Andrews Manufacturing Company, Eugene, Oregon.

27. M. P. Silverman and G. M. Cushman, "Voice of the Dragon: The Rotating Corrugated Resonator," *Eur. J. Phys.* 10 (1989): 298–304.

28. M. P. Silverman, "The Vortex Tube: A Violation of the Second Law?" *Eur. J. Phys.* 3 (1982): 88–92.

29. Edward M. Purcell, *Electricity and Magnetism* (New York: McGraw-Hill, 1985), pp. 143–44.

30. Quoted in J. Roger and P. McWilliams, *Life 101* (Los Angeles: Prelude Press, 1991), p. 330.

31. M. P. Silverman, "Self-directed Learning." Note 4.

32. J. Roger and P. McWilliams, *Life 101*, p. vi.

Why Brazil Nuts Are on Top:
Physics and the Art of Writing

Reading maketh a full man; conference a ready man;
and writing an exact man.
(*Sir Francis Bacon*)

I WAS A journalist before I was a professional scientist, and the love of language has always remained with me. At colleges and universities where I taught science, I was quite possibly the only instructor—of any faculty, including English—who bothered to point out on students' papers that "data" takes a plural verb or that "whom" (not "who") follows a preposition. To some this will seem pedantic carping. But to one raised to speak standard English and to enjoy its rich possibilities for expression, fractured language is received by the ear like misplaced notes in a symphony, or by the eye like errant signs in Maxwell's equations.

It is not, however, grammatical or orthographical peccadillos that alarm me when I read students' papers. Rather, it is the wholesale ignorance of general language structure, the impoverished vocabulary, the illogical trains of thought, the pervasive disharmony, inelegance, dysfunction—the clear distaste for writing and for the written word. This is alarming, but not surprising. No one ever convinced these students in their previous science courses that scientists, including physicists, need to know how to write.

Some years ago the institution at which I was teaching created an in-house journal entitled "The Write Stuff," for which the introductory issue was to focus on the importance of writing in each contributor's profession. I was asked to write on behalf of physics. The issue was published, but the editors decided not to include physics. Perhaps it was truly want of space—or perhaps it was that not even the faculty who comprised the editorial staff saw any real reason to emphasize writing skills to people whose lot in life was to construct apparatus and solve equations. Regrettably, the image of physics even in a liberal arts setting is often that of a narrowly technical discipline. In any event, whatever the cause, the glaring absence only reinforced the pervasive sentiment that for physicists and physics students writing is at best an unnecessary indulgence, not a vital skill.

I disagree. The following is an adaptation of that unpublished essay.

There was a time, perhaps, when an Isaac Newton, driven partly by the desire for solitude and an aversion to criticism, could ignore the outside world and carry on his work—alone and for his personal satisfaction—in the quiet of his private rooms. Genius that he was, Newton largely developed for himself what principles and experimental methods he needed to study the phenomena that interested him. Little inclined to share his findings, he was even less disposed to record them for public scrutiny. Were it not for his importunate friend Halley (of comet fame), it is quite possible that his epochal legacy, *Principia*, would not have seen the light of day.

Much has changed since the seventeenth and eighteenth centuries when the principal physicists and mathematicians of Europe knew one another, if not personally, then by written correspondence; together, all could have convened in one lecture hall. The natural philosophers of the day may have beaten a path to Newton's door in their pursuit of knowledge, but there are few such eminent and sought-after scientists, living or dead, who fall in the same category. With some hundreds of thousands of scientists the world over, it is unlikely that the thoughts and activities, however brilliant, of one who fails to record them in writing will survive long enough to be of use to anyone else.

It should be clear without further elaboration that original ideas and their careful execution are necessary for successful achievement in physics as in any other endeavor. But these alone are not sufficient. As an important part of our cultural heritage, physics involves the accumulation and sharing of knowledge. What is valid and valued is retained; what is false is modified or eventually discarded. Although scientists may disseminate the results of their research orally at scientific conferences or through informal discourse with colleagues, any physicist desirous of making some more permanent contribution to science must be able to write. And not just write, but write well. In fact, throughout his or her career, a physicist will almost certainly need to correspond with at least the first two of the following groups of people: (1) employers, supervisors, or sponsors, (2) scientific peers, and (3) nonscientist fellow citizens.

The need to write begins long before the scientific work begins. Indeed, without an articulate and cogently presented proposal justifying an intended project, the work may never begin at all. Physics research is expensive. Tchaikovsky may have had Madame von Meck to support him while he created masterpieces of music, but the venerable practice of patronage by the aristocracy never quite took root in physics, especially in America. An American physicist hopeful of financial aid will probably have to submit to the National Science Foundation (NSF) in writing—and frequently in decuple—a detailed description of the intended research.

A good proposal, by the way, is not merely a set of step-by-step technical instructions along the lines of (1) First I turn on the power supplies and detector, (2) next I orient the laser and polarizers, (3) then I record the signal as a function of . . . , and so on. I have evaluated many grant proposals over the years—not a few similar to the preceding—and can state with assurance that no submission, however technically feasible, is likely to get much attention or support unless the rationale for the project is made crystal-clear in acceptable written English at the outset. This is true whether judgment is made by an NSF panel, the governing board of a national laboratory, or the director of an industrial research laboratory. Physicists who want to see the fulfillment of their ideas must not only be familiar with the pertinent principles of physics, but also have the pursuasive skills of an attorney submitting a case before a jury—which, in effect, is exactly what they are doing.

Assuming that an investigation is accomplished successfully, the researcher desirous of receiving due recognition must then write a report of the work sufficiently interesting and complete that other scientists would want to read it and be able, if necessary, to confirm it. The twentieth-century image of a solitary, if not disequilibrated, scientist performing experiments in the secrecy of his basement may be appealing to science fiction readers and filmgoers (ever see *Back to the Future*?), but it is a far cry from the reality of physics as a professional discipline. Recognition ordinarily comes from publication in refereed journals. But with hundreds of thousands of physics papers published each year in journals now so numerous that not even librarians can keep track of them, is anyone likely to see, let alone remember, a particular research article? That, again, may depend almost as much on the literary merits of the paper as on the value of the scientific contents.

I can recall with amusement the Sidney Harris cartoon of two scientists "talking" to one another. The bubble over the first scientist shows a lengthy string of equations with abstract mathematical symbols; the second responds in kind. Then they both burst into laughter. There are no words, and the implication is clear that science is a secret language for scientists alone. I have often perused physics papers in journals purportedly for the entire physics community that project a message similar to that cartoon—papers so highly specialized as to be opaque even to most physicists. Such papers amuse me but leave no permanent impression.

Many laymen and scientists alike—to the extent they think about the subject at all—implicitly believe that the laws of physics are somehow just "out there" like apples to be plucked from a tree of knowledge, and that it is essentially immaterial who actually does the plucking. After all, the laws are the laws. Would the Earth attract the Moon any differently if the law of gravity were discovered by someone named John Smith instead of Isaac Newton? Nevertheless, irrespective of by whom a discovery is made, how it is written up and presented to the physics community *can* influence strongly the subsequent impact the research will have. Consequence is not divorced from history.

Albert Einstein, whose name—thanks to shallow and sensationalist press coverage throughout his life—is virtually a household word symbolic, not merely of genius, but also of science abstract and incomprehensible, was actually a master expositor. I have read in the original German many of his published scientific papers and personal letters to other scientists and have never failed to be impressed by the aptness of his phrases, the clarity and visual imagery of his examples, and the overall beauty of his language.

Einstein's classic paper (originally in German) "On the Electrodynamics of Moving Bodies" exposes the theory of special relativity in its entirety.[1] Others, as well as Einstein, will later provide additional insights, but within its domain of validity (neglect of gravitation) the theory, as first published, is correct, complete, and self-consistent. Rarely in the history of science does a theory spring with such mature finality from the head of a single creator, like Athena from the head of Zeus.

In contrast to a physics style increasingly dominant today, the young Einstein's approach was marked by a fundamental simplicity in both its verbal exposition and mathematical formalism. Beginning by pointing out a disturbing paradox in a deceptively simple phenomenon (magnetic induction in a conducting wire), he proceeded to the root of the difficulty in a logical and general way by examining the concepts of space and time, in particular the question of simultaneity, an issue that scientists from the time of Newton to the beginning of the twentieth century evidently found too trivial to warrant analysis. Employing a level of mathematics that demanded of the reader little more for the most part than basic algebra and elementary calculus, Einstein clarified the physical meaning of space and time in terms of operational procedures on moving clocks and measuring rods.

The reason for emphasizing here the expositional merits of Einstein's paper is that the presentation of these profound insights to the scientific community did not have to take place this way. Indeed, the basic mathematical relations were known to other scientists before Einstein. To one they represented a curious, but isolated, symmetry in the laws of electricity and magnetism. To another they implied an inexplicable force acting to compress objects in motion. To another, the relations served as the basis of a specific mechanical model of the recently discovered electron. It was only Einstein, however, who perceived the significance of his results in terms of general principles that transcended specific physical systems or the hypotheses of particular physical models. Our concept of the electron today differs radically from that of a century ago; yet Einstein's formulation of special relativity still remains intact with no known violations in its domain of applicability.

Though the laws of nature reflect an objective reality independent of their finder, it is nonetheless true that the one whose fortune it is to make a significant discovery can make as well a unique imprint on the work, its interpretation, its perceived generality, its aesthetic quality, its subsequent impact. In physics, the abilty to write counts—and not merely for the Newton or Einstein who may come along once every few centuries. Unfortunately, editors and reviewers do not always appreciate, and rarely encourage, the kind of creative writing that would make reading physics articles a pleasurable, as well as instructive, experience. But surprises do occur.

I recall an interesting article that appeared some years ago in a highly respected physics journal. Once a primary vehicle for publication of papers of immediate importance to all physicists, many of the articles that appear there now are often so narrowly focused and theoretically speculative that their very titles are expressed in neologisms comprehensible to only a small segment of the readership. Thus, in this very journal where one was most likely to read about the (experimentally unverifiable) beginning or end of the universe or a zoological garden of whimsically named (and probably nonexistent) elementary particles, I encountered a marvelously simple and counterintuitive paper about jostling nuts![2]

Shake a can of small particles in which embedded randomly in the interior are a few larger particles of greater density. The large particles on average move upward. *Move upward*, mind you—not downward so as to lower the total potential energy of the system.

I could easily have imagined that the authors, in order to be taken seriously by reviewers and editors, might have given their paper the sesquipedalian title "Monte Carlo Verification of Size Segregation of Granular Mixtures via a Nonequilibrium Local Geometric Mechanism in Contrast to Dislodged Metastable Configurations by Interstitial Filtration of Particles." But they didn't. Instead, to my pleasant surprise, the paper bore the title "Why the Brazil Nuts Are on Top." How such a lovely title—and instructive paper—passed the reviewers and editors is beyond me, but I am glad it did; I enjoyed reading it and learned from it. Over the years I have either forgotten, or never taken notice of, countless other articles—some, no doubt, containing interesting physics if only the contents were expressed intelligibly. But when good physics and clear and lively writing go together, the results are memorable.

Possibly the authors may have had to fight for the right to express their work creatively; that has been my own experience at times. Much of chapter 3 of this book originally comprised a paper of the same title ("How Deep Is the Ocean?/How High Is the Sky?"). An American editor rejected it, but the manuscript was readily accepted by a British journal and elicited, in fact, a warm letter of appreciation from Sir Brian Pippard, a don of Cambridge University and the honorary editor.

That a physics paper must often be phrased in intentionally obscure language to be acceptable is lamentable. It sustains an illusion that the principles of one branch of physics are even beyond the ken of physicists of other specializations, let alone an educated public. This unfortunate situation derives, I believe, from a fear of many scholars in all fields of science—and the "softer" the science, the stronger the fear—that what can be easily read and understood will be automatically scorned by one's peers as obvious and trivial. This fear is not unfounded.

Some years ago I had such an experience, humorous in retrospect, but certainly irritating at the time. Wondering about the properties of a strange kind of atom resembling helium but with two exotic particles (muons) instead of electrons orbiting the nucleus, I asked a number of distinguished nuclear and particle physicists at MIT, Harvard, Yale, and elsewhere for their opinions. In particular, I wanted to know whether the Pauli exclusion principle would lengthen, shorten, or have no effect on the lifetimes of the muons which, singly, decayed into other particles in about two microseconds.

No one, to my knowledge, had asked that question before; no theorist had studied this hypothetical atom; no experimentalist knew how to make it. Nevertheless, the question I asked was a fundamental one, but none of the physicists whom I queried, all of them "experts," could answer with any certainty. Of those willing to speculate, about 50% proposed that the Pauli principle lengthened the muon lifetime; about 50% decided oppositely. A few refused to take a stand.

More curious than ever, I sat down and performed the relevant calculation myself; it turned out not to be difficult. It also turned out to be surprising in view of the preceding "referendum": the Pauli principle had no effect at all. Thinking about the result in a more general way, I was able to discern a basic physical principle that substantiated the calculation and gave insight into the reason for the outcome.

I wrote up a short paper summarizing the analysis and emphasizing the general principle by which means insight, as well as calculation, could have yielded the answer. The paper was rejected. "Intuitively obvious," said the referee; "not in the least surprising," "straightforward exercise," could be "tackled by a third-year honours student," and so on. Had the paper consisted of pages of quantum field calculations (the Sidney Harris vision of physicists communicating) with little explanatory material, it would probably have been acclaimed as a groundbreaking study of a new atomic system.

Not one to be put off easily when he thinks he is right, I wrote a spirited rebuttal and asked to have it reviewed by the editorial board. It was a British journal, and British editors, I have found (no ethnocentrism intended), are usually fairly civil about adjudicating differences of opinion.

Certainly the results were obvious, I remonstrated, *after* they had been pointed out and explained. How obvious would they have been to the reviewer, however, were he presented beforehand with the same question put to the dozen or so nuclear and particle specialists? Would the reviewer have recognized at the outset the applicability of a basic principle that ultimately made the final result seem obvious? Moreover, of what relevance was the argument that the actual calculation turned out not to be difficult and could have been worked by a smart student? After all, the criterion for publication is not that an article be so mathematically recondite as to be incomprehensible to an undergraduate, but that it make an original and worthwhile contribution. Other points were raised as well, but the foregoing brief summary gives a general idea of the tenor of my response.

The editorial board apparently enjoyed the rebuttal. The honorary editor, writing for the board, said, "We all found your paper and reply well presented," but alas the paper was still turned down. The board discussed the matter "extensively," I was informed, after which the conclusion of the paper was ultimately declared to be obvious. Apparently the editors failed to appreciate the irony that what is clearly obvious hardly requires extensive discussion. In any event, the paper was submitted elsewhere, was readily accepted,[3] and, so far as I know, is still the seminal paper for its treatment of dimuonic atoms.

The board's response (as well as the subject matter of the paper) recalls to my mind an anecdote, probably apocryphal, about that infamous genius Wolfgang Pauli, whose trenchant criticism of his colleagues' research earned him the appellation from Paul Ehrenfest of "die Geissel Gottes" (The Scourge of God). The punch line contains the essential spirit of the story although I no longer recall where I first heard it related properly. In any event, Pauli, in the course of one of his lectures, wrote an equation on the blackboard and continued speaking as if the origin of the formula was entirely clear. Asked at last by one frustrated listener if the result was obvious, Pauli looked at the board a moment, scratched his head, and left the room. Upon returning a half hour later, he replied "Yes," and then continued the lecture with no further explanation or interruption.

I do not mean to persuade scientists and science students that they must resort to unintelligible exposition in order to have their work published. Indeed, just the contrary; such writing is a perversion of the purpose of scientific communication and should be resisted. In short, if the results of research are worth communicating, then they are worth the effort it takes to write in an attractive, readable, and edifying way—even if this goes against the trend toward pomposity and obfuscation as hallmarks of erudition.

Finally, beyond the solicitation of funds and the publication of research papers, there may come a time when a scientist feels compelled to express an opinion or advocate some course of action regarding substantive matters of general social concern. Perhaps it is the matter of proliferation of nuclear arms or hazards of nuclear energy, dissatisfaction with present methods of science teaching or allocation of federal funding, the dumping of industrial wastes or destruction of the ozone layer, or whatever. Regardless of the nature of the concern, a physicist will have little effect if he or she cannot address in clear, straightforward terms the people for whom the warnings are intended.

The Sidney Harris vision of a physicist's style of conversation is sometimes too close to reality. Niels Bohr, one of the architects of the nuclear age, was granted a single audience with the beleaguered Churchill during the darker days of the Second World War. Bohr had hoped to convince Churchill of the need to look upon nuclear power as a global resource for peace and not as a purely military matter to be kept secret for as long as possible. Perhaps it was the inauspicious timing of such a meeting; perhaps it was the soft-spoken meandering circumlocution of Bohr's style of speech. In any event the conversation was not a success, and an opportunity to influence subsequent policy regarding control over nuclear arms was lost.

Bohr was downcast when he left Downing Street and related to his son how he had almost been scolded by Churchill.[4] He then began what eventually was to become his famous Open Letter to the United Nations. But, alas, this lengthy document suffered a similar fate. I am a physicist with the technical background to understand Bohr and with views largely sympathetic to his, but I can hardly get through Bohr's letter without falling asleep; I can easily imagine what impact it must have had on U.N. representatives.

By contrast, I. I. Rabi, another physics Nobelist deeply involved in national affairs, was a man who wasted no words. What his speech may have lacked in rhetorical polish was compensated for by an almost surgical directness. Like Bohr with Churchill, Rabi had an audience (after the war) with Eisenhower. At issue was the critical matter of convincing the president of the importance of establishing a presidential scientific advisory committee. In an age of push-button obliteration of mankind, the need for the American chief executive to have rapid, personal scientific advice was perceived—by physicists in particular—as dire.

In the words of Hans Bethe, an advisory committee participant, "Rabi presented our . . . case to the President so concisely and so convincingly, six major points in one hour, and Eisenhower immediately understood. . . . He immediately said to his adjutant, 'You see that this is done.'"[5]

Two physicists, two world leaders, two critical decisions, and two diametrically opposite results. One can never know for certain the reasons for the different outcomes, but surely ability to communicate must have played a significant role. Bohr, raised in the intellectually refined household of a Danish professor, was a mogul of abstract thinking in the world of academic physics; but Rabi, raised in poverty on the streets of New York City, could talk to the president with the same directness as to his grocer.

My own experiences as a physicist called upon to address acute social issues have convinced me of the extreme importance of being able to write simply and nontechnically to people without special scientific training. As an invited speaker at an international conference designed to improve science education, I listened at the outset to a succession of elaborate programs stressing one or another philosophical framework, or historical event, or pedagogical device for inculcating scientific facts and calculational techniques. I listened to the historians recapitulate great historical periods to other historians. I heard philosophers philosophize eruditely to other philosophers. I did not hear much said to the general audience, which consisted to a large degree of science teachers (from elementary school upwards) needing deeper insight into what is essential to successful science teaching. Few among the speakers were scientists who engaged in scientific research or taught some fundamental natural science.

When at last the principal plenary session arrived, and it was my turn to speak, I approached the podium apprehensively. I had no special doctrines to espouse; it was not within my province to discuss the logistics (let alone the merits) of integrating Aristotle's metaphysics or the history of phlogiston into an introductory course where students were already struggling to understand the most elementary scientific concepts. And surely the audience would not have been interested in a lecture on quantum mechanics.

However, as one who had been exploring physics and teaching it for some three decades, I knew at least what made science interesting to *me*. And, as one who was schooling his own children (who did not suffer boredom quietly), I had learned though trial and error that the real desire to learn science grows from the same seed in anyone.

Without jargon I pointed out as directly as I could that science derives from human curiosity; that children are born with an innate curiosity; that children free to explore their environment, free to inquire without fear of ridicule and without an externally imposed agenda unrelated to their own personally meaningful questions will likely retain this essential curiosity; that the will to learn finds its source and its reward in its own exercise. Philosophy, history, computer software, and the whole bag of pedagogical tricks will not lead to lasting scientific interest in those whose capacity to wonder and freedom to explore have already been stifled.

When the paper was finished, I thought at first that surely the presentation was a disaster, for all I had done was point out observations so basic, so true, and in some ways so trivial, that no other conference speaker addressed them. But I learned soon afterward through the many short and private conversations in the corridors and gardens of the host facility that my basic message had gotten through and was appreciated. Subtle and abstruse distinctions between various pedagogical schemes may have passed over the heads of science teachers in need of practical advice, but each listener could comprehend straight talk about what makes science personally interesting and worth relating to others. The thoughts on science education that I summarized at the conference were published in full[6] and have in succeeding years brought from all over the world appreciative letters and requests for further information and advice.

The vast outpouring of scientific papers that floods our libraries is subject, I believe, to a physical law similar in effect to that which governs the jostled nuts. The dried-out ones of limited scientific interest, directed to the narrowest readership, and written in pedantic jargon will sink to the bottom and be forgotten. The "Brazil nuts," by contrast, weighty with scientific significance and expressed with clarity and élan, will rise to the top and be remembered much longer.

The art of writing in physics is more than the mere description of experimental procedure or the cataloging of data; it is the opportunity for self-expression and influence. Physics and engineering students whom I encounter today often regard the need to write with as much enthusiasm as the need to use a slide rule. And yet writing is a vital scientific skill. It is the means by which scientists secure for themselves the fruits of recognition, share their ideas with both their contemporaries and posterity, and work for the betterment of society.

NOTES

1. A. Einstein, "Zur Elektrodynamik bewegter Körper," *Annalen der Physik* 17 (1905): 891–921.

2. A. Rosato, K. J. Strandburg, F. Prinz, and R. H. Swendsen, "Why the Brazil Nuts Are on Top: Size Segregation of Particulate Matter by Shaking," *Phys. Rev. Lett.* 58 (1987): 1038–40.

3. M. P. Silverman, "The Lifetime of the Dimuon Atom," *Il Nuovo Cimento* 2 D (1983): 848–52.

4. See Aage Bohr, "The War Years and The Prospects Raised by the Atomic Weapons," in *Niels Bohr: His Life and Work as Seen by His Friends and Colleagues*, ed. S. Rozental (New York: Wiley Interscience, 1967), pp. 191–214.

5. See J. S. Rigden, *Rabi: Scientist and Citizen* (New York: Basic Books, 1987), pp. 247–49. Quotation on p. 249.

6. M. P. Silverman, "Two Sides of Wonder: Philosophical Keys to the Motivation of Science Learning," *Synthèse* 80 (1989): 43–61.

CHAPTER 17

What Does It Take. . . ?

To live in the presence of great truths and eternal
laws—that is what keeps a man patient when the
world ignores him and calm and unspoiled when the
world praises him.
(*Honoré Balzac*)

IN THE MAIN I have had a long and productive career in science. I have had the
satisfaction of realizing my childhood aspiration of being a physicist. I have had
the freedom to pursue many lines of inquiry that interested me and the pleasure
of "now and then finding a smoother pebble or a prettier shell than ordinary,"
to use Newton's metaphor for scientific discovery. In the course of my work I
have traveled widely, lived abroad frequently, and enjoyed the friendship and
collaboration of many scientific colleagues. Although my work has brought
me into contact with industry and government, by far the greater part of it took
place at colleges and universities—and the frequent contact with young people
has both inspired me and kept me young. Were I to choose a career over again,
I would probably choose the same one.

In an earlier book of nontechnical essays, *And Yet It Moves*,[1] I wrote of some
of my experiences as a scientist. The book was received favorably by reviewers,
and I have always been particularly fond of one perceptive and sensitive review
that bore the title "It's a Wonderful Life!"[2] It is, however, a bittersweet title, for
a wonderful life, which I have indeed led in some ways, has been by no means
an easy or privileged life—and there are useful lessons in it worth sharing.

As a physicist and teacher I am often called upon to give talks, and over the
years I have addressed many audiences whose listeners ranged from schoolchildren to ordinary townspeople to graduate students and professional scientists.
I particularly enjoy speaking with children, for they are still fresh, receptive to
the curiosities of nature, not yet molded by regimentation, and usually bursting
with questions. Why did you become a physicist? What do physicists do? Do
physicists like to do other things than physics? Were you smart in school? Is
physics hard? What do you have to know to be a physicist? And on and on.

The more mature audiences of high school and college age, as well as graduate
students, also ask questions, although fewer and of a more pragmatic bent. Can
I be a physicist if I am not very good in mathematics? Can I be a physicist if I
am not very good in the laboratory? If I were to study physics, could I get a job?

Who would employ me? How much could I earn? Is there much competition among physicists?

I try to respond to questions honestly, although reservedly, for time is usually short, and I do not want my replies to be either misleadingly optimistic or discouraging. This is particularly true of one question that almost always arises irrespective of the maturity and background of the audience. Schoolchildren who may have only just heard of the subject, and graduate students soon to receive a Ph.D. in it, ask the question. Worded a hundred different ways, it is still the same question: What does it take to be a scientist—or, more specifically, a physicist?

When asked, I reply: *unquenchable curiosity and resolute determination.* How trite these words may sound, but they have implications that are far from obvious. Often I have looked afterward into the eyes of an inquisitor, clearly happy to hear that mathematical virtuosity or experimental dexterity are not the key criteria. "I am very interested in science," he or she might volunteer, "and I know I would not give up if the work were hard." Good for you, I smile in return and move on to the next question, knowing sadly that under the circumstances a more complete reply was hardly possible and probably inappropriate. Yet the question is a profound and difficult one, and many physics students, postdoctoral researchers, and teachers might have trained more happily and productively for a different career if they only understood beforehand what depth of commitment the answer demanded.

"Anyone who works in science," wrote forensic anthropologist William Maples,

> knows the dull desperation and sharp anxiety of the early days in one's career. Few of us do not look back on those pinched, scraping times without a secret shudder, followed by a pang of relief that they are past. The miserable pay and financial woes; the long nights of study and the battles against sleep; the frightful hurdles of examinations; the climactic defense of one's doctoral dissertation; the hissing, malignant envy that is the curse of university life at all times and in all places; the constant struggle to get published, to win tenure, to carve out a niche and be recognized in one's field—all these torments are well known in Academe, and have been known to drive some people mad, even to suicide.[3]

Only someone intimately familiar with the social and psychological dimensions of science could write so graphic a description. I can almost see Dr. Maples looking over Fresnel's shoulder as the latter wrote his brother of his deep longings and frustrations, or auditing from an upper balcony the rancorous debates of the Paris Academy. The "hissing, malignant envy" that may have prompted Biot to sabotage the career of Arago or induced Arago to venge himself with Fresnel as his weapon are by no means mere entertaining episodes of wilder times gone by. Science has evolved radically over the past two centuries, but the obstacles to being a scientist have changed little.

I know of one young assistant professor returning from a leave of absence to find his laboratory empty—the product of several years of grant writing, equipment development, setup, and use all gone except for a few shelves of chemicals that lined the room. The disagreeable event, perpetrated by another faculty member, took place in the critical period before his tenure review, and as academic institutions are frequently concerned more with institutional harmony than with justice, the outcome was not difficult to predict. Deprived of his apparatus and ultimately his employment, the one untenured member of the department was "thanked" by the university for his services and sent on his way. He was determined to remain a physicist, however, and in the end did quite well for himself, although the painful thoughts of those uncertain days, when both the life of his career and the economic survival of his family were in jeopardy, are still vivid recollections. I know—for I was that young scientist.

Acts of malice, however, are not the roughest grindstone that can wear away a scientist's self-esteem and emotional drive. More discouraging still is to find one's whole field of endeavor—physics—of little interest, value, or priority within the institution where he or she is employed or among the public at large. In recent years the severe retrenchment and attempted elimination of entire physics departments at American institutions of higher learning have sent shock waves through the physics community. But even where outright suppression has not yet occurred, many a physics program may still experience lingering and distressing neglect.

In his autobiography, Nobelist Luis Alvarez wrote "I can't think of anything more rewarding than being a full professor at a great university."[4] There are relatively few "great" universities in the world, however, and most physicists, either by circumstance or by choice, will not end up in one. I have during my career worked at three and greatly appreciated the virtually limitless resources, both material and human, to which scholars there had access. It was not difficult to be a physicist then, in a climate of high regard and vibrant activity. A brief walk brought me to libraries replete with all the information I desired. A knock on an office door put me in contact with an authority in any sought-out field, however narrow. Laboratories and supporting workshops lacked for nothing. Departmental secretaries and scientific illustrators provided ready help in the preparation of course materials and research manuscripts. It was a golden age. But I have also come to know a very different environment.

Imagine, instead, pursuing physics where there is no machine shop or electronics shop or professional technicians to operate them. There is no librarian to look after physics books and journals—and physics reference materials are transferred to an inaccessible, ill-lighted, deskless subterranean holding room of the main library, or left in vertical stacks in a commons room used for classes and social gatherings. There is one harried secretary shared with other departments, and fewer physics faculty than fingers on one hand. Housed in the basement of the religion and philosophy building ("metaphysics" building), physics laboratories and storage rooms are inadequately ventilated, dehumidified, or secured, and expensive apparatus suffers water damage, vandalism, and theft. Classroom or laboratory instruction occupies most days, leaving a few scattered moments for research. There are no postdoctoral assistants nor graduate students nor for years at a time even undergraduate students to participate in physics research projects.

"Who could do physics there?" asked one of my European colleagues incredulously, comparing my working conditions to those of scientists in the former Soviet Union. I could—but *there* is not some far-flung corner of the globe; it is *here* in the United States. And there are assuredly other trained physicists working, not at great universities, but in elementary schools, middle schools, secondary schools, community colleges, four-year colleges, and teaching-oriented universities for whom the preceding paragraph will re-create, with varying degrees of similitude, a familiar picture of low status and neglect.

To remain an active scientist in such circumstances is far from easy; it requires resolute determination not to give up altogether. I have never yielded; the inner drive to be a scientist, to interrogate nature and wrest answers to many questions, is too strong. Equally important, in my own eyes my justification to teach physics is drawn from my activities as a working scientist. In the concluding chapter to *And Yet It Moves*, in which I discussed the attributes of successful science instruction, I wrote, "Teachers must, themselves, be motivated and inspired to read avidly and regularly in order to learn the lessons of the past and to keep abreast of the present and to undertake their own investigations, however modest in scope or means, in order to teach with confidence based on personal experience." This was not advice offered gratuitously from a position of privilege, but a hard-won lesson from a physicist who has struggled to remain an active scientist and worthy teacher amid difficult circumstances.

The likelihood of succeeding as a scientist under conditions of severely parceled time and inadequate resources depends critically on the projects one chooses to investigate. It is at this point where a wide-ranging curiosity can mean the difference between a satisfying career and a life of frustration. When physics students desirous of pursuing scientific careers ask me if they should seek a position at one place or another, I ask them in return: If you found nothing there but a sandpile, could you become interested in the physics of sand? They smile, thinking I am in jest. (The irony of the remark, however, is that the physics of granular materials—sand—has actually become one of the "hot" topics of contemporary research.)

As a rule, I never consciously sought after "hot" topics. It would have been pointless for anyone in my circumstances to attempt to compete with well-funded, well-staffed research groups at major research universities or in industry. The tragedy for many young Ph.D.'s, fresh from the womb of their professor's laboratory, is to have little curiosity about the physics outside their doctoral theses. With wide interests, however, there are wide choices.

In setting my own scientific course over the years, I have usually followed three simple guidelines. A problem must be (1) conceptually interesting, (2) theoretically tractable (with little more than pencil, paper, and a desktop computer), and (3) experimentally testable (preferably in my own laboratory). Whether others are interested in the matter or not has never influenced me. The prospect of becoming an expert in some single field has never attracted me. Indeed, it was in approaching an unfamiliar part of physics for the first time that I often came across the most interesting phenomena to investigate. I have little patience with philosophical meanderings that lead nowhere—that is, to no experimental resolution. (Quantum physics, an area to which I have given much thought,[5] is regrettably beset with issues of this kind.) Although I can appreciate the mind-stretching exotica that characterize what I call the "physics of the superlative state"—the smallest distance, the highest energy, the greatest density, the first and last moments of the universe—I have had little inclination to seek out problems in these areas, for they lie, I suspect, beyond the reach of experiment, certainly beyond any experiment I can perform. To conceive of an idea and be able to test it myself—*that* holds the greatest attraction for me.

In retrospect, the three simple criteria have served me well. The problems that fascinated me became of interest to others too, and I received many requests to collaborate. Besides being a gratifying validation of my own scientific acumen, these invitations and their subsequent realizations established close professional ties and lasting friendships that relieved the isolation of my immediate surroundings. How glad I am that I did not succumb to the manifold pressures that discouraged research, for I now have numerous colleagues whom I would otherwise never have known.

My penchant for selecting the unusual and unfashionable to study had another, somewhat surprising, consequence as well: It transformed me, the perpetual student, into an expert. According to another review of *And Yet It Moves*, "Here is a book to delight any physicist with the time and sense to enjoy a guided tour of some of the most fascinating byways in physics. If you're like me, you may be familiar with some of the topics covered in this collection of essays, but others you may not even have heard of. By contrast, the author, Mark Silverman, is an expert on all of them."[6]

Hardly, to be sure—but many journal editors seem to think so, for I have been receiving at times in excess of a manuscript a week to review. There is again a certain irony in this situation. Journals once reluctant to publish my own papers are continually soliciting my judgment of the suitability of others'. It is more than ironical, for it is also a poignant reminder of another severe obstacle that many an aspiring young scientist may expect to encounter: The difficulty of being published is inversely related to the prominence of the institution with which one is affiliated. Regrettably, there are reviewers and editors who simply cannot imagine original and creative research issuing from institutions they deem scientifically inferior. But research is not done by institutions, departments, foundations, or companies; it is done by individual people—and people with scientific ability can emerge from even the most unlikely places.

Most reviewers, in my experience, are fair and helpful, and if a paper is worthy of publication, it will eventually see the light of day (although not necessarily in the author's journal of choice). Meetings and conferences of professional societies also provide opportunities for presenting one's work, as well as facilitate contact with other scientists with similar interests. The essential is never to become demoralized, never to give in to self-pity; nothing of lasting value was ever achieved except by those who dared believe that something inside them was superior to circumstance. Those who have confidence in their work and the determination to pursue it will eventually receive what recognition they are due.

Finally, to be a physicist requires an almost adamantine vigilance against the dissipation of one's time by unproductive obligations. People often ask me how I have been able to teach a full load of college and university courses during the day, return home to teach elementary school through high school courses to my own children in the evening, and still maintain an active, multifaceted program of physics research. It is not easy, but it can be done. It requires will and organization.

On the blackboard in the office of one of my research colleagues is scrawled the admonition: "If you cannot say NO, you are not in control of your time." Time is a physicist's most precious resource, for to what avail are reference books and apparatus without it? As a member of a scientific community, I have tried throughout my career to return a measure of service for the satisfactions of being a physicist. I have willingly reviewed hundreds of papers for many editors; I have served on editorial boards for science publishers; I have put my time at the disposition of the National Science Foundation, the National Research Council, and other agencies, both domestic and foreign, to examine grant requests and graduate fellowship awards; I have voluntarily helped local schools improve their science curricula. Called upon in a capacity such that my time would be usefully spent, I have tried, whenever possible, to say "yes."

But I have also learned to say "no." In the academic world especially, the proliferation of faculty committees is an appalling and senseless drain on the time that ought to be devoted to teaching and scholarship. For every hint or shadow of an issue, however frivolous, there is a committee. There is a committee to form committees. There are committees to oversee committees. The possibilities are endless—but one's productive time is not. The primary function served by many academic committees in my own experience has been to provide a social forum for its members; it hardly mattered what person was on what committee—any warm, willing body would have sufficed. In such cases I say "no"; I will keep those spare moments to myself and be a physicist.

Those who wish to know what it takes to be a physicist must also consider whether the effort is worth it. There is no pat answer; it is a matter of personal attitude and expectations. The dearth of physics-related jobs, the preferential inclination of industry to hire engineers, the talk of closing down major national laboratories, the elimination of academic physics programs, and the pervasive sentiment among many students and much of the general public that physics is neither interesting nor relevant have generated much soul-searching within the physics community. There are suggestions for limiting graduate enrollments and restricting the number of physics Ph.D.'s, or for teaching physics in explicitly application-oriented ways that would help graduates to market themselves more effectively. These are all serious matters, but I find myself very much out of sympathy with the entire tenor of discussion that focuses narrowly on physics as an employable occupation.

Many years ago, with no family encumbrances and with little thought of future gainful employment, I heedlessly embarked upon the study of physics as a passion. I would do physics irrespective of my livelihood—at home, if necessary, like Rayleigh in his barn or Maxwell in his kitchen. Physics was not a remunerative affair as much as a way of experiencing life, an activity that fostered not only the capacity to reason but the enjoyment of beauty and the inculcation of worthwhile values. As much as art, music, and literature, physics formed a part of my cultural foundations. In the panorama of history, I saw the undertakings of physicists, no less than those of courageous mariners, lift from humanity the veils of mystery, ignorance, and superstition. I saw in the global extent of the physics community a stronger antidote to prejudicial stereotypes of human ability and intelligence than in all the fruitless programs of social engineering. Had I never earned a cent as a physicist, my studies of physics would not have been wasted.

As it turned out, I was fortunate and could survive economically by teaching physics. The conditions under which I worked were on the whole far from optimal, but it was attitude, and never these conditions, by which I defined myself as a physicist. Perhaps there is some truth in the exclamation of my ebullient ellipsometric colleague quoted in chapter 9: Physics *is* like a religion. It is founded on the belief that there is reason in nature which the human mind is capable of fathoming. The articles of faith are that the natural world is fundamentally interesting, that the effort to understand it is worthy of pursuit, and that the pursuit will be made despite all obstacles.

It is a demanding religion, but initiation is open to all.

NOTES

1. M. P. Silverman, *And Yet It Moves: Strange Systems and Subtle Questions in Physics* (New York: Cambridge University Press, 1993).

2. S. Stenholm, "It's a Wonderful Life," *Physics World* (December 1973): 54.

3. W. R. Maples and M. Browning, *Dead Men Do Tell Tales* (New York: Doubleday, 1994), pp. 20–21.

4. L. W. Alvarez, *Alvarez* (New York: Basic Books, 1987), p. 280.

5. M. P. Silverman, *More Than One Mystery: Explorations of Quantum Interference* (New York: Springer-Verlag, 1995).

6. J. R. Taylor, book review in *Am. J. Phys.* 62 (1994): 671–72. Quotation on p. 671.

ABOUT THE AUTHOR

Mark P. Silverman is Professor of Physics at Trinity College, Connecticut. He is the author of *And Yet It Moves* and *More Than One Mystery*.